# Reconstruction, Replication and Re-enactment in the Humanities and Social Sciences

*Edited by*
*Sven Dupré, Anna Harris,*
*Julia Kursell, Patricia Lulof and*
*Maartje Stols-Witlox*



Routledge
Taylor & Francis Group
LONDON AND NEW YORK

This book was realized with funding from the ARTECHNE project, a European Research Council (ERC) Consolidator Grant awarded to Sven Dupré under the European Union's Horizon 2020 research and innovation program (grant agreement No. 648718). Anna Harris's research contribution is funded by the European Research Council (ERC) under the European Union's Horizon 2020 research and innovation program (grant agreement No. 678390).

First published in 2020 by Amsterdam University Press Ltd.

Published 2025 by Routledge
4 Park Square, Milton Park, Abingdon, Oxon OX14 4RN
605 Third Avenue, New York, NY 10158

*Routledge is an imprint of the Taylor & Francis Group, an informa business*



ISBN: 9789463728003 (hbk)
ISBN: 9781041184973 (pbk)
ISBN: 9781003702375 (ebk)

NUR 694

Cover illustration: Anthea Walsh: Beady Eye, detail. Photo: Anthea Walsh, used with permission.

Cover design: Coördesign, Leiden

DOI: 10.5117/9789463728003



For Product Safety Concerns and Information please contact our EU representative:
GPSR@taylorandfrancis.com
Taylor & Francis Verlag GmbH, Kaufingerstraße 24, 80331 München, Germany

# Reconstruction, Replication and Re-enactment in the Humanities and Social Sciences

# Table of Contents

# Acknowledgements

The idea for this book was conceived during a NIAS-Lorentz Workshop in Leiden in June 2017, and it was further developed in an authors' meeting at the University of Amsterdam in December 2018. We wish to thank all participants and hands-on session leaders who were involved in the NIAS-Lorentz workshop and kindly acknowledge the important contributions which the Lorentz Center, and its dedicated staff, made to this book. We would also like to thank the Rijksmuseum Boerhaave and the National Museum of Antiquities in Leiden, the Organ Park and the Atelier Building in Amsterdam, and the 4D Lab and the Conservation and Restoration Program at the University of Amsterdam, for generously hosting us. The realization of both meetings, and of this book, received further support from the ARTECHNE project, a European Research Council (ERC) Consolidator Grant awarded to Sven Dupré under the European Union's Horizon 2020 research and innovation program (grant agreement No. 648718). Our editor at Amsterdam University Press, Erika Gaffney, has guided us through the publication process with tact and efficiency. Jill Briggeman and Marleen Schans (ARTECHNE) have given invaluable editorial support without which this volume would not have been possible.

The Editors
February 2020

# Introduction

*Sven Dupré, Anna Harris, Julia Kursell, Patricia Lulof, Maartje Stols-Witlox*

**Abstract**

Performative methods are playing an increasingly prominent role in research into historical production processes, materials, and bodily knowledge and sensory skills, and in forms of education and public engagement in classrooms and museums. Such methods, which we refer to as Reconstruction, Re-enactment, Replication, Reproduction and Re-working (RRR), are used across fields in the humanities and social sciences, from history of science and technology, to archaeology, art history, conservation, musicology and anthropology, among other disciplines. There is much to learn from interdisciplinary methodological reflection. RRR raises issues of truthfulness and accuracy, draws attention to process and performance as well as practices of documentation and facilitates communication with broader publics.

**Keywords:** truthfulness, accuracy, process, public, replica, ephemerality

Performative methods are playing an increasingly prominent role in research into historical production processes, materials, and bodily knowledge and sensory skills, and in forms of education and public engagement in classrooms and museums. Such methods, which we refer to in this book as Reconstruction, Re-enactment, Replication, Reproduction and Re-working (RRR), are used across fields in the humanities and social sciences, from history of science and technology, to archeology, art history, conservation, musicology and anthropology, among many other disciplines.

Throughout this book, a variety of terms are used in relation to performative methods. While the title focuses on Reconstruction, Re-enactment and Replication, these three are only a small selection of all the 'Re'-terms used to

Dupré, S., A. Harris, J. Kursell, P. Lulof, M. Stols-Witlox (eds.), *Reconstruction, Replication and Re-enactment in the Humanities and Social Sciences*. Taylor & Francis Group
DOI 10.5117/9789463728003_INTRO

describe activities in which researchers perform (past) practices.[1] Disciplines have acquired preferences for particular Re-terms. Reconstruction is the term of choice in conservation and restoration and in reference to digital or virtual reconstructions in archaeology; the 'replication method' is the point of reference in the history of science; and re-enactment seems to be the more common term in musicology and anthropology. The ways in which the Re-terms are used in the introduction and the different chapters of this book bear traces of these rich historiographical traditions. This also means that the use of specific Re-terms is not always consistent across the different chapters.

Despite the growing interest in performative approaches across disciplines, so far, reflection upon these RRR methods has largely remained within the disciplines. Yet, there is much to learn from interdisciplinary methodological reflection, especially with an eye towards considering more productively the generative potential of the challenges that come with the use of performative methods, challenges which are common to more than one discipline. In this book, an interdisciplinary group of authors bring their experiences of performative practices within their discipline in conversation with RRR methods in other disciplines. This, we suggest, offers deeper reflection on performative methods both within and across the disciplines. RRR raises issues of truthfulness and accuracy, draws attention to process and performance, as well as practices of documentation and facilitates communication with broader publics. We explore these themes which run through the chapters later in the Introduction, but first we look at how RRR methods have developed in archaeology, history of science and technology, musicology, and conservation and restoration, fields where RRR has a long tradition, as well as in anthropology, where it has been adopted more recently. While there are many more disciplines in which RRR methods have come to play a role, the mentioned disciplines have in common that their histories of using performative methods show traces of interdisciplinary cross-fertilisation upon which we can build.

1 Interdisciplinary reflection on the 're-' prefix is gaining momentum, see *An Errant Glossary*, ed. by Holzhey and Wedemeyer.

## Histories of RRR in Conservation, Musicology, Archaeology, History of Science, and Anthropology

Ever since people started restoring Cultural Heritage objects, attempts to heal damage have resulted in some kind of reconstruction with the aim of remaking lost areas or even complete objects.[2] Such reconstructions predate reconstructing as a research method, to investigate past or otherwise unknown artisanal practices. Traditionally, such restoration-reconstructions were executed by artisans or artists working from a crafts or artistic background.[3] The gradual emergence of the restoration profession in the nineteenth and early twentieth century led to the development of professional guidelines, in which the field formulated ethical codes discussing the limitations and boundaries of restorative actions, as discussed in the chapter by Maartje Stols-Witlox in this volume.[4] The earliest evidence of reconstructions investigating methods and materials of past artists can be traced back to the eighteenth century. Jilleen Nadolny et al. placed the beginnings of reconstruction methodologies in the 1750s, when French connoisseur Count Caylus was investigating ancient methods of encaustic painting through experiments with historical recipes combined with scientific analyses, the results of which were published with co-author and medical doctor Majault as *Mémoire sur la Peinture a l'Encaustique et la Peinture a la Cire* in 1755.[5] Elsewhere, Nadolny recounted how excavations of Ancient sites like Pompeii fuelled enthusiasm for Antique painting and efforts to understand its techniques aiming to find more stable alternatives to the painting materials in use at the time. Interest in discovering the early history of oil painting, more in particular in the 'lost' paint binding medium of Jan van Eyck, played an important role in the publication of historical recipe books and manuscripts throughout the nineteenth century,

2 Thomas, in 'Restoration or Renovation', writes about Renaissance restorations and provides examples of restoration treatments that included far-reaching reconstruction or repainting. Campbell, 'The Conservation of Netherlandish Paintings', pp. 20–21, discusses a case in 1436 Tournai, when a wall painting in a building was copied on paper by a painter before demolition of the original, in order to have a record of the painting in case the decision would be made to reconstruct the paintings 'as they were before' (quote from Campbell).

3 See for example Brayer, *Conservation in the Nineteenth Century*; Sitwell and Staniforth, *Studies in the History of Painting Restoration*; and Conti, *History of the Restoration*, on the role of artists and artisans in early restorations.

4 The first conservation education programmes were set up in the 1930s and 1940s, first with the Courtauld Institute, followed slightly later by programmes in Austria (Vienna), Germany (Munich) and in Italy (Rome). See Nadolny, 'A History of Scientific Examination', p. 340.

5 Nadolny et al., 'Art Technological Source Research', p. 6.

for instance for two authors whose seminal works on historical materials are nowadays still consulted, Mrs. Mary P. Merrifield and Sir Charles Lock Eastlake.[6] Both are known to have tested the methods described in these historical sources through reconstruction.[7]

The interest in historical painting techniques in Germany led to the publication of additional historical sources in the series *Historische Quellenschriften für Kunstgeschichte und Kunsttechnik des Mittelalters und der Renaissance* (Vienna, Braumüller), and by German painter and art academy professor Ernst Berger, published between 1893 and 1919. Berger himself experimented with the preparation of historical paint recipes. The German periodical *Technische Mitteilungen* (1884-1944), established by chemist Adolph Wilhelm Keim in Munich, aimed to provide artists with reliable technical information and included articles describing reconstructions of historical methods.

Daniel V. Thompson (1902-1980), art historian, conservator, professor and chemical engineer, emphasized the role of craftsmanship in arts. He noted in his *Materials and Techniques of Medieval Painting* that 'technique means materials and tools in action', and described technical study of paintings including 'the recognition of those systematic methods which combine taste and knowledge and competence'.[8] Thompson's interest in craftsmanship extended to reconstructions of Medieval techniques on the basis of historical recipe sources and the study of paintings.[9]

Interest in early oil painting practice has not dwindled, for instance with the art historian Pim Brinkman's (1993) research in to Van Eyck's binding medium and the Impact of Oil project (2007-2015), both employing reconstructions based on historical recipes as well as on scientific data from painting investigations.[10] The combination of historical source research, painting investigation and reconstructions characterises the current approach to reconstruction within conservation and technical art history.[11] Ever since

6 Eastlake, *Materials for a History*; Merrifield, *Medieval and Renaissance Treatises.*

7 For example, Merrifield, *Medieval and Renaissance Treatises*, pp. liv-lv, ccxxxv, in her introduction discusses reconstructions of enamel paints as well as oil processing recipes.

8 Thompson, *Materials and Techniques*, p. 19.

9 Thompson discussed his approach in interviews. Oral History interview with Daniel Varney Thompson, 25 September, 1974 / November 2, 1976. https://www.aaa.si.edu/download_pdf_transcript/ajax?record_id=edanmdm-AAADCD_oh_212102, Checked on February 2nd, 2019.

10 See about the Impact of Oil project (Netherlands Organisation for Scientific Research, 2008-2013): https://www.nwo.nl/en/research-and-results/research-projects/i/19/3019.html. Checked on February 2nd, 2019.

11 The Art Technological Source Research Study Group, a working group within the International Council of Museums Committee for Conservation (ICOM-CC), brings together researchers

the introduction of the concept of 'historical accuracy' in reconstructions by Leslie Carlyle in the early 1990s, increasing emphasis has been placed on sourcing materials that are deemed appropriate for the time period studied.[12] More recently, other terms have been suggested to describe reconstructions that aim to approximate historical practices, such as 'historically informed' by Spike Bucklow and 'historically appropriate' by Carlyle.[13]

This terminological discussion has a direct equivalent in current RRR practices in the study of music. The historiography of such practices is intertwined with the peculiar role of performativity in music. Music cannot subsist without enactment of some kind. Musicology – in the roughly 150 years of its existence – has attached the prefix 're' to this enactment. Most often, musical re-enactments have been building on practitioners' foregoing calls and attempts to reconstruct the music of earlier times. The first performance of Johann Sebastian Bach's *Matthäus-Passion* after Bach's death by Felix Mendelssohn Bartholdy in 1829 is generally seen as a starting point. Mendelssohn's revival of Bach's music soon merged with canon formation, eventually resulting in a notion of classical music that coincided with eighteenth- and nineteenth-century composition.[14]

The limits of the canon were put into question by the young field of musicology. The founding figures of musicology, Philipp Spitta, Friedrich Chrysander and Guido Adler introduced Baroque music into the canon and eventually extended the scope to pre-modern composers as well. Their main occupation was to provide editions as reference objects for the discipline. Written musical notation became the focus of musicological scholarship, by the same token. Yet, the gap between the notating systems also drew attention to the fact that music had been performed under quite different and constantly changing circumstances. Musicologists felt the need to also bring to life the growing corpus of recovered written notation in musical performance. Adler's fundamental paper 'Scope, Method and Aim of Musicology' (1885) urged scholars and musicians to collaborate in the

within the arts domain who study historical sources, often in combination with/through reconstruction. http://www.icom-cc.org/21/working-groups/art-technological-source-research/ Checked on February 4th, 2019. Art history is a relative latecomer to the adoption of performative methods, but the discipline can build upon sustained methodological reflection which considers re-enactment as part of the toolbox of art history since Erwin Panofsky's 'Art History as a Humanistic Discipline' (1938). See Davis, 'Art History, Re-Enactment'.

12 See Carlyle's chapter in this volume.

13 Bucklow in Wrapson et al., *In Artists' Footsteps*, p. 26. See Carlyle's chapter in this volume on 'historically appropriate'.

14 See, e.g. Applegate, *Bach in Berlin*.

'restoration, re-working and performance' of historical works.[15] A parallel endeavour to place historical performance within music scholarship was Hugo Riemann's reviving of the *Collegium Musicum* tradition at Leipzig University after 1900. Many other universities followed the example, including Columbia University, in whose *Collegium* the most prominent critic of the later 'authentic music' movement, Richard Taruskin, played viola da gamba for many years.[16] The most influential example of this institutionalisation of reviving music was the foundation of the *Schola Cantorum Basiliensis* in 1933. Multiple nodes in the emerging network of musicians and scholars lead from these institutions to the Early Music movement that took shape after World War II in the Western world.[17]

If these activities mainly focused on the relationship between musical notation and performance, collecting and replicating historical instruments forms another thread in RRR and music. An early example is the cembalo designed for Wanda Landowska by the French company Pleyel in 1903. This instrument featured a strong bass register, unknown in the original instruments, but necessary for performance and recording in her times.[18] By that time, historical instruments had become collector's items. Instrument collections were reorganised and opened to the public. The growing knowledge about instruments reached beyond the community of musicians and scholars to buyers and forgers. The interested readership could be referred to the catalogues of famous collections, such as that of the *Gesellschaft der Musikfreunde* in Vienna, the *Musée instrumental du conservatoire royal de musique* in Brussels or the *Deutsches Museum* in Munich.[19]

Overseers of museum collections in particular held a fierce discussion over whether the instruments should be preserved in their current state or restored for playing. Curt Sachs, director of the Berlin collection, suggested integrating recordings into displays and engaged Landowska to play on a harpsichord that had come to the collection from the Bach family. Her phonographic cylinder recording of Bach's *Italian Concerto* from 1908 was

15 Adler in Mugglestone, 'Guido Adler's "The Scope"', pp. 1-21, quote on p. 17, translation slightly changed by Julia Kursell. On the history of musicology, see Auhagen et al., *Musikwissenschaft 1900-1930*, and more specifically on Adler: Martin Eybl, 'Guido Adler, die Denkmäler der Tonkunst in Österreich'.

16 Cf. Taruskin, *Text and Act*.

17 For example, Adler's disciples Josef Mertin and later Munich ordinarius Rudolf von Ficker, and the musicians Gustav and Marie Leonhard, Nicolaus and Alice Harnoncourt, or Paul O'dette. New light on the female participants in this endeavor has been shed by Mimi Mitchell, 'The Revival of the Baroque Violin', Unpublished PhD dissertation, University of Amsterdam (2019).

18 Eigeldinger, *Wanda Landowska*.

19 See, for instance, Ruth-Sommer, *Alte Musikinstrumente*.

meant to be played on site, but the wax cylinder recording proved too frail for this purpose and the idea was picked up only much later in a systematic way.[20]

After 1945 the notion of 'early music' came into use for explorations of a pre-classical repertoire. This was backed up by the formation of a lively community of musicians, listeners, scholars and instrument builders who fed their findings into the emerging market for digital recordings. The dry studio sound of the 1980s meant rethinking questions such as tempo and timbre in different terms than those of the nineteenth-century concert hall, if not necessarily more appropriate ones than the period practices. Today's notion of RRR rediscovers musicianship in new ways, connecting with the material turn in the social sciences and humanities.[21] Not only is the beginning of the movement nowadays an object of reconstruction under the header of 'historically informed performance practice', listening habits have also changed and keep changing dramatically, as the contributions in this volume will discuss.

As in conservation and restoration and music and musicology, RRR has a long and prominent history of use in the discipline of archaeology. However, while experimental archaeology has roots in the nineteenth century when archaeologists attempted to recreate the technologies of the past, it was only in 1979 that John Coles published a book on experimental archaeology, thus establishing the discipline of the same name.[22] In more recent years, Roeland Paardekooper and the EXARC network have done much discipline-formation work, from compiling bibliographies and documenting experimental archaeological projects to establishing experimental archaeology publication venues and writing histories of the discipline.[23]

In a special issue of *World Archaeology* on experimental archaeology, Alan Outram, professor of archaeological science, stated that one typical characteristic of experimental archaeology is that it is done in the field. He differentiated between experimental archaeology and experiments or tests in the laboratory, while specifying how they relate: 'A gulf is left between such laboratory work and how such processes may have been achieved in the past, with a limited range of materials, technologies and a lesser control upon the environment. Experimental archaeology comes into its own at this point. What has been learned in the lab can now be taken further; hypotheses

20 See Elste, *Meilensteine der Bach-Interpretation*.

21 See, e.g. the frequent references to music in Tim Ingold's *Making. Anthropology, Archeology, Art, and Architecture* (New York and London: Routledge, 2013).

22 Coles, *Experimental Archaeology*; Outram, 'Introduction to Experimental Archaeology', pp. 1-4; Reeves Flores and Paardekooper, *Experiments Past*.

23 https://www.exarc.net. Checked on March 31st, 2019.

can be tested in a range of environmental conditions that aim to reflect ... "actualistic" scenarios.'[24] To emphasise this open-air (as in, outside the controlled environment of the lab) character of experimental archaeology, Outram spoke of 'actualistic' experiments. The use of the term 'actualistic' also recognises the much-discussed issue with the 're'-prefix which misleadingly suggests that archaeologists can reconstruct the past exactly.

The aims and types of RRR methods in archaeology are diverse: from trying out archaeological techniques in actualistic scenarios and experimental investigations into formation processes of the archaeological record, to constructions testing the design of a building and research into technological processes and the *chaîne opératoire*. Especially in the latter categories of experimental archaeology, digital techniques have had a major impact. The latest development of data visualisation is the introduction of computer-generated techniques.[25] Three-dimensional models (3D), virtual reality (VR) and augmented reality (AR) animations are increasingly being used in reconstruction processes. These technologies also attempt to simulate ancient building techniques and visualise decisions taken by the archaeologists. Thereby, they allow archaeologists to explore processes that would otherwise remain hidden, or to answer questions that have never before been asked.[26] Archaeologists started to apply digital tools and 3D modelling in the early 1980s. Already in the 1990s they discussed the value and credibility of 3D modelling.[27] Today, many archaeologists experiment with the newest digital technologies for reconstruction.[28] These developments also embrace more disciplines, because the reconstruction of environments, architecture as well as material culture deals with (digital) data and artefacts from art history, history and heritage and memory studies.[29]

24 Outram, 'Introduction to Experimental Archaeology', p. 2.

25 On the history of archaeology: Stiebing, *Uncovering the Past.* An excellent overview of the history of reconstructing archaeological phenomena is provided by Piccoli, 'Visualizing Antiquity', pp. 225-59 and Moser, 'Archaeological Visualisation', pp.192-233.

26 Important publications that discuss the potential of 3D modelling: Hermon, 'Scientific Method', pp. 13-22; and for VR: Nick, 'Documenting and Validating', pp. 254-273; on the added value for ancient architecture: Lulof et al., 'The Art of Reconstruction', pp. 333-337

27 For an excellent overview of the early rise of 3D modelling and its reception in archaeology: Lanjouw, 'Discussing the Obvious', pp. 1-13; on the resistance to this methodology in humanities: Ratto, 'CSE as Epistemic', pp. 567-587.

28 Since 2000, these new digital technologies have been unanimously embraced in archaeology; see Barceló et al., *Virtual Reality.* For discussion of their value and deception, see Posluschny et al., *Layers of Perception.*

29 On the use of 3D modelling withina wide range of disciplines: Opgenhaffen and Sepers, '3D Modelling', pp. 411-414 and Lury et al., *Routledge Handbook of Interdisciplinary Research Methods.*

Precisely to safeguard its methodology as scientific, experimental archaeologists have long distanced themselves from recreational and educational evocations of past life and technology by re-enactment groups dressed up in period costume.[30] However, in more recent years, some RRR projects have embraced closer connections to these practices. One example is the Iron Age Village in Sagnlander Lejre, a Historical-Archaeological Experimental Centre in Denmark, which was studied in the 1990s by an anthropologist who participated in the life of the Iron Age Village.[31] In such projects, volunteer and citizen non-archaeologists are involved in experimental archaeological research on equal terms with professional archaeologists.

Given its focus on past technologies, it is no surprise that in the twenty-first century experimental archaeology has connected with the history of technology and science.[32] In the history of the latter discipline, re-working, the reproduction or re-enactment of experiments is a methodological practice closely tied up with the issue of real versus thought experiment. A central question with which historians of science in the 1960s were concerned was whether Galileo performed experiments or whether his reported experiments were rather thought experiments. According to Alexandre Koyré, the imperfection of Galileo's instruments and experimental procedures did not enable any real experiments to have occurred. In the 1960s and 1970s, Tom Settle and James MacLachlan used experimental reproduction to show that Galileo did perform real experiments.[33]

Soon the question shifted from whether Galileo performed experiments, to how he acquired the skills to do experiments. Ever since, opening the black box of skill (or tacit or gestural knowledge) has been a central concern of experimental history of science, as conceived by Otto Sibum in the 1990s. Inspired by Falk Riess' addition of a laboratory course using replicas of historical scientific instruments to the physics teacher training programme at the University of Oldenburg, historian of science Sibum started reworking historical experiments using carefully reconstructed instruments.[34] While the history of experimental history of science is closely intertwined with science education, for Sibum, experimental history of science should also be self-reflective: experimental historians of science should be aware of the limitations of the reproductions of historical experiments. Experimental

30 Outram, 'Introduction to Experimental Archaeology', p. 3.
31 Holten, 'Engaging Experiments'
32 Stauberman, *Reconstructions*.
33 MacLachlan, 'Galileo's Experiments' and Settle, 'An Experiment in the History of Science'.
34 Sibum, 'Reworking the Mechanical Value'.

reproductions generate historical possibilities and impossibilities, Sibum argued, but they are not able to tell what 'really' happened.[35] The rise of experimental history of science occurred at the time that the material turn took hold in the discipline, and historians of science became interested in instruments and the material culture of science more broadly.

History of science appears to have converged towards 're-working' as the term of preference to refer to performative methods in recent years, the term placing emphasis on process rather than product.[36] Nevertheless, experimental history of science has, and presumably will continue to use a wide variety of Re-terminologies. The only term which is considered misleading is 'replication', precisely because of its associations with its present-day scientific usage for repetition of an experiment to check or confirm the validity of prior results. The aims of historians of science and technology are quite different: 'When historians rework or reproduce a process or an experiment as a historiographical tool they are not replicating in these scientific [...] senses, but are instead seeking fresh historical information.'[37]

Sibum coined the term 'experimental history of science' in the 1990s, inspired by experimental archaeology, pointing to an interdisciplinary awareness of the use of performative methods. An important difference between archaeology and other disciplines (history of science, musicology, conservation) is the type of sources which are used in conducting research. Textual sources (such as recipes, musical scores, and experimental accounts, including lab books) are much more widely available in conservation, musicology and history of science than in archaeology; sometimes in the absence of instruments and experimental apparatus they are the only remains of past scientific and artistic practices. These fields require a close reading and engagement with both material culture and historical text sources. Performative engagement with texts has made researchers in these fields attentive to issues of language and questions of interpretation of terminology.[38] In recent years, this has resulted in projects in the history of science which take texts as their object of study, cooperating with conservators.

One example is Pamela Smith's Making and Knowing Project at Columbia University, which centres on a French, sixteenth-century manuscript in the Bibliothèque Nationale de France. Here, reconstruction serves the

35 Sibum, 'Experimental History of Science'.

36 Fors et al., 'From the Library to the Laboratory'.

37 Fors et al., 'From the Library to the Laboratory', p. 9.

38 By reconstructing alchemical practices, Lawrence Principe showed that the metaphorical language of alchemy (such as the 'Tree of the Philosophers') has its roots in actual experience in the laboratory. See Principe, 'Apparatus and Reproducibility'.

elucidation of historical texts, from the names of materials or processes to the distillation of taxonomies and how historical actors conceived of materials informing the processes of writing.[39] For Smith, reconstruction is a method of critical reading, assisting the understanding of the text and giving insight into the composition of the manuscript and the motivations and reasons behind writing. The treatment of texts as 'paper technologies', considered as embedded in material practices of reading and writing in the arts, is also the point of departure of the ARTECHNE project led by Sven Dupré. Adding yet another interdisciplinary reflection on performative methodology to the equation, inspired by the ethnographic experiment, ARTECHNE re-works recipes not only to reflect on the relationship between texts and practice, but more importantly, to gain insight into what artisans do with texts.[40] This brings us to the history of RRR in the discipline of anthropology.

Re-enactment methodologies have only more recently gained recognition in the work of ethnographers due to a number of developments in fieldwork methodologies. One of these is the increasing popularity of visual anthropology, whereby ethnographers retrace routes and revisit localities with cameras in hand. Sarah Pink and Kerstin Leder Mackley,[41] for example, argue that video re-enactments, whereby interlocutors are asked to re-enact their practices before a video camera, are a way to research and collaboratively apprehend, with research participants, aspects of their everyday life that are usually hidden. The Brazilian artist ethnographer Veridiana Zurita[42] explores this beautifully in her work in the Amazon, as she films families re-enacting scenes from their favourite soap operas; the insights into their lives are generated from the ways in which they adapt and add flourish to the scripts. Zurita's work highlights creative ways in which ethnographic fieldwork and artistic research are finding synergy; re-enactment methodologies are an experimental space where these approaches meet.

This performative turn in ethnographic methods can also be found in anthropological links with methods used in historical disciplines, especially oral history. For example, Paul Geissler and Ann Kelly[43] explored the methodological and theoretical potential of re-enactment during their ethnography of a research station in Tanzania to probe relationships between materiality, memory, place and affect. Geissler and Kelly used the

39 Smith, 'Historians in the Laboratory'.
40 Mann et al., 'Mixing Methods'.
41 Pink and Leder Mackley, 'Re-enactment Methodologies', pp. 146-151
42 More information on the project can be found on Zurita's website: http://www.veridianazurita.com/index.php?/projects/televizinho/. Checked on April 4th, 2019.
43 Geissler, *Traces of the Future.*

now inactive field station as a site to bring together retired African and European workers to re-enact experiments and laboratory procedures. The ethnographers also re-enacted scientific method sections of papers written by researchers from the field station. They found that these performative methods rendered habitual movements, and the emotional labour of the scientific work conducted there into a productive conversation between the past and present.

As the music and language scholar Vanessa Agnew stated, 're-enactment's emancipatory gesture is to allow participants to select their own past in reaction to a conflicted present. Paradoxically, it is the very ahistoricity of re-enactment that is the precondition for its engagement with historical subject matter'.[44] The purpose of these re-enactments in Tanzania, Geissler and Kelley assert, was not so much to add tacit, embodied and relational dimensions of past events, as they suggest is the goal of re-enactment in other fields, but rather to 'shed light on the ethical and political tensions in our relationship with the past, and the experience of loss and decay – and to provoke and sense unexpected constellations, excess that moved beyond the script'.[45] These experiments are part of anthropology's shift since the 1980s from representation to presentation. This move opened up now common tensions in anthropology between authenticity and artifice, representation and desire, past and present.

As well as exploring re-enactment methodologies, anthropologists have also, perhaps more commonly, taken practices of re-enactment as a focus of their work. For example, Michaela Schäuble,[46] also using film and photography as a research tool, examines re-enactment and religious performances as sites of revitalising and negotiating tradition, heritage and cultural identity in Southern Italy. The anthropologist Lucy Suchman[47] researches the re-enactment and simulation of war in the Middle East with US soldiers, looking, for example, at the role of Hollywood in re-enacting war scenes for training, while Katherine Johnson reflects upon historiographic methods through her ethnographic study of a Jane Austen Festival.[48] Petra Tjitske Kalshoven[49] has written previously about Indian hobbyism, or Indianism in Europe, a practice that developed out of a strong fascination with Native American life in the eighteenth- and nineteenth centuries. The

44 Agnew, 'What is Re-enactment?', p. 328.
45 Geissler, *Traces of the Future*, p. 153.
46 Schäuble, 'Living History?', pp. 1-19.
47 Suchman, 'Configuring the Other', pp. 1-36.
48 Johnson, 'Rethinking (Re)doing', pp. 193-206.
49 Kalshoven, *Crafting the Indian*.

hobbyists themselves used ethnographies to base their re-enactments of North American Indian lifeworks, and in a move typical of anthropologists, Kalshoven draws upon the practices of her informants to reflect upon her own practices as a researcher.

Thus, the four main contributions of anthropology, as a relative latecomer among the disciplines to the topic of RRR, could be considered as the following. Firstly, the field contributes a great deal of reflexivity to RRR research with considered attention to researchers' own practices of researching, and offers insights into how the work of the researched can reflect upon the work of researchers themselves. Secondly, anthropologists help to expand the repertoire of fieldnote making and documentation of relevance to RRR methodologies across disciplines. Ethnographers are finding exciting new ways to extend fieldnote practices, using visual methods, or revitalising methods from other fields such as psychogeography and walking (see Jo Vergunst's chapter in this volume). Thirdly, anthropology is a field which is experiencing a close alignment with artists with regards to thinking through crossovers between artistic research and ethnographic fieldwork, including playful experiments with RRR. Finally, anthropology explores productive tensions between past and present realities through re-enactment, looking not at how to bring the 'past to life', but rather how performances may disrupt temporalities normally taken for granted.

## This book

The disciplinary historiographies of RRR methods reveal what motivates researchers in different fields to introduce performative aspects, and provide insight into the way researchers have tried to come to terms with the limitations and problems associated with attempts to reconstruct environments, objects or processes. It also becomes evident that the emerging interdisciplinary dialogue on RRR methods has already started to enrich the potential of RRR methods in the fields of conservation, archaeology, history of science and anthropology. To foster this dialogue, in 2015, we organised a Netherlands Institute for Advanced Science (NIAS)-Lorentz workshop in Leiden, bringing together practitioners using performative methods across disciplines in the humanities and social sciences. The workshop involved lectures as well as hands-on laboratory workshops, walks and artistic experiments. This book builds on the collective, bodily and sensory experiences of the workshop. It is furthermore, to the best of our knowledge, the first book to bring together interdisciplinary reflections upon RRR methods.

This is an important step in the development of performative methods in research in the humanities and social sciences.

In this introduction, we highlight four themes which run through all of the chapters and could serve as the basis of future reflection: (1) Truthfulness and accuracy; (2) 'Re'-terminology: process or product; (3) Communication and the public; and (4) Documenting RRR practices and the challenge of ephemerality.

## Truthfulness and accuracy

One of the starting points of the RRR practitioners in this book is their conviction that they cannot fully recreate the past, as our situatedness in place and time as well as our mindsets as researchers make it impossible to relive the past. In some fields, such as conservation for example, characteristic qualities of the object of study are assumed to have become lost with time and distance, as a result of ageing and degradation, or because of limited documentation. Words and images are considered insufficient to describe ephemeral aspects, variously referred to as 'tacit', 'implicit', 'gestural' or 'embodied' knowledge. Moreover, the traces we ourselves leave, our own 'footsteps' as researchers, increase this distance. The anthropologist Kalshoven puts this nicely when, in reference to the taxidermists she studies, she maintains that 'perfection is never really achieved: for the expert taxidermist, every mounted specimen is a rehearsal for the next one, making the practitioner move from one prototype to the next, which he or she hopes to be an improvement on the previous one'. The open-endedness of the process seems to capture a fundamental characteristic of RRR practices.

Carlyle uses the metaphor of the 'bull's eye', the centre of a target such as used in darts or archery, to clarify her approach to the 'accuracy' of reconstructions of historical paint recipes. Comparisons with actual historical paintings find their place in the centre of the 'bull's eye', and serve as a main point of reference for reconstructions that aim to be as historically accurate as possible, while acknowledging that accuracy can at best be approached. Archaeologists Patricia Lulof and Jill Hilditch add another aspect to this issue, drawing attention to the dimension of time. While a painting was (often) created in a relatively short time, archaeological sites may have grown and changed over centuries. Hence, a clear 'target' in time may not exist.

Interestingly, while some fields struggle to define and bridge the distance to the past, Kalshoven emphasises its positive effects: while 'the act or thing being replicated (or reconstructed or re-enacted) will never be quite the same as the emulated act or thing on which it was supposedly modelled', this situation creates 'opportunities to undermine or critique underlying models and ideologies'. Others, like the historian of science Sibum, emphasise that 'it is important to understand that reworking past experiments by means of performance is not an attempt to find out how it really was'. It is evident that instead of an inherent weakness, the impossibility to replicate the past is seen as constructive. RRR offers researchers the opportunity to create the conditions that allow them to raise interesting questions about past processes or objects of study. In musicology, sound is an extremely elusive target. In their chapter on historical organs, Peter Peters and Julia Kursell therefore focus on learning the processes involved, instead of imitating 'site-specific' solutions, while Hans Fidom emphasises the importance of *process* reconstruction to signal his reservations regarding the possibility of recovering or replicating a past sound concept. The aim of process reconstruction, according to Fidom, 'was not to build yet another "Bach-organ" but to conceive a sound world that would fit historically informed performances of Bach's music as well as inspire making new music'.

The choice for a specific Re-term is directly connected to reflections on temporalities of RRR research and on its historicity, recognising the impossibility of walking in the streets of the past. Leslie Carlyle opposes historically accurate reconstruction to an older pedagogical practice in conservation to which she refers as 'reproduction', that is, the pedagogical practice of copying (or replicating, one might say) the image of a painting to understand how it was created. Instead, 'reconstruction' uses historically appropriate materials to 'produce historical models at the material level, not only in terms of surface appearances'. Nevertheless, historians of science continue the use the term 'reproduction' to draw 'attention to the final products and the means of their production, working to explain the underlying processes or reactions involved as well as the material circumstances that help shape these products'. While Fidom introduces the term 'process reconstruction', the anthropologist Jo Vergunst questions the use of the 're'-prefix in 're-enactment' alltogether to complicate the 'simple divide between originality and copying', opposing acting and enacting to 'mechanical replication'.

### 'Re'-terminology: process or product

The term 'process' is indeed central to RRR practices. Researchers employ RRR to learn about skills, workshop choreographies, and thus the workings of past processes. Similar to the *chaîne opératoire* as it is called in archaeology (see Hilditch in this book), Peters and Kursell discuss the importance of process reconstruction in the North German Organ Research Project: 'recovering the sound of historical organs required relearning artisanal building techniques that were displaced by modern building techniques over a century ago'.

As evident from Sibum's account of a re-enactment of the paddle-wheel experiment of James Joule, re-working can give insight into the operative knowledge and skill of this scientist, the complexity of the experiment, the likely set-up in the laboratory, and the choreography of actions shows that the experiment required an assistant; all aspects that did not make it into Joule's scientific publication. Sibum emphasises that 'only the performance of experiment can provide access to a working knowledge that existed in the scientific laboratories, artisanal workshops and studios of the past'.

In other disciplines, the balance might shift more towards product than process. Paintings conservator Stols-Witlox relates this shift to her profession, writing that 'the profound way in which our aim to be invisible in conservation interventions is integral to the profession, may in fact be why in conservation terminology to describe methods that replicate artistic processes equally focuses on the object'. This brings us to the importance of 'Re'-terminology. Roughly speaking, a distinction is evident between two groups of terms, those that relate to process and those that are result-driven (e.g. replication of an object or an architectural structure such as a house). This distinction in focus (process or product) relates to the aim of the research, but is also to a certain extent discipline-guided.

However, several authors also adopt other terms in addition to the more standard terminology used in their disciplines. For example, RRR methods in conservation typically have more object-focused aims. Carlyle uses 'reconstruction' in accordance with the tradition in conservation and restoration, but she also refers to re-enactment when she discusses the less commonly studied ways that oil paint behaves under the brush or palette knife during application. The archaeologist Lulof also conceives of digital reconstructions as incorporating 're-enacting' building processes. Fidom, Kursell and Peters adopt the term 'process reconstruction' to differentiate their approach from the more standard re-enactments in music. Process reconstruction is 'more than simply re-enacting historical craftmanship,' as they write; more also than the reconstruction of a material object – i.e. the musical instrument or the organ – as sound is also involved.

However, even when the focus is on scrutinising process or *chaîne opératoire*, performative methods can still entail the making of a (virtual) reconstruction of an object (on the basis of excavated archaeological remains as Lulof describes) or a replica of a scientific instrument by historians of physics Sibum and Peter Heering in their respective chapters in this book. While Sibum creates *replicas* of historical scientific instruments in order to *perform* past experiments, he insists on the term 're-working' and categorically objects to the use of 'replication' to refer to his method to avoid confusion with replication in science (as discussed above). He also prefers 'reworking' above 're-enactment', because he wants 'to indicate that every performance of an experiment is a unique event and certainly different from the original one. Hence, even if 'reworking' is striving for re-enactment of a past experimental performance the term 'reworking' ensures not to think that repetition, replication or re-enactment is possible. At the most, by means of performance of experiment we are able to reconstruct past events.'

Despite reservations towards 'replication', Kalshoven, among others, uses the term 'replication' which she conceives as 'the skilled act of copying a thing or a practice'. Rather than a reproduction of the past, she believes that 'replication' can be given a playful and future-oriented twist. 'After all,' Kalshoven reminds us, 'to replicate, from the Latin "again" and "to fold", is the practice of folding anew, once more, yet again.'

## Communication and the public

RRR methods have not only offered insights into research questions across the disciplines represented in this book, but they also offer a way in which to engage with broad audiences, whether in classrooms, museums or other public settings. Thijs Hagendijk et al.'s chapter offers one of the most explicit examples of the opportunities for RRR in the classroom. Their plea is for more hands-on history, supported by the rich insights from students into crucial yet often ignored scientific practices such as failure, improvisation, trial and error. When students made things with their hands – a single bead-lens microscope for example – they were able to be playful with science in a way that other kinds of teaching do not allow, and in a less hierarchical setting than is commonplace in universities. As Hagendijk, Heering, Principe and Dupré write, 'the power of RRR approaches in the classroom is to create collaborative learning environments in which students learn together and alongside their teachers'. Through RRR, students and teachers make historical knowledge together.

RRR in the museum is discussed by Stols-Witlox, who outlines the possibilities of digital reconstructions of art objects in the museum context, as does Lulof, through the example of the disputed reconstruction of a building which is on display at a local museum close to the archaeological site. Kalshoven studies taxidermists who made objects both for display in museums as well as for teaching.

Anthropologist Vergunst discusses many different audiences for the walks in his chapter: the RRR community, postgraduate students and the general public in Aberdeen. The material accumulated in RRR research can also lead onto public projects through the building of databases – databases of recipes for further research by others, in Carlyle's case, or databases for building reconstruction in Lulof's.

## Documenting RRR practices and the challenge of ephemerality

Sibum describes one of the strengths of RRR, writing that with a 'dynamic conception of embodied knowledge that is bound to the performative actions of the researcher or collective, we are able to grasp those cultural repertoires of action which are essential for the formation of this experimental knowledge but usually escape the historian's attention because these belong to different worlds of sense which are often described as non-verbal or oral knowledge traditions'. But how can we document these observations? As RRR scholars, we find ourselves facing the very same challenge of documenting embodied and material work that results from our practices, as faced by the scientists, artisans, musicians or other makers we study. How can we find ways to trace bodily engagements, material forms and hands-on learning?

In this respect, ethnographic note-taking practices are of particular interest. In his chapter, Vergunst discusses the imaginative possibilities of mapping, and the methods of psychogeography, next to fieldnotes in the traditional sense. He points to the maps that anthropologists make, but hardly publish in their fieldnotes. For Vergunst, these methods of documentation highlight how ethnography is a practice, not a methodology; it is a form of inquiry and observation and a way of going about the world that we are part of. Kalshoven offers a form of documentation that aligns with Vergunst's approach. In her ethnographic study of taxidermy, Kalshoven commissioned an artist to work with her on shared interests (see front cover). The final result was not only an intriguing way in which to share their research with others, but also one that helped Kalshoven explore the theoretical possibilities of her work.

Lulof addresses the need in RRR research to not only think critically about why we document, but also about the technological forms in which we do this, when there is so much on offer in the digital realm for cataloguing, creating databases and amalgamating materials. How can we leave room for doubt, annotation and change over time? Within the digital world of 3D modelling, scientific value has been doubted and fear of deception remains persistent, as structural uncertainty is difficult to visualise in digital models.[50] In practice, models generally give one version of the past instead of several interpretations; in this perception, digital models offer nothing more than the artist's watercolour-impressions of the past from the nineteenth century. The fear to deceive may have caused the focus on developing a common ground for 'good practices' within the field.[51] The focus on ways to visualise uncertainty characterises this trend.

In particular, the recent development of powerful visualisation platforms such as the three-dimensional Geographic Information System (3DGIS), and the introduction of digital acquisition tools provide opportunities to fully visualise and study in three dimensions (3D) the spatial and temporal relationships between the fragmented pieces of information detected in archaeological excavations. Very recently, these techniques have been connected to Unmanned Aircraft Systems (UAS, drones) and Lidar (Light detection and ranging) data, with the use of photogrammetry, to reconstruct archaeological sites, ancient built structures and city complexes in 3D and 4D; the fourth dimension being generally accepted as the 'element of transformation in time'.[52]

Together, the chapters in this book push for a more reflective approach to and use of RRR methodologies. Interdisciplinary reflection, as we engage in in this book, does not aim at developing a universal RRR method, one design that fits all disciplines. Instead, the gains of the interdisciplinary cross-fertilisation practiced in this book are first and foremost disciplinary, that is, specific and tailored for each discipline. By comparing methods across disciplines, the disciplines find inspiration and critical distance to reflect upon, and then further develop their own RRR practices. Through this interdisciplinary comparison of RRR methods, practitioners have access to pools of knowledge and experience in other disciplines that may help them deal with issues they likewise encounter; comparisons with RRR practices in other fields also offer the opportunity of making RRR

50 Reilly, 'Towards a Virtual Archaeology', discusses the 'pretty picture' trap as well as suggests how to visualise uncertainty; see Barceló et al., *Virtual Reality*, for more on this subject.

51 Frischer, *Beyond Illustration*. This period of doubt lasted from 1997 to 2008. Thinking about best practices, however, took away most concerns.

52 See Dell'Unto, 'The Use of 3D Models', pp. 51-58, Dell'Unto, 'Using 3D GIS', pp. 305-323 and Waagen, 'New Technology', pp. 11-20, for the most recent discussions on the use of GIS, 3D modelling and photogrammetry in studying excavations and ancient architecture.

practitioners aware of sometimes silent assumptions about RRR design in their respective disciplines so as to make room for more explicit reflection.

Which new insights has interdisciplinary cross-fertilisation generated, and how is this reflected in the chapters of this book? We have already mentioned four themes, but here finally, we want to highlight two issues as a starting point for future RRR research. First, this book has brought together process and product-oriented RRR practitioners. As we have seen, disciplines have acquired preferences for these two orientations, which is reflected in the typical 're'-terms which they have adopted. Interdisciplinary cross-fertilisation in this book has led to the insight that acting subjects and produced objects are always present in RRR research across disciplines and regardless of the process or product orientation, yet in variant constellations. The anthropologist Kalshoven, struck by the intense focus on materials in other disciplines, was attentive to materials in her work on replication in taxidermic practices, and has adopted the term 'reworking', introduced by historian of science Sibum, where anthropology more typically opts for 're-enactment'. Conversely, the conservator Carlyle, a strong advocate of intellectual rigour regarding material choices and whose home discipline typically uses the term 'reconstruction', adopts in addition the term 're-enactment' in her chapter to channel attention to skill and performance. Hopefully, this book contributes to making explicit the variant constellations of human agents and produced objects providing fruitful avenues for RRR research design.

Secondly, interdisciplinary cross-fertilisation also highlights creativity as a central aspect of RRR research. Research, perhaps foremost in the disciplines of anthropology and music, which are less historically inclined, complicates the temporalities of RRR and the distinction between original and replica or copy. While the aspirational qualities of historical accuracy are already widely recognised, it seems appropriate for this book to take the next step and to move away from the assumption that re-doing implies a lack of creativity. Indeed, only through interdisciplinary cross-fertilisation have the historical disciplines in this book become less blind to creativity. This recognition will allow RRR researchers to compare past and current creative practices rather than opposing the research agendas of historians and less historically inclined disciplines. There are many more gaps and threads this book leaves dangling and which will need further exploration in the future. We hope that this book inspires methodological innovations in creating performative approaches that question how to explore productively the tensions between past and present, and consider how the materiality and sociality of repetition in its many variations informs particular ways of making and knowing.

## Bibliography


Agnew, Vanessa, 'What is Re-enactment?', *Criticism*, 46:3 (2004), pp. 327-339.

Applegate, Celia, *Bach in Berlin. Nation and Culture in Mendelssohn's Revival of the St. Matthew Passion* (Ithaca and London: Cornell UP, 2005).

Auhagen, Wolfgang, Wolfgang Hirschmann and Tomi Mäkelä (eds.), *Musikwissenschaft 1900-1930. Zur Institutionalisierung und Legitimierung einer jungen akademischen Disziplin* (Hildesheim: Olms, 2014).

Barceló, Juan A., and others (eds.), *Virtual Reality in Archaeology* (Oxford: Archaeopress, 2000).

Brajer, Isabelle, *Conservation in the Nineteenth Century* (London: Archetype, 2013).

Brinkman, Pim, *Het geheim van Van Eyck. Aantekeningen bij de uitvinding van het olieverven* (Zwolle: Waanders, 1993).

Campbell, Lorne, 'The Conservation of Netherlandish Paintings in the Fifteenth and Sixteenth Centuries', *Studies in the History of Painting Restoration*, ed. by Christine Sitwell and Sarah Staniforth (London: Archetype, 1998), pp. 15-26.

Coles, John M., *Experimental Archaeology* (London: Academic Press, 1979).

Conti, Alessandro, *History of the Restoration and Conservation of Works of Art* (Amsterdam: Butterworth-Heinemann, 2007).

Davis, Whitney, 'Art History, Re-Enactment, and the Idiographic Stance', in *Michael Baxandall, Vision and the Work of Words*, ed. by Peter Mack and Robert Williams (Farnham/Burlington: Ashgate, 2015), pp. 69-89.

Dell'Unto, Nicolò, 'The Use of 3D Models for Intra-site Investigation in Archaeology', in *3D Recording and Modelling in Archaeology and Cultural Heritage*, ed. by Fabio Remondino and Stefano Campana (Oxford: Archaeopress, 2014), pp. 151-158.

Dell'Unto, Nicolò, 'Using 3D GIS Platforms to Analyse and Interpret the Past', in *Digital Methods and Remote Sensing in Archaeology. Archaeology in the Age of Sensing*, ed. by Maurizio Forte and Stefano Campana (New York: Springer Publishing, 2016), pp. 305-323.

Eastlake, Charles Locke, *Materials for a History of Oil Painting* (London: Longman, Brown, Green and Longmans, 1847), 2 Vols. [Reprint: *Methods and Materials of Painting of the Great Schools and Masters*, 2 vols. (New York: Dover, 1960)].

Eigeldinger, Jean-Jacques, *Wanda Landowska et la renaissance de la musique ancienne* (Arles: Actes Sud, 2011).

Elste, Martin, *Meilensteine der Bach Interpretation 1750-2000. Eine Werkgeschichte im Wandel* (Stuttgart: Bärenreiter, 2000).

Eybl, Martin, 'Guido Adler, die Denkmäler der Tonkunst in Österreich und die Anfänge der Musikwissenschaft an der Universitat Wien: Konzeptze und Perspectiven', in *Musikwissenschaft 1900-1930: zur Institutionalisierung und Legitimierung einer jungen akademischen Disziplin*, ed. by Wolfgang Auhagen,

Wolfgang Hirschmann and Tomi Mäkelä (Hildesheim: Georg Olms Verlag, 2017), pp.250-261.

Fors, Hjalmar, Lawrence M. Principe, and H. Otto Sibum, 'From the Library to the Laboratory and Back Again: Experiment as a Tool for Historians of Science', *Ambix*, 63 (2016), pp. 85-97.

Frischer, Bernard, and Anastasia Dakouri-Hild, *Beyond Illustration, 2D and 4D Digital Technologies as Tools for Discovery in Archaeology* (Oxford: BAR International Series 1805, 2008).

Hermon, Sorin, 'Scientific Method, *chaîne opératoire* and Visualization: 3D Modelling as a Research Tool in Archaeology', in *Paradata and Transparency in Virtual Heritage*, ed. by Anna Bentkowska-Kafel and Hugh Denard (Farnham: Ashgate, 2012), pp. 13-22.

Holten, Lars, 'Engaging Experiments: From Silent Cultural Heritage to Active Social Memory', in *Experiments Past: Histories of Experimental Archaeology*, ed. by Jodi Reeves Flores and Roeland Paardekooper (Leiden: Sidestone Press, 2014), pp. 269-283.

Johnson, Katherine, 'Rethinking (Re)doing: Historical Re-enactment and/as Historiography', *Rethinking History*, 19 (2015), pp. 193-206.

Kalshoven, Petra Tjitske, *Crafting the Indian: Knowledge, Desire, and Play in Indianist Reenactment* (New York: Berghahn Books, 2012).

Lanjouw, Tijm, 'Discussing the Obvious or Defending the Contested: Why Are we Still Discussing the 'Scientific Value' of 3D Applications in Archaeology?', in *The Three Dimensions of Archaeology, Proceedings of the XVII UISPP World Congress, 1–7 September 2014, Burgos, Spain*, ed. by Hans Kamermans et al. (Oxford: Archaeopress, 2015), pp. 1-13.

Lulof, Patricia, and others, 'The Art of Reconstruction. Documenting the Process of 4D Modeling, Some Preliminary Results', in *Proceedings of the 2013 Digital Heritage International Congress (DigitalHeritage)*, ed. by Alonzo C. Addison et al. (Marseille: IEEE Xplore, 2013), Vol. 2, pp. 333-337.

MacLachlan, James, 'Galileo's Experiments with Pendulums: Real and Imaginary', *Annals of Science*, 33 (1976), pp. 173-185.

Mann, Anna M., and others, 'Mixing Methods, Tasting Fingers: Notes on an Ethnographic Experiment', *HAU: Journal of Ethnographic Theory*, 1 (2011), pp. 221-243.

Merrifield, Mary P., *Medieval and Renaissance Treatises on the Arts of Painting. Original Texts with English Translations*, 2 Vols. (London: Murray, 1849) [Reprint 1967, 2 volumes reprinted as 1, Mineola, New York: Dover, 1999].

Moser, Stephanie, 'Archaeological Visualisation: Early Artefact Illustration and the Birth of the Archaeological Image', in *Archaeological Theory Today* (revised edition), ed. by Ian Hodder (Cambridge: Polity, 2012), pp. 292–322.

Mugglestone, Erica, 'Guido Adler's "The Scope, Method, and Aim of Musicology" (1885): An English Translation with an Historico-Analytical Commentary', *Yearbook for Traditional Music*, 13 (1981), pp. 1-21.

Nadolny, Jilleen, 'A History of Early Scientific Examination and Analysis of Painting Materials, ca. 1780 to the Mid-Twentieth Century', in *The Conservation of Easel Paintings*, ed. by Joyce Hill Stoner and Rebecca Rushfield (London: Routledge, 2012), pp. 336-340.

Nadolny, Jilleen, Mark Clarke, Erma Hermens, Ann Massing and Leslie Carlyle, 'Art Technological Source Research: Documentary Sources on European Painting to the Twentieth Century, with Appendices I-VII', in *The Conservation of Easel Paintings*, ed. by Joyce Hill Stoner and Rebecca Rushfield (London: Routledge, 2012), pp. 3-32.

Opgenhaffen, Loes and Maarten H. Sepers '3D Modelling: Crossing Traditional Boundaries Between Different Research Areas', in *The International Archives of the Photogrammetry, Remote Sensing and Spatial Information Sciences,* Vol. XL-5/W4 (2015), pp. 411-414.

Outram, Allen K., 'Introduction to Experimental Archaeology', *World Archaeology*, 40 (2008), pp. 1-7.

Piccoli, Chiara, *Visualizing Citiscapes of Classical Antiquity: From Early Modern Reconstruction Drawings to Digital 3D Models* (Oxford: Archaeopress, 2018), pp. 6-49.

Pink, Sarah, and Kerstin Leder Mackley, 'Re-enactment Methodologies for Everyday Life Research: Art Therapy Insights for Video Ethnography', *Visual Studies*, 29 (2014), pp. 146-154:

Posluschny, Alex, and others (eds), *Layers of Perception*, Proceedings of the 35th International Conference on Computer Applications and Quantitative Methods in Archaeology, CAA (Bonn: Rudolf Habelt, 2008).

Principe, Lawrence M., 'Apparatus and Reproducibility in Alchemy', in *Instruments and Experimentation in the History of* Chemistry, ed. by L. Holmes and T. H. Levere (Cambridge, MA: The MIT Press, 2000), pp. 55-74.

Ratto, Matt, 'CSE as Epistemic Technologies: Computer Modelling and Disciplinary Difference in the Humanities', in *Handbook of Research on Computational Science and Engineering: Theory and Practice*, ed. by Joanna Leng and Wes Sharrock (Hershey: IGI Global, 2012), pp. 567-587.

*RE-. An Errant Glossary*, ed. by Christoph F.E. Holzhey and Arnd Wedemeyer (Berlin: ICI Berlin Press, 2019).

Reilly, Paul, 'Towards a Virtual Archaeology', in *Computer Applications and Quantitative Methods in Archaeology* (CAA 90), ed. by Sebastian Rahtz and Kris Lockyear (Oxford: Tempus Reparatum, 1991), pp. 132-139.

*Routledge Handbook of Interdisciplinary Research Methods*, ed. by Celia Lury et al. (London: Routledge, 2018).

Ruth-Sommer, Hermann, *Alte Musikinstrumente. Ein Leitfaden für Sammler* (Berlin: Richard Carl Schmidt & Co, 1916).

Ryan, Nick, 'Documenting and Validating Virtual Archaeology', *Giornale di Archeologia e Calcolatori*, 12 (2001), pp. 254-273.

Schäuble, Michaela., 'Living History? Reenacting the Past and Promoting "Tradition" in the Dalmatian Hinterland', *Nationalities Papers*, 47 (2017), pp. 198-216.

Settle, Thomas, 'An Experiment in the History of Science', *Science*, 133 (1961), pp. 19-23.

Sibum, H. Otto, 'Experimental History of Science', in *Museums of Modern Science*, ed. by Svante Lindqvist (Canton, MA: Science History Publications, 2000), pp. 77-86.

Sibum, H. Otto, 'Reworking the Mechanical Value of Heat: Instruments of Precision and Gestures of Accuracy in Early Victorian England', *Studies in the History and Philosophy of Science*, 26 (1995), pp. 73-106.

Smith, Pamela H., and The Making and Knowing Project, 'Historians in the Laboratory: Reconstruction of Renaissance Art and Technology in the Making and Knowing Project', *Art History*, 39 (2016), pp. 210-233.

Staubermann, Klaus, *Reconstructions Recreating Science and Technology of the Past* (Edinburgh: NMSE – Publishing Ltd, 2011)

Stiebing, William H., *Uncovering the Past: A History of Archaeology* (Oxford: Oxford University Press, 1993).

Suchman, Lucy, 'Configuring the Other: Sensing War through Immersive Simulation', *Catalyst: Feminism, Theory, Technoscience*, 2 (2016), pp. 1-36.

Taruskin, Richard, *Text and Act. Essays on Music and Performance* (Oxford: Oxford University Press, 1995).

Thomas, Anabel, 'Restoration or Renovation: Remuneration and Expectation in Renaissance 'acconciatura'', in *Studies in the History of Painting Restoration*, ed. by Christine Sitwell and Sarah Staniforth (London: Archetype, 1998), pp. 1-14.

Thompson, Daniel V., *Materials and Techniques of Medieval Painting* (London, 1938) [unabridged and unaltered republication London: Dover, 1956].

Waagen, Jitte, 'New Technology and Archaeological Practice. Improving the Primary Archaeological Recording Process in Excavation by Means of UAS Photogrammetry', *Journal of Archaeological Science*, 10 (2018), pp. 11-20.

Wenzel Geissler, Paul, and Ann Kelly, 'Kidewu's Return: Reenacting Mid 20th-century Science', in *Traces of the Future: An Archaeology of Medical Science in Africa*, ed. by Paul Wenzel Geissler, Guillaume Lachenal, John Manton and Noemi Tousignant (Bristol, UK: Intellect, The University of Chicago Press, 2016), pp. 150-153.

Wrapson, Lucy, Jenny Rose, Rose Miller and Spike Bucklow, *In Artists' Footsteps: The Reconstruction of Pigments and Paintings. Studies in Honour of Renate Woudhuysen-Keller* (London: Archetype, 2012).

## About the Authors

**Sven Dupré** is Professor of History of Art, Science and Technology at Utrecht University and the University of Amsterdam. He directs the project 'Technique in the Arts: Concepts, Practices, Expertise, 1500-1950' (ARTECHNE), supported by a European Research Council (ERC) Consolidator Grant. The project undertakes the experimental reconstruction of historical recipes to open the black box of the transmission of technique in the visual and decorative arts.

**Anna Harris** is Associate Professor of the Social Studies of Medicine at Maastricht University. She studies medical practices, working at the intersections of science and technology studies and anthropology. She currently leads the ERC project 'Making Clinical Sense: A comparative anthropology of medical techno-perception'. This sensory study of craft, skill and learning brings together historians and anthropologists.

**Julia Kursell** is Professor of Musicology at the University of Amsterdam. Before coming there, she was a research fellow that the Max Planck Institute for the History of Science and at Bauhaus University Weimar. She is one of the founding editors of the journal *History of Humanities* and one of the directors of the Vossius Center for the History of Humanities and Sciences at the University of Amsterdam. She is the author of two books: *Schallkunst. Eine Literaturgeschichte der Musik in der frühen russischen Avantgarde* (Sagner 2003) and *Epistemologie des Hörens: Helmholtz' physiologische Grundlegung der Musiktheorie* (Brill 2018).

**Patricia Lulof** is Associate Professor of Mediterranean Archaeology at the University of Amsterdam. She specializes in Pre-Roman Archaeology, Ancient Architecture, and (digital) building techniques. She is scientific director of the Interdisciplinary 4D Research Lab of the Faculty of Humanities at the University of Amsterdam, founded in 2012; and was PI of 'Archaeology of Architecture' (2013-2014, Faculty Research Priority Area Creative Industry-Digital Humanities), 'Biographies of Buildings. Virtual Futures for our Cultural Past' (2014-2015, Netherlands Institute for Advanced Study), the KIEM/NWO project 'Living in the Past. Reconstructing Etruscan Houses in Acquarossa' (2017-2018), and 'Van Klooster tot Kwartier' (2019-2020, Amsterdam University Fund).

**Maartje Stols-Witlox** is Associate Professor Conservation and Restoration at the University of Amsterdam. She is an art historian and paintings conservator; she specializes in the investigation of historical artists' recipes, with particular emphasis on reconstructions with 'historically informed' materials. She is project leader of the NWO Free Competition Humanities project 'Down to the Ground, a historical, visual and scientific analysis of coloured grounds in Netherlandish paintings, 1550-1650, core-team member of the 'Irradiation Passport for Art' (seed money, NICAS/NWO) and is affiliated to the ARTECHNE-project (Utrecht University).

# 1. Replication as a Play on Categories: The Case of Taxidermy

*Petra Tjitske Kalshoven*

**Abstract**
The epistemic and moral frameworks that underlie practices of replication may differ significantly between and even within communities of practice. This makes replication intellectually exciting as a lens that brings human category making into sharp focus. Drawing on ethnographic examples from fieldwork with taxidermists and artist-taxidermists and paying close attention to material and ideological turns in their endeavours, I argue that the case of taxidermy reveals replication to be a constant play with categories in a bid to create prototypes, and critiques of prototypes, that express and shape human perspectives on nature. Replication models a past for the future: It looks ahead, rather than back, reworking as well as reflecting the world.

**Keywords:** Category making, prototyping, modelling, naturalization, saliency, taxidermy

'Les prodiges de Monsieur Renkin nous enchantent car ils se jouent des distinctions et des catégories aristotéliciennes.'[1]

'A key process in the continual becoming of natural objects [...] is that like other museum things, they are *naturalized*. Naturalization is what happens when categories or objects cease to be strange, and as such is an integral part of classification.'[2]

1 Della Noce-Bertozzi, 'L'homme de chambre', p. 80.
2 Alberti, 'Constructing Nature', p. 83.

Dupré, S., A. Harris, J. Kursell, P. Lulof, M. Stols-Witlox (eds.), *Reconstruction, Replication and Re-enactment in the Humanities and Social Sciences*. Taylor & Francis Group
DOI 10.5117/9789463728003_CH01

## Introduction

Humans are classifiers. Any analytical enquiry, any situational assessment, any sensory impression hinges on the subject's ability to make distinctions and thus on category making. Because humans are classifiers par excellence, paying attention to practices of classification is a prime anthropological concern. Making distinctions through classification is not just a straightforward process of making sense of otherwise overwhelming sensory impressions by slotting them into familiar frameworks ('naturalizing'), it is also a strategic and rhetorical process of seeking to manipulate such impressions in specific ways, or of withholding judgment, for example when one indulges in an illusion. Playing with categories and taxonomies, and taking pleasure in this play, is what makes us distinctly human.

In this chapter, I consider a practice that is predicated on this human propensity for category making and breaking and that exercises this propensity through the use of replication. Replication, which I conceive of as the skilled act of copying a thing or a practice, transfers the copied thing or practice into a new category (the replica) whilst striving to make it mirror, make it appear to belong to, or critique, its original category. This play with categories, steeped in moral implications, makes replication a delightfully human preoccupation.[3] The particular practice of replication I discuss here is the profession and craft of taxidermy, which is particularly intriguing because the shift from one category (live animal) into the other (stuffed skin; animal-object; scientific specimen; artwork) involves, actually requires, death. This shift is precisely what taxidermy generally seeks to gloss over: In an, at first sight, ironic and yet profound way, taxidermy is about replication of 'life'.

Taxidermy literally means skin arrangement (from the Greek 'taxis', order/arrangement, and 'derma', skin). It involves the taking apart and subsequent rearranging of parts of dead animals in combination with man-made materials, usually, but not always, in such a way that they look lifelike again. Taxidermy blurs not only conceptual categories of life and death but also object categories, such as 'natural entity' versus 'piece of craftwork' or 'museum artefact'. Category transitions in taxidermy are not straightforwardly linear; taxidermied specimens transform in material, symbolic, and functional ways that articulate differently. What is

3 For an ethnographic analysis of 'Indianism', a practice of replication and re-enactment predicated on a deeply rooted European fascination with Native American cultures, see Kalshoven, *Crafting 'the Indian'* and Kalshoven, 'Moving in Time'.

more, besides shifting categories themselves, taxidermied specimens also contribute to the establishment of categories in specific historical and cultural contexts: taxidermy, through its products of replication, has played an important role in processes of scientific classification as it has helped to shape the taxonomies that came into being since the Enlightenment with the discovery and preservation of unfamiliar species outside of Europe.[4] And taxidermy still plays this role in museum contexts, in particular through the creation of so-called study skins.

Taxidermy, both as a practice and as a material phenomenon, has recently made a comeback, and has (re)gained prominence in museum and art contexts.[5] From 2011 onwards, intrigued by this development, I have carried out an ethnographic study among professional taxidermists and artist-taxidermists (that is, contemporary artists who use taxidermy in their art practice) in the UK, the Netherlands, and Belgium. An important anchoring point for the study has been my membership of the UK Guild of Taxidermists, where professionals and artists come together to share expertise and enhance their craft practice, whilst often differing in their perspective on the role of taxidermy, as I explain below.[6]

An anthropological analysis of taxidermy is of interest to studies of re-enactment, replication and reconstruction because it brings out the saliency of a practice that, in replicating 'life', both sheds light on, and troubles, human category making. Taxidermy naturalizes and denaturalizes; it familiarizes and defamiliarizes. Taxidermy intrigues because its appearance triggers a recognition of the real and holds a promise of something preserved *as is*. At the same time, this surface appearance is at odds with what hides inside; in the process of creating a taxidermied animal-object, called a 'mount' by practitioners, the skinned body is replaced by a model, a morphologically correct form or 'manikin' fabricated from wood wool, polyurethane, balsa, or foam. 'Is it real?' museum visitors will ask. 'What's inside?' (interview with museum taxidermist Rebecca Desjardins, Naturalis, Leiden, 31 August 2015). Troubling the boundaries between organic and inorganic, animate and inanimate, life and non-life, real and fake, taxidermied animal-objects

4 Star, 'Craft vs. Commodity'; Wonders, *Habitat Dioramas*; Poliquin, *The Breathless Zoo*; for a history of taxonomy, see Asma, *Stuffed Animals*.

5 See, e.g., Alberti, *The Afterlives of Animals*; Aloi, *Speculative Taxidermy*; Dalla Bernardina, 'Hymnes à la vie?'; Desmond, 'Displaying Death, Animating Life'; Eastoe, *The Art of Taxidermy*; Fuller, *Voodoo Salon*; Kalshoven, 'Gestures of Taxidermy'; Lange-Berndt, *Animal Art*; Michaud, 'De l'animal au trophée'; Milgrom, *Still Life*; Poliquin, *Ravishing Beasts* and Strivay, 'Taxidermies'.

6 See Kalshoven, 'Gestures of Taxidermy', for a more elaborate discussion of dynamics at the UK Guild.

have a physical indexicality to life that is perceived as being only skin-deep. Drawing on cognitive anthropologists Roy D'Andrade and Pascal Boyer's ideas on the prototypical and the counter-intuitive respectively, and on my ethnographic work, I argue that taxidermy, as a result of its particular materiality anchored in skills of replication that enable the transition from animal to animal-object, achieves saliency by triggering a variety of operative concepts in human categorizing that oscillate between the prototypical and the counter-intuitive in gauging what it means to be 'true to nature'.[7] Producing taxidermy (in French, *naturaliser*) both constitutes and disturbs acts of naturalization.

Taxidermists trouble our expectations of what it means to be 'true to nature', and in doing so they highlight the interest and potential of replication both as a practice and as an analytical concept. What is more, replication in taxidermy constitutes 'nature' in a manner that is future-orientated rather than looking back, thus contributing to studies of re-enactment, replication and reconstruction a novel perspective on temporalities. Being 'true to nature' is a key aspect in an influential piece written on taxidermy by anthropologist Donna Haraway in 1989, to which I turn first.

## 'Teddy Bear Patriarchy': Taxidermy as a gendered practice of naturalization

In her analysis of renowned American taxidermist Carl Akeley's practice in the 1920s, the heyday of scientific taxidermy, Haraway emphasized the gendered ideology behind Akeley's displays as well as his construction of a specific, idealized representation of nature. In a team effort with dozens of skilled collaborators, Akeley created what became known as the Akeley African Hall in the American Museum of Natural History in New York City, a museum dedicated, as Haraway argues, to manhood personified by

7 Similar questions have been explored in the history of science, e.g., by Lorraine Daston and Peter Galison in a study of imagery through time, in which they grapple with the concept of 'objectivity' (Daston and Galison, 'The Image of Objectivity'). The authors discuss drawings of human and non-human animal bodies in scientific atlases dating from the eighteenth until the early twentieth century. Atlas makers sought to render nature accurately, guided by terms such as 'typical', 'ideal', 'characteristic', or 'average'. Over time, however, they differed as to which specimen (or series of specimens) would be most representative of the species actually found in nature. Daston and Galison find that objectivity, glossed as 'being true to nature', does not have the same meanings or implications throughout the period they study. In other words, objectivity is not a stable concept over time.

president, hunter, and conservationist Theodore Roosevelt. Specimens in the African Hall were (and are) arranged in dioramas as 'actors in a morality play' performing in habitat groups.[8] Akeley sought to collect the 'typical specimen, group, or scene', where 'typical' implied perfection.[9] In Haraway's compelling analysis, Akeley's gorilla group becomes the focal point. It was the imposing silverback that exemplified this kind of perfection, the virile adult male that had been killed in a heroic stand-off with man, to be resurrected as a nurturing emblem of nature for urban museum visitors (Figure 1.1).

In the hunting encounter, '[p]erfection was known by natural kinship; type, kind, and kin mutually and seminally defined each other'.[10] Analogies, Haraway suggests, could be drawn between the perfect gorilla and the perfect human family, with the silverback reminding visitors of what it meant to be a natural man. The perfect specimen in the encounter had to be an adult male of perfect proportions and appropriate virility. This imposing male was then combined with a female and a young specimen to form the perfect, and perfectly representational, family unit. 'Perfection', Haraway writes, 'inhered in the animal itself, but the fullest meanings of perfection inhered in the meeting of animal and man, the moment of perfect vision, of rebirth. Taxidermy was the craft of remembering this perfect experience [...] Memory was an art of reproduction.'[11] The choice of words is interesting here – rather than 'memorizing' or 'celebrating', Haraway writes 'remembering', and one is tempted to read this as 're-membering', putting the limbs of the dead animal back together again – or perhaps also reassembling members of a particular category in the perfectly replicated habitat of the diorama. Even though she discusses the process of *making* in taxidermy only briefly, taxidermy shines through as a practice of reconstruction, of resurrection even, where death is overcome.

Haraway inverses the usual story of taxidermy replicating nature: 'Akeley "typified" nature, *made nature true to type*. Taxidermy was about the single story, about nature's unity, the unblemished type specimen.'[12] In her account, taxidermy mounts become both models of and models for nature. Rather than recreating what was there, replication models the future, and in doing

8 Haraway, 'Teddy Bear Patriarchy', p. 29.
9 Haraway, 'Teddy Bear Patriarchy', p. 40.
10 Haraway, 'Teddy Bear Patriarchy', p. 41.
11 Haraway, 'Teddy Bear Patriarchy', p. 41.
12 Haraway, 'Teddy Bear Patriarchy', p. 31; italics mine. Haraway's use of the term 'type specimen' deviates from what biologists mean by it. Rather than an ideal type, a type specimen is the first individual specimen to which the existence of a species and its scientific name are attached (personal communication Becky Desjardins, Naturalis; Carol Davies, Kendal Museum).

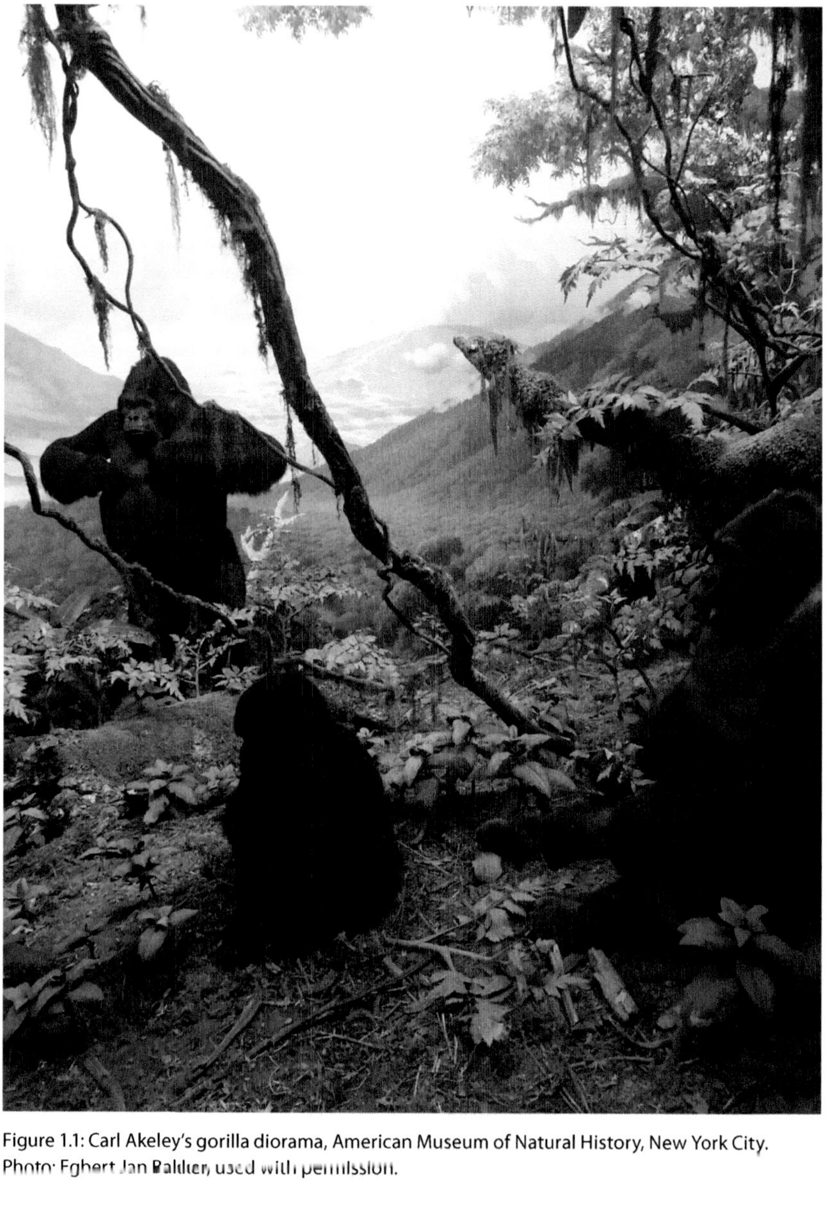

Figure 1.1: Carl Akeley's gorilla diorama, American Museum of Natural History, New York City. Photo: Egbert Jan Bakker, used with permission.

so categorizes this future as natural. In Haraway's analysis, the Natural History Museum was involved in an attempt to cure human decadence through its displays, with 'exhibition, eugenics, and conservation' serving to protect a manhood that was exercised by WASPs in outdoorsy pursuits[13]

13 Haraway, 'Teddy Bear Patriarchy', p. 55-56.

– with a disturbing tale of institutional and casual racism lurking beneath. Haraway's account of early twentieth-century taxidermy, then, is a morality tale of men in the roles of hunters and taxidermists killing, preserving, and displaying, in an ideologically motivated quest for perfection, replicas of nature that came to stand for nature reconfigured as a political space. In Haraway's analysis, taxidermy constructs (rather than imitates) nature; it becomes a model for culture as well as nature, with perfection as the key to success as it underpins the rhetoric of the emblematic encounter, in which the hunter and the hunted animal become commensurate in their perfection. Once this story is accepted, the artefacts on display have become naturalized (see Alberti, opening quotation) as they cease to be strange, with the animal family a familiar presence acting out a fable about humanity and the dominating role of the virile man. Through these naturalized artefacts, culture is naturalized, seems to be Haraway's message.

## Category making: Saliency of the prototypical

If a piece of taxidermy comes to stand for nature, and becomes normative for nature, then taxidermy as a process of replication plays with temporalities; it is not only about recreating past life, it is not only about creating timeless, generalized types, it is also about creating prototypes for the future. I suggest that thinking about replication in terms of prototypes enriches the operational realm of replication from both a temporal and a conceptual angle.

Some twenty years ago, cognitive anthropologist Roy D'Andrade engaged with and defined the noun 'prototype' to think through matters of classification. He discussed psychologist Eleanor Rosch's use of prototypes in tests designed to examine the way in which people classify living beings such as trees or birds. Rosch used 'prototypes' in the sense of 'the clearest cases of membership defined operationally by people's judgments of goodness of membership in the category'.[14] It appeared from Rosch's tests that prototypicality in the case of birds was largely determined by taxonomic relations among birds; a bird was considered prototypical when it was closely related to a large number of other species of birds, sharing many attributes with them, thus constituting a 'best example' in a category.[15]

14 Quoted in D'Andrade, *The Development*, p. 118.
15 D'Andrade, *The Development*, p. 119-120.

Drawing on Rosch, D'Andrade defined a prototype as 'a typical example of a type of object capable of being held in working memory, often with many properties "chunked" together to form a rich, specific image'.[16] The 'chunking together' refers to the tendency of the human cognitive system to group features and attributes together so that they are turned into 'object-like things'. This example, then, has saliency. According to D'Andrade's definition, a sparrow would be a more prototypical bird than a penguin because a sparrow shares more features with other birds than a penguin does. In Haraway's analysis, the dynamics are different as a specific specimen, a gorilla, comes to stand for a species, the gorilla, but it also comes to stand more generally for a force of nature – exemplary nature, leading the way for culture. From this perspective, Akeley's silverback is not a prototype but rather an ideal type. It is better, more imposing than D'Andrade's sparrow; its features are extraordinary rather than shared; they are perfect. Subsequently, the larger category, nature, becomes equated with that ideal type (Haraway's 'nature made true to type'), and as a result, this ideal type is assumed to be the prototype. The silverback exemplifying natural perfection in Haraway's analysis has been turned into a strategic model for manhood, and each group in an Akeley diorama becomes an ideal type standing for a species. It is as if Haraway reveals that the public is duped into mistaking the ideal type for the prototype, nature, which then becomes the prototype for culture. The taxidermist achieves a category mistake by creating a powerful illusion, and it is skilful replication that enables such an illusion to come about.

## Teddy Bear Patriarchy in the aftermath

Haraway's story about the ideologies and narratives that are built into science has been deeply influential in scholarly circles. But what do Akeley's dioramas narrate nowadays, almost a century since his hunting expeditions? The dioramas still feature in the American Museum of Natural History and are considered masterful and mesmerizing recreations; during my conversations with taxidermists in the United Kingdom, Belgium and the Netherlands, several discussion partners mentioned Akeley's work with reverence. Nowadays, the museum seeks to mobilize the dioramas in conveying a story with a solidly environmental message. In 2010, the museum's wildlife artist and curator Stephen C. Quinn traced Akeley's footsteps to the Democratic Republic of Congo's Virunga National Park

16 D'Andrade, *The Development*, p. 179.

to find and recreate the scene in which the gorilla diorama is set. He told *The New York Times* that 'the whole purpose of this expedition is to use art to tell an environmental story'. '[T]he dioramas in the museum', he said, 'should be used to nurture environmental awareness', crediting Akeley with the mountain gorilla's survival.[17] Implied here is the claim that by killing gorillas for the diorama, Akeley made gorilla lives continue, not only by making specific gorillas live on as mounted specimens but by actually breathing life into a species threatened with extinction, made visible and 'real' for generations of museum visitors in an encounter with 'nature'. Through the materiality of taxidermy, with its strong connotations of the real, and a harnessing of the symbolic and rhetorical potential of the gorilla as humanity's beleaguered kin, categories of life and death are troubled and stretched.

This potential that taxidermied specimens have as narrative tools has been more generally recognized and embraced by curators from a post-colonial perspective and is part of a recent regeneration of taxidermy in museum environments.[18] Trophy heads are dusted off, brought out of the stores, and made to perform stories about imperialist conquest or extinction of species and natural habitats. Replicas of nature-under-threat, they have become actors in a different kind of ideology that implies a different kind of categorizing in which doubts are thrown on humankind's position at the top of a long-established taxonomy.

From my recent ethnography, it appears that Haraway's 'single story' about taxidermy in the 1920s has become a multiple one. Whilst taking on new life in the hallowed space of the natural history museum, taxidermy has also wandered out of it and found a niche in contemporary art, where taxidermy is very much in vogue. Contemporary artists use taxidermy to express political statements about human – animal relations in the Anthropocene, associated with the destructive human impact on the environment. As I will suggest, artists, whilst embracing the craft of taxidermy, seek to make nature true to a different type than Akeley did according to Haraway's analysis. These artists also make nature true to a different type than most

17 See: https://cityroom.blogs.nytimes.com/2010/11/21/aross-the-world-revisiting-a-familiar-scene (checked on 16 October 2018), and https://cityroom.blogs.nytimes.com/2011/01/02/from-gorilla-diorama-to-its-inspiration-in-africa (checked on 16 October 2018).

18 See Poliquin, *The Breathless Zoo*, for taxidermy's shapeshifting potential, and Alberti, *The Afterlives of Animals*, for animal-object histories. For a thoughtful discussion on Haraway's analysis from a contemporary viewpoint, see https://americanscienceblog.com/2016/01/22/new-feature-americanscience-re-reads-the-classics-or-teddy-bear-patriarchy-in-2016 (checked on 16 October 2018).

professional taxidermists do nowadays. The skilled craft of replicating nature as precisely as possible still exists in private, often one-man studios where degrees of perfection articulate with the demands of specific commissions. Different groups of practitioners have different agendas when it comes to category making and prototyping, which articulate with different stances to replication as a craft. By paying attention to such differences, we can gain insights into contemporary perceptions of 'nature' and contemporary politics of naturalizing. Importantly, these insights can also tell us something about replication as a strategy for change, through its creation of prototypes, rather than a strategy for preservation. Having taken my cue from Haraway, I discuss examples of three different, yet interrelated realms of replication in early twenty-first-century taxidermy, namely: 1. lifelike taxidermy; 2. scientific museum taxidermy; and 3. contemporary art.

## Contemporary lifelike taxidermy: 'True to nature'

Nowadays, professional taxidermists typically work for a range of clients with different requirements: a hunter commissioning a trophy; a grieving individual hoping to have a pet revived; an artist working on an installation; or a museum seeking to enliven its collections. The UK Guild of Taxidermists, of which I became a member during my fieldwork, makes explicit that practising taxidermy is all about a play on categories; its motto is 'creating the illusion of life'. Creating an illusion implies knowing how to play on category mistakes; this is the *métier* of the taxidermist, and it follows clear, categorical rules in achieving this illusion. The craft of taxidermy involves different skill sets, including a solid knowledge of morphology, anatomy, and animal behaviour.[19] Taxidermists also learn to engage in empathic, bodily understandings of the animals they replicate through what I have called morphological approximation.[20] The more straightforward part of the trade is a mastery of materials (both organic and inorganic) and tools in the transfer from dead animal to lifelike mount. Some taxidermists insist on taking elaborate measurements of the dead animal, which are then replicated on the mount that they construct. In this case, an abstraction is made from nature, turning a body and its morphology into a series of measurements. This is called the engineering approach. Others prefer an approach considered

19 For close engagement with the craft of taxidermy, see Patchett, 'Witnessing Craft' and Patchett, 'The Taxidermist's Apprentice'.
20 Kalshoven, 'Gestures of Taxidermy'.

more holistic, paying attention to proportions whilst sculpting shapes rather than reproducing precise measurements. These approaches partly correlate with the taxidermist's choice of materials used in creating the so-called form or manikin that sits inside a taxidermied animal,[21] materials that lend themselves to measuring and shaping to varying extents.

Most clients ask for a realistic rendering of the animal they commission to be mounted. Practitioners themselves gloss realistic as 'lifelike' and speak of extending an animal's life through their skill. A mounted specimen (that is, a piece of taxidermy) must be true to nature; nature is firmly seen as the model for emulation. So, what does it mean to be 'true to nature' in contemporary taxidermy? In my discussions with taxidermists, I found that almost a century after Akeley crafted his masterful pieces, perfection is still an important principle guiding professional taxidermists. In my ethnography, it became evident that a prototypical specimen for twenty-first-century taxidermists to look for *and* to aspire to is an unblemished, healthy, good-looking specimen. When taxidermists find or receive an animal for mounting, the first thing they will do is check its condition by feeling for broken limbs and assessing the quality of feathers and fur in order to determine whether it is worthwhile to mount it; working on a damaged specimen is generally seen as a waste of time because of the effort needed to make it look perfect, if at all possible. This is in particular the case with taxidermists who participate in competitions; judges, I was told, are not forgiving of asymmetry[22] or a crooked neck or messy feathers because that could point to a mistake on the taxidermist's part rather than a mistake in nature[23] (Figure 1.2 and Plate 1.1). The effort that a taxidermist puts into a specimen for a competition is not the same effort that goes into a commission for an agreed-upon price. Professional taxidermy, then, is about categorizing in a quite pragmatic sense. Firstly, what is worthy of preservation and then, which preserved specimen makes it into what category: museum quality, educational purpose, collector's item, private commission, eBay, junk shop. Professional taxidermists, then, approach their practice strategically, but

21 Kalshoven, 'Nature's Inner Workings'.

22 But contrast a conversation at a NatSCA (Natural Sciences Collections Association) workshop, Lancashire Conservation Studios, Preston, 1 April 2015: 'Is asymmetry something to be avoided?' I asked. 'No', said Jack, a widely admired bird taxidermist, 'I like asymmetry'. The symmetrical looks 'dead, even stuffed!' amateur taxidermist Isobel added. Yet they acknowledged that judges may mark down asymmetry.

23 This concern with aesthetics and perfection works in strikingly similar ways in competitions involving living animals, such as shows of domestic duck species (interview with artist and skeleton builder Morag Jones, 25 November 2014).

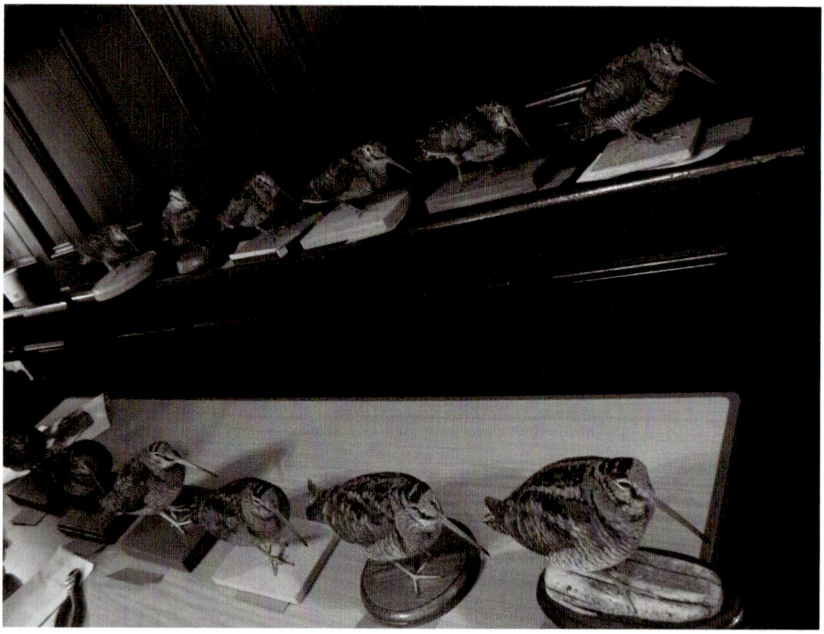

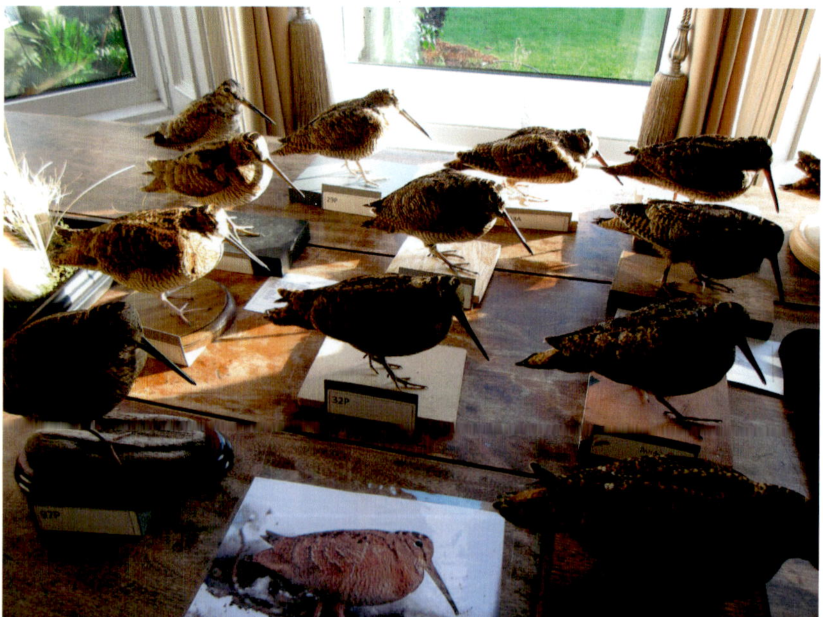


Figure 1.2 & Plate 1.1: Competing for perfection, replicating a photographic image: Woodcock Avian Challenge, UK Guild of Taxidermists, Annual Convention 2016, Derbyshire, England. Photo: author.

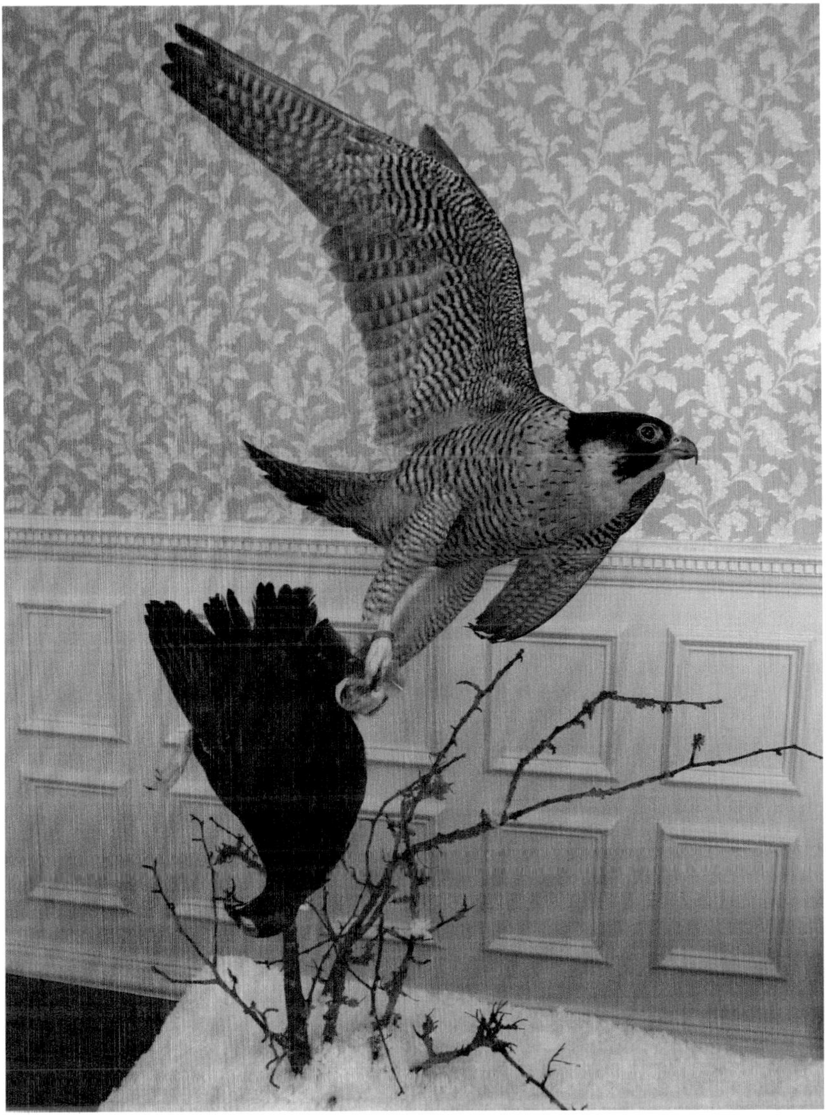

Figure 1.3: Perfect nature: Mike Gadd's award-winning Peregrine Falcon with Red Grouse, UK Guild of Taxidermists, Annual Convention 2014, Derbyshire, England. Photo: author.

those considered the top practitioners, when given sufficient time, ideally strive for 'lifelike' renderings of nature that are better than the real thing. Being 'true to nature' implies making a mount look as vigorous as possible. So, specimens, and nature through these, are made true to a type that exudes health and liveliness (Figure 1.3). In this endeavour, perfection is never really achieved (cf. Stols-Witlox, this volume, on imperfect copies in art conservation and restoration). For the expert taxidermist, every mounted specimen is a rehearsal for the next one, making the practitioner move from one prototype to the next, which he or she hopes to be an improvement on the previous one. And yet, perhaps ironically, to strive for the prototypical, in this case the perfect model of nature, taxidermists must take nature apart, insert artifice, and they must work on the margins between different categories, turning a flaccid skin into a lively mount. This is so because creating mounts involves the use of materials that must replace perishable parts on the dead animal, and usually the skeleton gets fully or almost fully replaced as well. This means that the taxidermist uses materials that are unrelated to the living animal to achieve a natural, lively look.

This play on categories can go even further, as one of my discussion partners, Belgian taxidermist and model maker Pierre-Yves Renkin, explained to me (conversation near Namur in April 2015). Renkin is a French speaker, so when we spoke, he used the French term for doing taxidermy and creating a mount, which is *naturaliser* (the less formal, and rather derogatory term is *empailler*, fill up with straw, stuff). As with the English term *to naturalize*, this can mean to give someone another nation's nationality or to make an animal or a plant at home somewhere else, but in French it also means to create a natural-looking taxidermied mount.[24] Sometimes, however, Renkin told me, to achieve a natural look, one needs to *dénaturer* ('denature'). To denature, he explained, means to play tricks on one's audience by using materials from one species to achieve a convincing look in a specimen of another species; to achieve an illusion, as he put it: to mislead or confuse the onlooker. In a striking example, he mentioned how he had made a miniature horse head for a TV commercial. To achieve this, he had used the head of a rabbit with its long ears, took the rabbit skull out, and replaced it with a model he had sculpted for a horse head, which then got covered in the rabbit skin with the ears. In this process, the rabbit had been denatured to make it look like something else. Renkin showed me other examples of this process, where he had used stork feet for a pink flamingo, or emu feet for a dodo (for extinct species such as the dodo or great auk, a patchwork of parts of other species has to be sourced to arrive at a convincing 'replica').[25] In a contribution to a volume celebrating Renkin's work, the model maker is likened to a mystic, to a

24 https://www.littre.org/definition/naturaliser (checked on 16 October 2018): 'Introduire dans un pays et y faire prospérer' […] 'Préparer les dépouilles des animaux, pour leur rendre l'aspect naturel. J'ai porté chez le fourreur une belle peau de renard, pour en faire une descente de lit ; la tête sera naturalisée'.

25 For Renkin's and other model makers' work on great auk models, see Kalshoven, 'Piecing Together the Extinct Great Auk'.

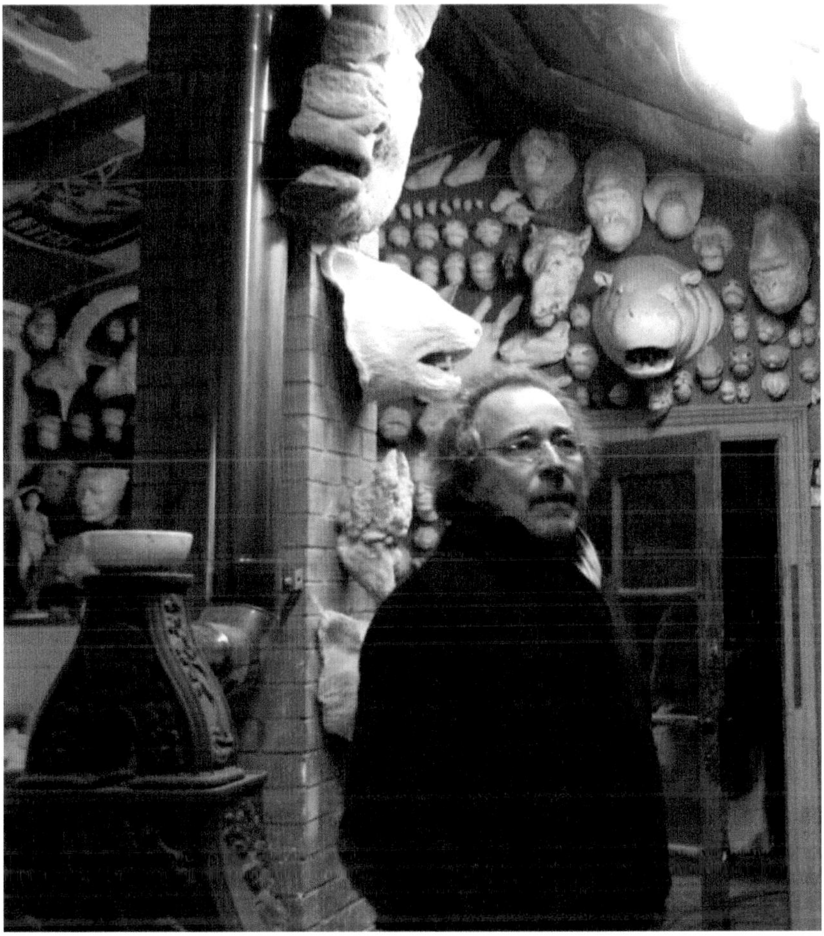

Figure 1.4: Model maker Pierre-Yves Renkin surrounded by his collection of plaster moulds, in his studio near Namur, Belgium, 2016. Photo: author.

shaman and a demiurge: 'il communique avec les esprits de la nature qui, alliés au sien, inspirent ses recréations ou ses créations d'animaux [...] Les prodiges de Monsieur Renkin nous enchantent car ils se jouent des distinctions et des catégories aristotéliciennes'[26] (Figure 1.4). In harnessing matter in a bid to denature rather than naturalize, an additional category mistake (on top of the play with life and death) is achieved, mesmerizing onlookers. Replication takes devious turns to convince an audience of an underlying original existing 'in nature'. True to nature, in this case, becomes true to nature as imagined by an exceptionally skilled human playing a crafty Creator.

26 Della Noce-Bertozzi, 'L'homme de chambre', p. 80.

## Scientific perspective: Study skins

Some professional taxidermists work for natural history museums to create lifelike mounts, including replicas of extinct species. Such mounts are used for didactic displays, increasingly so, as I explained above, to convey messages about the environment and humanity's role in it. Full mounts may be commissioned for study purposes as well,[27] but there is also another kind of skin preparation that goes on in museums which is considered the more scientific kind, and which has a long history. I watched this process of preparation at Naturalis Biodiversity Center in Leiden, the Netherlands, demonstrated by curator and taxidermist Rebecca Desjardins (31 August 2015). In the Live Science space of Naturalis, Desjardins prepared a so-called study skin of a black redstart for the centre's scientific collection, while the public could look on and ask questions. A study skin is a preserved bird skin that is not mounted to look 'real' but simply filled with cotton wool or tow, remaining relatively flat and easy to store for scientific reference of colour, size, and feathers. From what Desjardins explained to me, I understood the study skin to be a record of a natural specimen that comes about through a stylized process of replication. While she manipulated the dead bird, about a dozen visitors to the natural history collections walked into the live science space, where seats were provided as well as a screen magnifying Desjardins's hands. Examining the bird, she said, there is no visible moult – in song birds, moulting occurs symmetrically (unlike in sea birds); if moulting is asymmetrical on a song bird, this may be caused by regrowth after an attack. There is a bit of fat on the body, she found, which means the bird was in good health. Desjardins filled out a form with measurements she took from the bird. It is odd, she remarked in an aside to me, how conventions differ. In North Carolina, where she had worked in a small museum as a bird collection curator, the under-wing measurement would be taken from the natural curvature of the wing. In Naturalis, received practice was to measure the wing stretched flat onto a horizontal plane, which she suspected would lead to slightly different outcomes.

Part of her work closely resembled full taxidermy in that she stripped the carcass of its skin. Unlike in display taxidermy, however, where the skinless body is left undisturbed before being discarded, she probed further inwards, extracting the bird's last meal and exposing its tiny testes, indicative of its

27 NatSCA workshop, 1 April 2015: Jack had just finished one bird using five different carcasses, he told us. 'Well', said curator Claire, 'that is very important to know when commissioning – it would disqualify such a specimen from DNA sampling'.

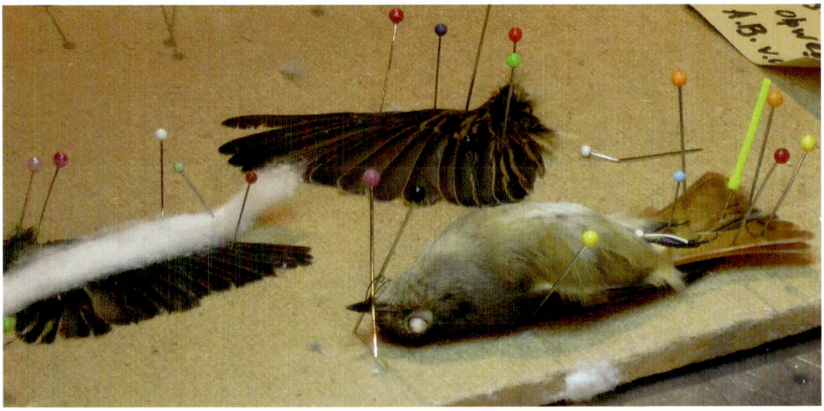

Plate 1.2: Black Redstart, study skin prepared by conservator Becky Desjardins, Naturalis, Leiden, the Netherlands, August 2015. Photo: author.

state before the breeding season. She then proceeded to cut off one wing from the specimen, explaining that she would add it to the study skin all spread out with the feathers neatly visible, aiming to 'maximize every animal'. From the spread wing, ornithologists would be able to read the bird's age.

The skin was folded back over the left-in skull, a wooden stick inserted to stabilize the simple shape, cotton wool added for stuffing and rolled into small balls to replace the eyes, and then the skin was sewn up presenting the familiar surface of feathers again, which the curator fluffed up to hide a second, smaller piece of wood that she added in lieu of the cut-off wing (Plate 1.2). Connected to the remaining, folded wing by a string through the bird's breast, this small stick provided support to the remaining wing should anyone in future try to pick up the bird by its wing tip. As she pinned the black redstart onto a cork surface for drying, Desjardins took care to fan out the tail feathers to enhance visibility. Her Dutch colleague, she remarked, preferred to leave the tail closed. From each specimen, she would take a DNA sample that was kept in a tiny container filled with formaldehyde. Experts compare strands against international databases, she explained. Collection series of study skins make it possible to study change over time. Sparrows, for example, were bigger a hundred years ago, perhaps because they had to do more flying in between places, looking for grains on farms.

Desjardins's practice privileged content-rich surface volumes over morphologically correct representation.[28] In fact, most of the bird's materiality

28 For taxidermy's entanglements with surface and volume, see Kalshoven, 'Re-Animating Skin'.

was captured in measurements and matter readable to a specific, scientific, audience only. Turned into a study skin, the black redstart was immobilized in a manner optimized for scientific readability, allowing access to different amplitudes for bits of information – one wing folded, the other wing spread. Its real-life morphology, however, needed to be imagined from the measurement sheet. 'My teachers in the United States', she said, 'emphasized the importance of accurate measurements to such an extent that, if done properly, these should allow you to dispense of the body altogether.' 'Why did she bother', I asked, 'to fill and stitch up the skin at all?' 'Well', she said, 'because it still helped to convey the idea of a real specimen, and it made study skins easier to handle.'

With study skins, then, specimens are made readable and materially accessible for expert consultation; they are not made to achieve an illusion or to celebrate nature's vigour or beauty, but rather to unlock nature's measurable laws. Nature, here, is made to measure. But not all modes of measuring are valued the same, as Desjardins noted with some surprise comparing her experience in different museum settings. Replication in the preparation of study skins is a matter of engagement with materials through stylized abstraction, remnants turned into records for future knowledge making and yet quite evocative of a living bird's 'natural' behaviour and potential, both individually and collectively.

## Contemporary art: Perturbing the lifelike

Scientific practice in natural history museums and the modes of classification it upholds have come under scrutiny not only from scholars such as Haraway but also from contemporary artists intent on uncovering the institutional politics underlying taxidermy practice.[29] In direct interplay with the renaissance of mounted specimens in museum spaces, taxidermy is decidedly in vogue in contemporary art. Contemporary art has embraced taxidermy as a mode and material of expression on the topic of human – animal relations, in particular over the past two decades, and tends to do

29 This critical stance is particularly evident in studies in which academics collaborate with an artist, e.g., Snaebjörnsdóttir and Wilson, *Nanoq*, on an exhibition reuniting stuffed polar bears; Patchett and Foster, 'Repair Work', which reflects on remnants of the extinct blue antelope; and Patchett, Foster, and Lorimer, 'The Biogeographies of a Hollow-Eyed Harrier'. In the latter piece, a museum study skin is taken back to the bird's habitat as part of an art project that explores and critiques wildlife management. For a fascinating analysis of taxidermy's renaissance in art circles, highlighting its materiality and ontological ambiguity, see Aloi, *Speculative Taxidermy*.

so through interventions in museum and gallery spaces. Contemporary artists, while embracing the craft of taxidermy, seek to make nature true to a different type than professional taxidermists, following in the footsteps of Akeley, tend to do. This occurs through different approaches to category making, and in particular to the constitution of new kinds of prototypes.

Some artists make their mounts themselves. Others use discarded pieces of taxidermy or commission mounts from professional taxidermists for inclusion in their artworks. Amongst artists who embrace taxidermy there is a tendency to trouble the lifelike, realistic approach for which professional taxidermists aim. Some artists do so by working with poor specimens or trophies stripped naked.[30] Others make hybrid fantasy creatures[31] or emphasize death or decay, or injury at the hands of humans, rather than perfect-looking life[32] (Figure 1.5 and Plate 1.3). The lifelike approach in taxidermy is what we have become used to seeing; the kind that has become naturalized in museum settings from the heyday of Akeley's dioramas onwards. The contemporary artistic approach is predicated on troubling our expectations both of taxidermy and of 'natural' life. It is predicated on showing what is counter-intuitive rather than prototypical or ideal.

Contemporary artistic engagement with taxidermy is varied and extensive. I will give one example of the defamiliarization underlying such engagement from a collaboration with textile artist Anthea Walsh.[33] As part of my research on taxidermy I commissioned work from her for a small exhibition at Kendal Museum in the English Lake District (Autumn 2013). The exhibition was designed as an intervention in the important collections of Victorian taxidermy on display at the museum. In realizing *Approximating Animals*, her title for the exhibition, Walsh made several art pieces combining embroidery and taxidermy. The taxidermy in these pieces came from parts of a green woodpecker that the artist had acquired as an old decommissioned museum specimen. She dismantled the bird and used various bits of it to create separate pieces framed by embroidery hoops. She made feathers and down seamlessly merge into stitches of embroidery, combining bird remnants and human patterning. The materials used included 'a combination

30 See Baker on 'botched taxidermy' in *The Postmodern Animal* and '"You Kill Things to Look at Them"'.

31 For German artist Thomas Gruenfeld's 'Misfits', see Dillon, 'Curiosity', p. 110, for 'Siamese twins' created by Dutch artists Les Deux Garçons, see Borg and Smets, *Les Deux Garçons*.

32 Consider Damien Hirst's dead animal installations, created by a team of taxidermists including UK Guild member Emily Martin, which sparked the artistic interest in taxidermy in the UK; or Polly Morgan's installations, see Morgan, *Psychopomps*.

33 For other examples from my ethnography, see Kalshoven, 'Gestures of Taxidermy'.

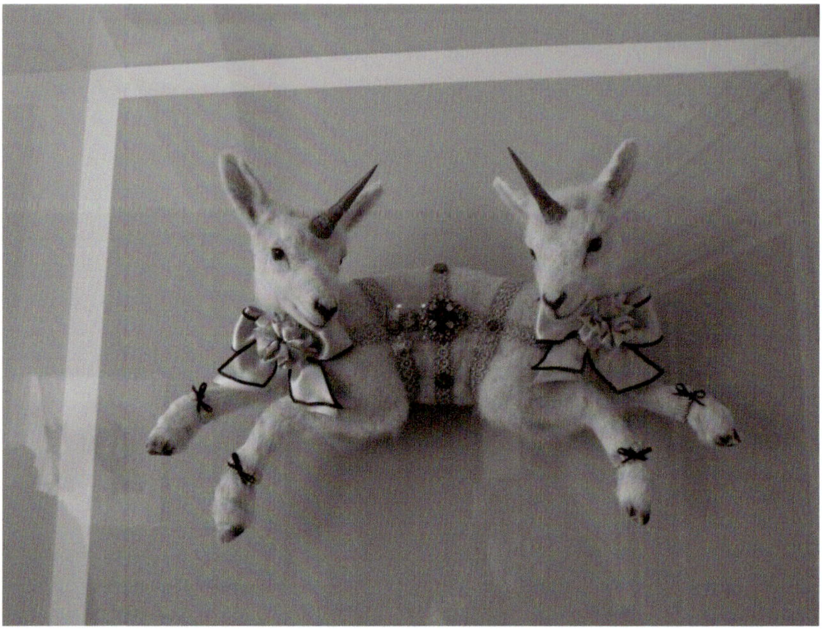

Figure 1.5: Taxidermy sculpture by Dutch artists Les Deux Garçons. Exhibition *PURE*, Museum aan het Vrijthof, Maastricht, April 2016. Photo: author.

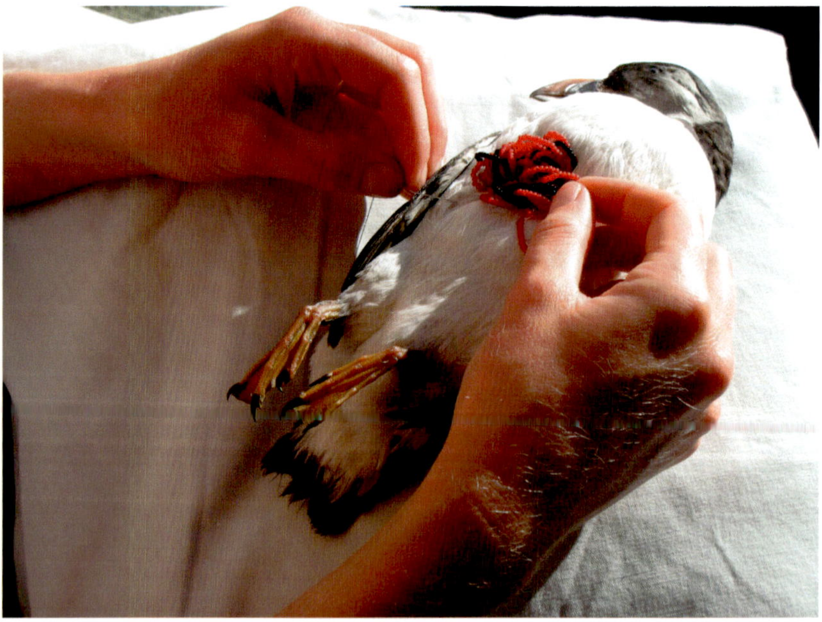

Plate 1.3: Artist Fiona Dean with a puffin using red beads to make it look as if the puffin bleeds, Edinburgh, July 2013. Photo: author.

of natural and synthetic in the embroidery threads, the sequins and beads [...] The ready-made embroidery hoops [are plastic], with a metal clasp at the top. The embroidery is on dyed natural linen (the dye is synthetic...) and the wool felt under the linen supporting the structure is industrial felt and probably has some synthetic content in the mixture of fibres' (Anthea Walsh, personal communication). The artist used the conventional, flat surface of the embroidery frame to present her work, to which the bird parts gave some relief. Threads unfurled and the woodpecker seemed to have exploded rather than been given volume. And yet there remained something very dynamic about the bird. The bird-ness was recognizable without the skin having been stretched over a correctly shaped manikin. Perhaps it was the intertwining of human-made and 'natural' materials that made her work so intriguing. What struck me was how the dismantled bird, displayed in several embroidery hoops, conveyed a real, almost defiant, sense of life, perhaps more strikingly so than a perfectly mounted 'realistic' specimen. This was particularly the case in the piece that contained the bird's expressive head, with one foot jutting out of the confines of the embroidery hoop. It was as if the trophy was kicking back (Plate 1.4).

In the comment book we left in the museum, some visitors expressed wonder about the artworks' puzzling life force, whilst others protested against what they considered a lack of respect for the bird, literally robbed of its integrity: the only taxidermy that is acceptable, one visitor remarked, is the realistic, natural history kind. In Walsh's work, replication of life showed an interesting twist in that it did not aim for a realistic (lifelike, dead, or injured) reconstruction of a woodpecker, nor did it distil abstracted, standardized features of the live animal. Rather, its mode of replication resided in exploring human – animal analogies through the use of materials that mimicked one another: stitches resembling feathers, and feathers merging into stitches (Plate 1.5).

## Category making: Saliency of shifting prototypes

The stir caused by Walsh's and other contemporary artistic approaches to replicating nature reminded me of a thought-provoking article by Pascal Boyer on human categorizing, which draws on the fields of evolutionary psychology and cognitive anthropology to discuss mechanisms involved in the selection and transmission of religious concepts.[34] Boyer's main

34 Boyer, 'Functional Origins of Religious Concepts'.

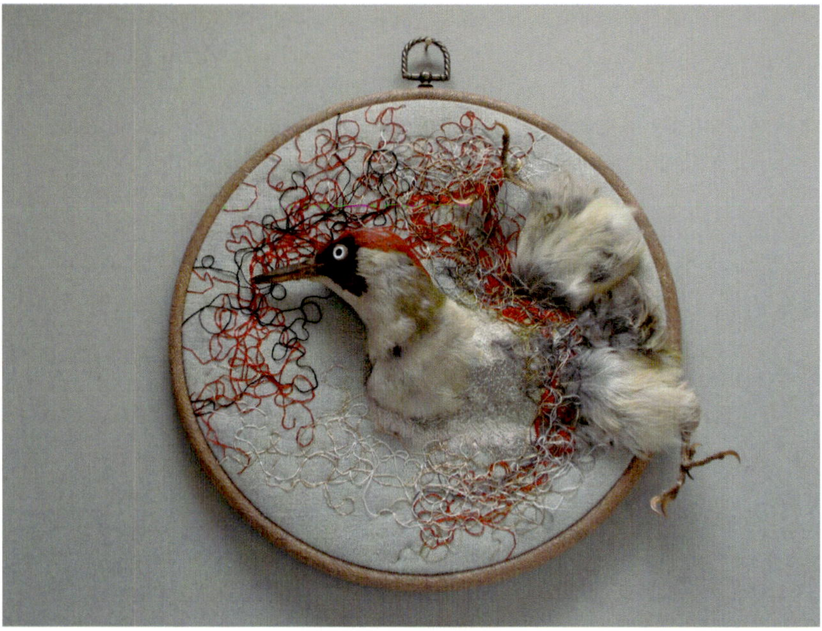

Plate 1.4: Anthea Walsh: Unfurl mixed media 30cm x 30cm. Photo: Anthea Walsh, used with permission.

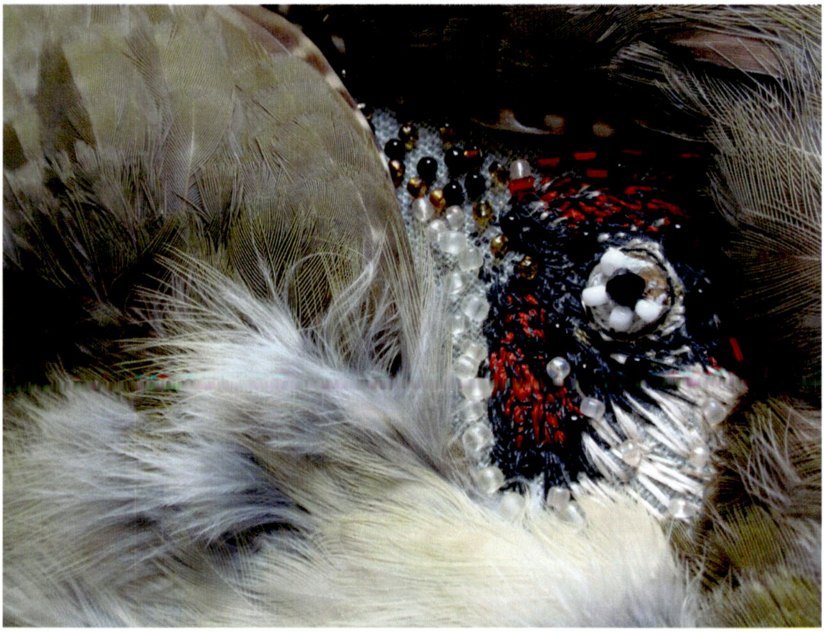

Plate 1.5: Anthea Walsh: Beady Eye, detail. Photo: Anthea Walsh, used with permission.

question is: Why is it that some supernatural concepts acquire widespread popularity, whereas others do not? With taxidermy, supernatural concepts are perhaps not of primary concern (*pace* Della Noce-Bertozzi, above), but because taxidermy works on the borders between life and death and between living being and artefact, Boyer's experiment may still provide a useful analogy. According to Boyer, the human mind is equipped with an intuitive ontology, which he defines as 'a set of expectations about the kinds of things to be found in the world'.[35] Supernatural concepts activate ontological categories whilst violating intuitive expectations. For example, a statue of the Virgin would activate the category 'artefact'; however, the belief that the statue may speak or think violates what is normally associated with the category 'artefact'.[36] This makes the concept a salient one. Boyer distinguishes five ontological categories: person, animal, plant, artefact, and natural object.[37] He cites experiments measuring recall performance as evidence for the importance of saliency as a determining factor in the transmission of religious concepts.[38] Only those concepts that are recalled may subsequently be transmitted. According to Boyer's theory, the counter-intuitive (that which does *not* fit the pattern; that which does *not* match the other items on the list) is more memorable and thus more likely to last in human imagination. Remembering, then, shapes the future.

I suggest that, in the case of taxidermy, the saliency of prototypicality highlighted in D'Andrade's account finds a complement in Boyer's discussion of the saliency of the non-prototypical. Taxidermy is a play with classification in that it combines saliencies of the prototypical and of the counter-intuitive in different, socially competitive ways. Lifelike taxidermy is salient because of the realism of the prototype it seeks to perpetuate through faithful replication-through-artifice. It is also salient precisely because of the artifice that sustains its replication; it is something else from what it purports to be, which provokes, alienates, or intrigues the human observer. Complicating this tension between the prototypical and the counter-intuitive is contemporary artistic taxidermy, which seeks to rework (cf. Sibum, this volume) a different kind of nature, which explicitly includes

35 Boyer, 'Functional Origins of Religious Concepts', p. 196.

36 Boyer, 'Functional Origins of Religious Concepts', p. 197.

37 Boyer, 'Functional Origins of Religious Concepts', p. 198. The question arises whether such 'intuitive categories' are indeed part of the universal intuitive ontology with which the human mind, according to Boyer, is equipped or whether there is something 'cultural' about these very categories. What *is* clear, at least in a Western context, is that, in the case of taxidermy, categories such as animal, artefact, and natural object become quite unstable.

38 Boyer, 'Functional Origins of Religious Concepts', p. 199ff.

human (destructive) intervention, or rather, it seeks to *rework taxidermy* with a provocative twist that makes it salient compared to the kind of taxidermy with which we have become familiar in museum environments.

Contemporary art's critical take on taxidermy, however, is becoming rather recognizable, almost prototypical, perhaps even constituting a new form of 'naturalization'. Ironically, in artistic critiques of the human impact on nature, nature is 'made true to type' as well. Contemporary art seeks to break frames yet constitutes new ones. What is considered prototypical shifts over time, and human conceptions of nature (as programmatic for culture; as perfect; as measurable; as under threat) shift along, as reflected in, and co-constructed by, taxidermy.

Important in these processes of familiarization and defamiliarization is the sheer materiality of stuffed specimens as reworkings of what came before. Taxidermied mounts are salient, alluring, and repulsive because they feature remnants of animals that have obviously died, even if they are made to look quite lifelike. 'Nature' provides the materials for taxidermy and serves as its model for replication. In the different practices I discussed, taxidermy's relationship with nature turned out to be complex, and was articulated in verbs with 'nature' at its root. Taxidermy was naturalized (as in Alberti, opening citation); it was also argued to actively engage in naturalizing (as in Haraway's account); and it could be denatured (as transpired from the conversation with Renkin and from Della Noce-Bertozzi's remarks on his work). Taxidermy was made to abstract from nature (as in the case of study skins); and it denaturalized (making nature, and human intervention in it, seem strange or wrong, as in contemporary art). What is clear from the varied roles in which animal skins are seen to feature is that the single, ideological story that Haraway found in taxidermy has become a multi-vocal, messier affair. This has interesting implications for replication both as a practice and as an analytical concept. As a *métier* purported to restore a once-living creature, taxidermy creates prototypes that refer to a future rather than trying to emulate what is lost. In the present of practice, taxidermists and artist-taxidermists work according to and towards a variety of ideals that resonate with the past whilst creating collections and conceptual frameworks for the future. The case of taxidermy shows replication to be a constant, political and ideological, play with categories that moves back and forth in time as animal skins get stripped, used, and reused to make statements about natural worlds. It shows replication to be a skilled act of prototyping for the future. After all, to replicate, from the Latin 'again' and 'to fold', is the practice of folding anew, once more, yet again.

## Bibliography


Alberti, Samuel J.M.M., 'Constructing Nature behind Glass', *Museum and Society*, 6 (2008), pp. 73-97.

Alberti, Samuel J.M.M., *The Afterlives of Animals: A Museum Menagerie* (Charlottesville: University of Virginia Press, 2011).

Aloi, Giovanni, *Speculative Taxidermy: Natural History, Animal Surfaces, and Art in the Anthropocene* (New York: Columbia University Press, 2018).

Asma, Stephen T., *Stuffed Animals and Pickled Heads: The Culture and Evolution of Natural History Museums* (Oxford: Oxford University Press, 2001).

Baker, Steve, *The Postmodern Animal* (London: Reaktion Books, 2000).

Baker, Steve, '"You Kill Things to Look at Them": Animal Death in Contemporary Art', in *Killing Animals*, The Animal Studies Group (Urbana: University of Illinois Press, 2006), pp. 69-98.

Borg, Lucette ter, and Sandra Smets, *Les Deux Garçons – Hun Wereld – Their World* (Maartensdijk: B for Books BV, 2014).

Boyer, Pascal, 'Functional Origins of Religious Concepts: Ontological and Strategic Selection in Evolved Minds', *Journal of the Royal Anthropological Institute*, 6 (2000), pp. 195-214.

D'Andrade, Roy, *The Development of Cognitive Anthropology* (Cambridge: Cambridge University Press, 1997).

Dalla Bernardina, S., 'Hymnes à la vie? Sur l'engouement récent pour les bêtes naturalisées', *Terrain*, 60 (2013), pp. 56-73.

Daston, Lorraine, and Peter Galison, 'The Image of Objectivity', *Representations*, 40 (1992), pp. 81-128.

Della Noce-Bertozzi, Roger-Philippe, 'L'homme de chambre: voyage dans les mondes intérieurs ou système universel des cosmogonies de chambre', in *Le monde de Pierre-Yves Renkin*, ed. by Jean-Paul Heerbrant (Brussels: Centre Albert Marinus, 2011), pp. 77-109.

Desmond, Jane, 'Displaying Death, Animating Life: Changing Fictions of "Liveness" from Taxidermy to Animatronics', in *Representing Animals*, ed. by Nigel Rothfels (Bloomington: Indiana University Press, 2002), pp. 159-179.

Dillon, Brian, *Curiosity: Art and the Pleasures of Knowing* (London: Hayward Publishing, 2013).

Eastoe, Jane, *The Art of Taxidermy* (London: Pavilion Books, 2012).

Fuller, Errol, *Voodoo Salon* (Billingshurst: Summers Place Auctions, 2014).

Haraway, Donna, 'Teddy Bear Patriarchy: Taxidermy in the Garden of Eden, New York City, 1908-1936', in *Primate Visions: Gender, Race and Nature in the World of Modern Science* (London: Routledge, 1989), pp. 26-58.

Heerbrant, Jean-Paul, *Le monde de Pierre-Yves Renkin* (Brussels: Centre Albert Marinus, 2011).

Kalshoven, Petra Tjitske, *Crafting 'the Indian': Knowledge, Desire, and Play in Indianist Reenactment* (New York: Berghahn Books, 2012).

Kalshoven, Petra Tjitske, 'Moving in Time, Out of Step: Mimesis as Moral Breakdown in European Re-enactments of the North American Indian Woodland', *Journal of the Royal Anthropological Institute (N.S.)*, 21 (2015), pp. 561-578.

Kalshoven, Petra Tjitske, 'Gestures of Taxidermy: Morphological Approximation as Interspecies Affinity', *American Ethnologist*, 45 (2018), pp. 34-47.

Kalshoven, Petra Tjitske, 'Piecing Together the Extinct Great Auk: Techniques and Charms of Contiguity', *Environmental Humanities*, 10 (2018), pp. 150-170.

Kalshoven, Petra Tjitske, 'Re-Animating Skin: Probing the Surface in Taxidermic Practice', in *Surfaces: Transformations of Body, Materials and Earth*, ed. by Mike Anusas and Cristián Simonetti (London: Routledge, 2020), pp. 62-79.

Kalshoven, Petra Tjitske, 'Nature's Inner Workings: Manikins and the Lifelike in Contemporary Taxidermy', in *3D Modelling: Bodies and Buildings in Anthropology, Anatomy and Architecture*, ed. by Elizabeth Hallam (London: Routledge, 2020). Contracted for the series *Anthropological Studies of Creativity and Perception*, series editor Tim Ingold.

Lange-Berndt, Petra, *Animal Art. Präparierte Tiere in der Kunst, 1850–2000* (Munich: Verlag Silke Schreiber, 2009).

Michaud, Maxime, 'De l'animal au trophée: réification ou relation amoureuse?', *Anthropologie et Sociétés*, 39 (2015), pp. 165-178.

Milgrom, Melissa, *Still Life: Adventures in Taxidermy* (Boston: Houghton Mifflin Harcourt, 2010).

Morgan, Polly, *Psychopomps* (London: Haunch of Venison, 2010).

Patchett, Merle, and Kate Foster, 'Repair Work: Surfacing the Geographies of Dead Animals', *Museum and Society*, 6 (2008), pp. 98-122.

Patchett, Merle, Kate Foster, and Hayden Lorimer, 'The Biogeographies of a Hollow-Eyed Harrier', in *The Afterlives of Animals: A Museum Menagerie*, ed. by Samuel J.M.M. Alberti (Charlottesville: University of Virginia Press, 2011), pp. 110-133.

Patchett, Merle, 'Witnessing Craft: Employing Video Ethnography to Attend to the More-than-Human Craft Practices of Taxidermy', in *Video Methods: Social Science Research in Motion*, ed. by Charlotte Bates (London: Routledge, 2015), pp. 71-94.

Patchett, Merle, 'The Taxidermist's Apprentice: Stitching Together the Past and Present of a Craft Practice', *Cultural Geographies*, 23 (2016), pp. 401-419.

Poliquin, Rachel, 'The Matter and Meaning of Museum Taxidermy', *Museum and Society*, 6 (2008), pp. 123-134.

Poliquin, Rachel, *Ravishing Beasts: The Strangely Alluring World of Taxidermy* (Vancouver: Museum of Vancouver, 2009).

Poliquin, Rachel, *The Breathless Zoo: Taxidermy and the Cultures of Longing* (University Park, PA: The Pennsylvania State University Press, 2012).

Snaebjörnsdóttir, Bryndis, and Mark Wilson, *Nanoq: Flat Out and Bluesome. A Cultural Life of Polar Bears* (London: Black Dog Publishing, 2006).

Star, Susan Leigh, 'Craft vs. Commodity, Mess vs. Transcendence: How the Right Tool Became the Wrong One in the Case of Taxidermy and Natural History', in *The Right Tools for the Job: At Work in Twentieth-Century Life Sciences*, ed. by Adele E. Clarke and Joan H. Fujimura (Princeton: Princeton University Press, 1992), pp. 257-286.

Strivay, Lucienne, 'Taxidermies: le trouble du vivant', *Anthropologie et sociétés,* 39 (2015), pp. 251-268.

Wonders, Karen, *Habitat Dioramas: Illusions of Wilderness in Museums of Natural History* (Acta Universitatis Upsaliensis. Figura Nova Series 25, 1993).

## About the Author

**Petra Tsjitske Kalshoven** is a social anthropologist at the University of Manchester. Drawing on a background that combines the Humanities and Social Sciences (M.A. Classical Languages and Cultures, Leiden University; Ph.D. Cultural Anthropology, McGill University), she explores skilled manifestations of human curiosity, simulation, play, and rhetoric as these find expression in different practices, ranging from reenactment, museum practice, or taxidermy to nuclear decommissioning. Currently immersed in an ethnographic project around the UK's nuclear facilities in North West England, she pursues her interest in human expertise and the persuasive ways in which people engage specific materials and landscapes in shaping uncertain futures.

# 2. Bringing the Past to Life: Material Culture Production and Archaeological Practice

*Jill Hilditch*

**Abstract**
Performative methods in archaeology provide a valuable heuristic tool for investigating the many behaviours and interactions of both producers and consumers of material culture. Focusing on the potter's wheel at Bronze Age Akrotiri as a socially embedded performance of technical know-how, this chapter outlines an integrated approach to material engagement across three arenas of archaeological action – experiment, analysis, and visualisation – connected by an explicit engagement with the *chaîne opératoire* approach. An innovative tool-kit is presented for the investigation of this technology by the wider archaeological community. Given the large-scale regional and diachronic questions that the adoption and adaptation of ancient technologies can raise, a collective approach is proposed for the interpretation of the potter's wheel.

**Keywords:** material engagement, ceramic technology, experiment, analysis, visualisation

## Introduction

Archaeologists seek to understand the past, that is the people, objects and environments that have shaped, and been shaped by, human experience. Yet, we can never fully recreate it. These complex, multi-scalar interactions between humans and the material record were *performed* both on a daily basis and throughout millennia. This task is further complicated by the study of an incomplete record of partial material remains, often the only

Dupré, S., A. Harris, J. Kursell, P. Lulof, M. Stols-Witlox (eds.), *Reconstruction, Replication and Re-enactment in the Humanities and Social Sciences*. Taylor & Francis Group
DOI 10.5117/9789463728003_CH02

surviving tangible link to a myriad web of intangible human behaviours or practices. In the face of such challenges, archaeologists have developed theoretically-informed methodologies for investigating material culture. These methodologies integrate reconstruction, reproduction and replication (RRR) practices with considerations of human agency, material engagement, and practice theory to seek greater understanding of processes of production, networks of learning, identity construction, and technological transmission in the past.

This chapter first presents a brief overview of how performative methods and material engagement practice have developed at the intersection of archaeology, sociocultural anthropology and the natural sciences. The second part focuses on how current archaeological practice uses reconstruction and replication to shed new light on ancient technology, in particular the process of ceramic manufacture and the transmission of the potter's wheel in prehistory. The adoption and adaptation of the potter's wheel at Bronze Age Akrotiri (prehistoric Greece) is discussed from an integrated *chaîne opératoire* approach across three loci of activity: experiment, analysis and visualization.

## Performance and practice, now and then

Following the notion that all practices are, or were, performed within the broad arena of human activity, archaeology is a systematic attempt to characterize these performances in their original spatial and social contexts, as well as how they develop over time. The interaction between humans and material culture has been one of the defining relationships within modern archaeology; 'most archaeologists today see material culture as instrumental to how people create, experience, give meaning to, negotiate and transform their world'.[1] When face-to-face with material culture from the past, our curiosity invariably leads to the question; who made this and why did they make it? The performance, or act, of artefact production offers a unique opportunity to answer these questions, affording archaeologists a means by which to characterize the individuals who made, distributed and used these material artefacts. If we go on to consider the communities in which these individuals performed their craft, then we can also reach wider interpretations about the social practices and networks of interaction within which these communities operated. In short, tangible remains can

1 Dobres and Robb, 'Doing Agency', p. 161.

shed light on intangible aspects of past human societies. In the theoretical overview that follows, examples are given in relation to ceramic material culture.

## Bodily performance and social space

The physical and technical knowledge needed to produce material culture, to make a pot, knap a flint tool or cast an axe head, is learnt or transferred by watching and copying 'techniques' until their constituent gestures become habit. These habitual, unconscious actions were first discussed by the French sociologist Marcel Mauss who 'radically separated the phenomenon of techniques from nature, and embedded it within social reason and social practices'.[2] Concepts of bodily gestures and actions, and subsequent sequences of rhythmic gestures, were, therefore, crucial to Mauss as he considered the body an aspect of the self that we live *through*, not in. One of Mauss' rare students, André Leroi-Gourhan, developed this idea by viewing artefacts as extensions of the body, meaningfully constituted through the results of sequences of gestures applied to material.[3] It was Leroi-Gourhan who coined the term *chaîne opératoire*, or operational sequence, which has led to an explicitly technological approach to studying material culture, 'reinsert[ing] technical activities into their gestural, spatial and temporal dimensions'.[4] In embracing this tradition, archaeological research into material culture acknowledges that all learning, past and present, takes place within a social context, hence making technical acts simultaneously social acts.[5] In effect, material engagement allows us to conceive past acts of production as 'meaningful and socially negotiated set[s] of material-based practices'.[6]

Technical or technological choice is a fundamental concept that allows archaeologists to uncover socially 'embedded' technological activities at both individual and group level.[7] Technological choices and social practice are closely related as 'techniques are first and foremost social productions, any technique is always the physical rendering of mental schemas learned

2 Mauss, *Sociologie* and Schlanger, 'Techniques', p. 23.

3 Leroi-Gourhan, *Gesture*.

4 Schlanger, 'Mindful Technology', p. 147.

5 Leroi-Gourhan, *Gesture*; Dobres and Hoffman, 'Social Agency'; Van der Leeuw et al., 'Technical Traditions'; Van der Leeuw, 'Giving the Potter'; Lemonnier, 'Introduction' and Dobres, 'Of Great and Small'.

6 Dobres and Hoffman, 'Social Agency', p. 213.

7 Lemonnier, 'Introduction'.

through tradition'.[8] Ethnographic observation within anthropology has been invaluable in leading us towards an understanding that technical actions for ancient people in pre-industrial societies could be quite 'elastic', given variable raw material properties in combination with other techniques.[9] As a result, technical strategies such as in pot-making have several solutions, or alternatives, and 'every pot is the unique result of a series of choices between alternative techniques'.[10] This applies not only to ceramic manufacture but to all prehistoric 'technicians', their products and their techniques. Even stone tool production, despite being a subtractive rather than additive process, is not entirely linear in character and the artisan could implement a number of different production strategies.[11] So, within material culture production, it is the choices that are crucial in determining the nature and shape of a product, its effectiveness and its life expectancy.[12]

Technical choices also relate to the wider issue of social agency and Bourdieu's *habitus*, i.e. to what degree individuals have the power, knowledge or freedom, to operate outside the socially constituted system of cognitive and motivating structures within their society.[13] Bourdieu defined practice as 'the product of processes, which are neither wholly conscious nor wholly unconscious, rooted in an ongoing process of learning which begins in childhood and through which actors know – without knowing – the right thing to do'.[14] Technical behaviours are not then solely determined by the social (or symbolic) context of production, nor entirely by material or environmental factors: rather it is the goal of technological studies to understand the micro-scale social processes behind the microscale prehistoric technical gestures.[15]

## Reconstructing the archaeological framework of 'choice'

The initial step in inferring social context and meaning from material culture is to envisage technical choices as stages within Leroi-Gourhan's gestural sequence, or *chaîne opératoire*. The individual stages can be reconstructed

8 Lemonnier, 'Introduction', p. 3.

9 Van der Leeuw, 'Giving the Potter', p. 239; Mahias, 'Pottery Techniques' and Gosselain, 'Social and Technical Identity'.

10 Sillar and Tite, 'The Challenge', p. 3.

11 Edmonds, 'Description', p. 57.

12 Van der Leeuw, 'Giving the Potter', p. 241.

13 Bourdieu, *The Logic* and Dobres and Hoffman, 'Social Agency', p. 222.

14 Jenkins, *Key Sociologists*, p. 44.

15 Dobres and Hoffman, 'Social Agency', p. 226.

using detailed materials analyses to identify the technical behaviours involved in producing an artefact. Ceramic vessels are conducive to technological reconstruction because they have an observable structure that can be related to their operational sequences. By studying ceramic production technologically, it is possible to reconstruct the choices of potters by their behaviours within the gestural sequence, i.e. raw material procurement and processing, vessel forming, surface treatment and decoration, as well as firing conditions. The benefit of the *chaîne opératoire* as a 'middle range interpretative methodology' then lies in the ability to combine empirically-grounded analytic methods with a robust anthropological theory of social reproduction concerned with everyday practice, embodiment and identity, which embeds technicians into social collectives.[16]

Sander van der Leeuw has made a cross-cultural analysis of the *chaînes opératoires* of a number of pottery-making traditions to highlight the 'conceptual anchors' (shape, parts of a pot and the specific sequence of manufacture) and 'executive functions' (tools and techniques, rotating and support of vessel) that are both more difficult to manipulate than the raw materials.[17] It is these factors within ancient material culture production, examined within a social context, that have the greatest amount of information to give scholars. By studying material artefacts, scholars can recognize and then juxtapose these 'gestural traces of norms and variants' within the *chaîne opératoire* to reveal the degree of freedom technicians experienced within their social contexts.[18] We can also go further still by acknowledging that the cognitive correlates of tradition and innovation can be accessed by identifying how the *chaîne opératoire* is modified to introduce new products.[19] Even though the basic nature of the product can remain constant over a period of time, the *chaîne opératoire* does not necessarily follow the same patterns.[20] It is possible to reveal profound changes in the character of organization within production, either in structures or choreography, which can also be used to interpret the social context of artefact production, distribution and consumption. In other words, it is possible to 'pinpoint and define small coherent combinations of technical features that correspond directly to given social groups'.[21]

16 Dobres and Robb, 'Doing Agency'.

17 Van der Leeuw, 'Giving the Potter'.

18 Dobres, *Technology and Social Agency*, p. 155.

19 van der Leeuw, 'Giving the Potter', p. 141.

20 Edmonds, 'Description'.

21 Mahias, 'Pottery Techniques', p. 170; also Gosselain, 'Social and Technical Identity', and Lemonnier, 'Introduction', who goes on to associate technical choices with social identity and difference.

The *chaîne opératoire* is consequently a rigorous framework for structuring the approach to the social aspect of the overall context of production, i.e. environmental and technological constraints, economic and subsistence base, social and political organization and the ideology or belief systems of the people making the choices. The intentionality and decision-making involved within a particular *chaîne opératoire* can, however, be elucidated without tying those gestural sequences to a particular embodied individual.[22] This framework can also allow scholars to glimpse the social meanings behind material culture production, by highlighting various states of 'belonging' and 'not-belonging' within small coherent social groups.

**Social scale and communities of practice**

Determining socially meaningful groups as a scale of analysis is a primary concern for anthropologists, sociologists, ethnographers and archaeologists alike.[23] There are problems of establishing which attributes should be used to define groups, how visible such group boundaries may be and also to what extent these boundaries may overlap spatially and temporally.[24] With respect to material culture production, perhaps the most valuable of social collectives is that of Etienne Wenger's 'community of practice', 'the social configurations in which our enterprises are defined as worth pursuing and our participation is recognizable as competence'.[25] Using earlier work with Jean Lave on 'situated learning',[26] a theory of learning as a form of participation, Wenger goes on to state that 'social participation is not just in local events of engagement in certain activities with certain people but a process of being active participants in the practices of social communities and constructing identities in relation to these communities'.[27] By participating or engaging in local activities and interactions, learning 'reproduces and transforms the social structure in which it takes place'.[28] In turn, collective learning reflects the social context in which we wish to participate and, therefore, the practices of a community (joint enterprise, mutual engagement, shared repertoire etc.) are always socially meaningful.

22 Hodder, 'Agency and Individuals', pp. 25-26.
23 Lock and Molyneaux, *Confronting Scale*.
24 See Stark, *The Archaeology*.
25 Wenger, *Communities of Practice*.
26 Lave and Wenger, *Situated Learning*.
27 Wenger, *Communities of Practice*, p. 4.
28 Wenger, *Communities of Practice*, p. 13.

As Wenger succinctly stated, 'in practice, manual activity is not thoughtless and mental activity is not disembodied'.[29]

The concept of a community of practice is extremely valuable to archaeological theory because the material and symbolic negotiation of membership within such a community *can* be traced in the archaeological record: the 'technical profile of [an individual's] products will resemble other members of those social groups'.[30] With respect to the production of pottery, this negotiation could manifest itself in clay paste recipe, formation or shaping technique, the tools used to decorate a vessel, the method or motif of painted decoration or even the firing conditions. This constant negotiation of meaning and identity (both as individual members and as a group) can occur over long or short periods of time, but all communities of practice are sustained by the introduction of new members who gain competency through situated learning. Identifying the performance of particular *chaînes opératoires* allows archaeologists to situate individuals within communities of practice, therefore allowing the social context of production, the gestures involved, and the learning of those gestures, to be enmeshed within a network of dynamic relations that can inform the long-term study of material culture.

An important point to emphasize is that the concept of a community of practice operating in the past should not be synonymous with previous definitions of archaeological 'cultures'. Many communities of practice may be operating within a large social group, either as specific differences in the performance and learning of a single technical task or with respect to the production and exchange of different materials. These different communities of practice may also operate at different scales of analysis, often linked to the physical or ideological properties of the material involved. From a ceramic perspective, forming techniques appear to be the most resistant to change as they are learnt through motor skill acquisition within social contexts, in contrast to decorative expression, or cooking methods, that may crosscut residential units as a result of marriage relations or other social processes.[31] This crosscutting of social groups by communities of practice creates contradictions and overlaps within material culture patterning, as different types and changing forms of community membership have the potential to influence or create formal variation within the operational sequence, even

29 Wenger, *Communities of Practice*, p. 45.

30 Jeffra, 'Experimental Approaches', p. 142.

31 Sassaman and Rudolphi, 'Communities of Practice', p. 407. See also Gosselain, 'Materializing Identities; Gosselain, 'Social and Technical Identity' and Van der Leeuw, 'Giving the Potter'.

after long periods of perceived stability that exhibit very little variation. In other words, a meso-scale perspective can provide evidence for similarities between intensive skill transmissions, as well as unconscious practices that can help us understand in greater depth 'the learning structures and mobility of the makers of the objects we analyse'.[32]

## Archaeology and RRR practices

The terms 'reconstruction', 're-enactment', and 'replication' are most strongly associated with archaeological practice through the subfield of experimental archaeology, where practitioners are still often (incorrectly) characterized as 're-creating activities, artefacts, structures and processes that happened in the past'.[33] This chapter seeks to expand this familiar arena of performative archaeological practice and instead emphasize a *chaîne opératoire* approach combining material analysis with experimental and digital archaeologies.[34] Such an approach starts during post-recovery artefact analysis, allowing archaeologists to reconstruct past production processes, moves through systematic experimental replication, and finishes in the public domain, by communicating the results in a way that visualizes ancient objects and technologies. These three components, analysis, experiment, and visualization, can be integrated using the *chaîne opératoire* approach to structure their investigative process (Figure 2.1). The rest of this chapter uses ceramic manufacture and technology to illustrate how archaeologists use material culture to reconstruct the crafts, technologies and social practices of the past.

Ceramics are ubiquitous in the global archaeological record. This is for many reasons, not least the availability of their raw materials, as clays can form almost anywhere in the natural landscape. Fired clay was used for a vast range of objects, decorations, and structures, and its durability and compositional stability during deposition has ensured the survival of these objects for generations of archaeologists to discover. As it turns out, this is wonderfully convenient because ceramics are ideally suited for investigating technological innovation. Few other material categories offer such potential

32 Broodbank, 'Minoanisation', p. 60.

33 Outram, 'Introduction', p. 2.

34 The Tracing the Potter's Wheel Project is funded by an NWO Vidi Grant (2016-2021). The project is directed by Dr. Jill Hilditch (PI), in collaboration with Dr. Caroline Jeffra (postdoctoral research fellow) and Loes Opgenhaffen (PhD researcher) at the University of Amsterdam (UvA). More information on project activities and results can be found at https://tracingthewheel.eu (checked on February 28th, 2019).

| CHAÎNE OPÉRATOIRE | ANALYSIS | EXPERIMENT | VISUALISATION |
|---|---|---|---|
| **raw material acquisition** | geological prospection/compositional analysis | clay pits/fills | photomicrographs of ceramic fabric and geological samples |
| **paste processing** | textural analysis (petrography/SEM); macroscopic analysis | temper/sediments within site | high resolution photography; ceramic fabric photomicrographs |
| **forming** | macrotrace analysis; textural analysis (petrography/SEM) | device/tools/specialised architecture | high resolution photography; video capture; 3D scanning |
| **surface treatment** | compositional analysis (petrography/SEM/XRF); macrotrace analysis | pigments/tools | high resolution photography; 3D scanning |
| **firing** | structural analysis (FTIR/petrography/SEM) | kiln architecture | ceramic fabric photomicrographs |
| FINISHED VESSEL | **Laboratory-based testing and replication for reconstructing the *chaîne opératoire*** | **Investigation of forming techniques and tools through hypothesis testing and replication** | **Innovative methods for experiencing the material and physical aspects of objects and technologies** |

Figure 2.1: Investigating the ceramic *chaîne opératoire* (summary of the methodology of the Tracing the Potter's Wheel project). Table: author.

for materializing the full range of social complexity within ancient societies. Ceramic objects provide evidence for eating, drinking, cooking, processing, storing, lighting, transporting and worshipping activities, to name a few, across multiple social and economic contexts. Developments in past activities and interactions are attested through changes in the ceramic objects, either in their shape or form, size, decoration, colour, skill of execution (finish quality) and even their production method. Yet, we do not just glean information on the potter or producer from such changes. Material culture production does not take place in a vacuum; the affordances and constraints of exchange mechanisms and the socially motivated desires of consumers also play a powerful role in how material culture changes through time. In this way, production/transformation and consumption acts are a window onto the social practices of both the potter and consumer and how both construct their identity in relation to their wider social worlds.

## Analysis

Once a ceramic object has been recovered from its depositional context, the analysis phase begins. There are many stages of analysis that take place, frequently in more than one location. Analysis often starts in the dusty, temporary storerooms of an excavation before leading to the decidedly dust-free environment of a scientific laboratory and finishing in a permanently dusty collection of a museum. Although typological and stylistic analyses are undertaken to provide archaeologists with relative dating sequences, this section discusses scientific or ceramic fabric analysis, more specifically the compositional and textural investigation of fired and unfired clay bodies. The central questions of scientific analysis concern composition and technology, both of which are crucial for identifying which technical choices were made at each stage of the *chaîne opératoire* for the object being investigated.

Usually carried out on fragmentary pieces of the original vessel, ceramic fabric analysis can identify the types of raw materials used in the production of a particular vessel or ware, or indeed an entire ceramic assemblage.[35] This in turn facilitates the characterization of the technological choices employed by the potter in the raw material acquisition, paste processing and even forming technique stages of the production sequence. It is important to establish whether vessels were locally produced or imported when found within an assemblage deposit at a site. Macroscopic analysis of the ceramic fabric using the naked eye or a small hand lens (up to x60) is the most cost-effective means of establishing compositional and technological groups within an assemblage. Visual differences in the colour and texture of the clay, as well as any inclusions present in the clay matrix, are used to build broadly compatible fabric groups. These groups are then sampled to carry out more detailed, but ultimately destructive, methods of microscopic and chemical (mineralogical or elemental) analysis.[36] The integration of compositional and textural information at macroscopic, microscopic and chemical level provides the most effective means of characterizing ceramic fabrics, and comparing them with available raw materials in the landscape.[37] Determining whether a ceramic assemblage was locally produced requires geological comparatives, either in the form of detailed geological maps of the area or replication experiments using geological sediments (sands, clays, etc.) collected from the vicinity of a site. Replication involves the processing of collected sediments into experimental briquettes to transform the raw materials into a replica ceramic paste. Analysing the geological sediments, experimental briquettes and the archaeological ceramics side-by-side can

35 After Shepard, *Ceramics for the Archaeologist*. See Hilditch, 'Ceramic Analysis', for an overview of current ceramic fabric research carried out within the Aegean.

36 X-ray fluorescence (XRF) and scanning electron microscopy with energy dispersive spectrometers (SEM-EDS) are now the most widely available techniques for compositional and structural analysis of the ceramic body and its various layers.

37 The importance of moving coherently between scales of analysis was first highlighted by Peacock, 'The Scientific Analysis'. For a case study highlighting how integration is achieved today, see Hilditch et al., 'Taking the Rough'.

indicate the specific technical choices made by ancient potters in the *chaîne opératoire* (Figure 2.1). Reconstruction of the ceramic *chaînes opératoires* used within a ceramic assemblage at a particular site forms the empirical foundation for investigating the variation within local pottery producing communities, allowing subsequently broader regional and temporal differences to be identified.

## Experiment

Expertise in the practical and material knowledge of a craftsperson emerges from material engagement, not just from a lifetime's experience of working with the material: It is 'a knowledge born of sensory perception and practical engagement, not of the mind with the material world but of the skilled practitioner participating in a world of materials'.[38] Material engagement is a central theme of experimental archaeology, a subdiscipline of archaeology that has been perceived as focusing its attention on 'reproducing former conditions and circumstances' or 'replicating past phenomena'.[39] The heart of experimental archaeology in fact lies in hypothesis-testing through experiment.[40] Academic context, and appropriate reference to the specialist literature, is also fundamental to the performance of successful experimental archaeology, and helps to differentiate the *experimental* from the *experiential*. This is an important distinction because re-enactment, as defined elsewhere in this volume, falls into this latter category, primarily in the form of 're-enactment groups, outdoor education and public presentation centres, and other demonstrations of past life and technology'.[41] Experiential activities are still important, however, as they provide a means for experimenters (archaeological specialists) to gain competence in the performance of particular activities, that is, the requisite skills or techniques of the experiment. They can also clearly and meaningfully convey the results of archaeological research to non-specialists. Experiential practices are necessary precursors then to successful experiments and effective ways of communicating them.

Experimental archaeology has historically focused on the technological questions, the 'how' and 'what' of material culture, yet has rarely engaged with the social questions, such as 'who' and 'why'.[42] To elucidate these social

38 Ingold, 'Materials against Materiality', p. 15.
39 A summary of the debate of the nature of experimental archaeology is given by Outram, 'Introduction', p. 2.
40 This point is emphasized by Outram, 'Introduction', and Marsh and Ferguson, 'Introduction'.
41 Outram, 'Introduction', p. 3, echoing the earlier work of Reynolds, 'The Nature of Experiment'.
42 Jeffra, 'Experimental Approaches', p. 141.

phenomena we need to turn our attention to the social contexts in which craftspeople learn and reproduce their crafts, or, in other words, we must return to the archaeological framework of 'choice', the *chaîne opératoire* approach. The choice of forming technique for ceramic vessels is one stage of the *chaîne opératoire* in which social phenomena play a significant role. This is due to the concept of salience within pottery-making techniques.[43] The choice of a potter to hand-build a vessel, or to use a wheel to finish a hand-built rough-out, or a wheel to shape a raw lump of clay into a vessel, is not known to the distributer or user of that finished vessel. Further still, the tools and gestures required to perform each forming technique also remain a mystery, if only looking at the finished vessel. Forming techniques represent knowledge that is not transmittable in the finished object but must be transmitted person-to-person.[44] Experimental archaeology is able to engage with the identification of forming techniques by reconstructing the sequence of actions that leave diagnostic macrotraces on the surfaces of ceramic vessels. The use of rotative kinetic energy (RKE) in the form of a tournette or a weighted wheel device leaves distinctive traces on the surface of a pot (rilling). If a hand-built coiled rough-out of a vessel is finished on a wheel, then other diagnostic traces can form that reveal the initial coiled structure in combination with the RKE (s-cracks and tension seams). Replication is embedded in the experimental process, because it is wrong to assume that the full range of potential diagnostic features will be present on each experimentally-produced vessel.

In practice, the strength of the experimental approach for identifying and tracing the transmission of the potter's wheel is the greater detail accessible in discussing *chaînes opératoires* diachronically. Moreover, the power of experimental archaeology lies in its ability to act as a bridge between data on the ground and higher level theories of social interaction. Experimental archaeology cannot be defined as solely replication or reproduction practice, but rather it is the interplay of these practices that provides a robust platform for considering socially meaningful questions through material engagement.

## Visualization

Communicating technological change has always been a difficult task. Archaeologists are trained to identify and interpret material culture changes to create the material profiles or signatures of specific cultural groups. As archaeology

43 Gosselain, 'Materializing Identities' and Gosselain, 'Social and Technical Identity'.

44 Gosselain, 'Materializing Identities.'

has developed from its nineteenth-century roots in antiquarianism, so have its methods of visualization: 2D drawings, technical illustrations, and photographs have slowly been supplemented and, perhaps in some cases, even supplanted by digital and 3D techniques.[45] New visual technologies such as 3D scanning and modelling offer huge potential for enhancing research strategies within archaeology.[46] Yet, currently, the acceptance of these technologies has mostly served to automate existing archaeological practice.[47] So although many archaeological objects now exist as 3D scans and can be viewed on a computer screen, their conventional classification systems remain the dominant means to describe and present these artefacts to specialists and non-specialists alike. Fortunately, there is a growing number of projects designed to address this disconnect between academic research and public awareness, and to promote ancient technological studies to contemporary society.[48]

As with the experiment and analysis components, visualization draws upon the *chaîne opératoire* approach to structure the process of acquiring data and interpreting the results. Experimentation with equipment is a crucial first step in establishing a workflow that other specialists can replicate. Issues of cost, portability, resolution and suitability to the task must also be considered when choosing hardware and software for the scanning and subsequent processing of the models. In many ways, the digital archaeologist develops a workflow that mirrors an ancient craftsperson's *chaîne opératoire* of technical tasks, although the end product is somewhat different. Whereas a potter produces a tangible, fired clay vessel, the digital archaeologist produces a virtual, high-resolution model of a fired clay vessel that can be rotated and manipulated by the viewer.[49] Close collaboration is required between the analyst, the experimenter and the visualizer, because a clear understanding of what is technologically important is required, i.e. what needs to be visualized and to what resolution. In the case of wheel-made pottery, the distinctive surface traces made by RKE must be visible to the viewer, alongside more traditional views of the shape profile and location on the vessel. Replication of the 3D scanning process is also essential if

45 See Molloy, 'Recent Advances', for a current overview of the use of 3D techniques for studying archaeological objects.

46 See also Lulof, this volume.

47 Opgenhaffen, *Visualising the Potter's Wheel*.

48 The Pottery Goes Public Project (2015-2018) is presented in Opgenhaffen, Revello Lami and Kisjes, 'Pottery Goes Public'. See also The Bridges Collection: https://thebridgescollection.wordpress.com (checked on February 28th, 2019) and the Digital Epigraphy and Archaeology Project: www.digitalepigraphy.org (checked on February 28th, 2019).

49 Molloy and Milić, 'Wonderful Things?', p. 111.

other specialists are to add their compatible models to an open access database of diagnostic wheel traces. The creation (or following) of technical manuals is an important step in the digital archaeologist's workflow that allows archaeological specialists to build a detailed, annotated online reference collection for the identification of wheel traces in their own ceramic assemblages. On the other hand, an open access data repository also allows non-specialists to follow the questions, methods and practices of archaeologists by accessing the dataset virtually, manipulating the scale and angle of models and asking their own questions of the material 'in hand'. This is a significant step forward towards challenging the traditional perception of ancient artefact and technology research within archaeology as being rather static in its presentation to the public.

## Tracing the potter's wheel in the Bronze Age Aegean

In the case of the Bronze Age Aegean (3100-1200 BC), terms such as Anatolianization, Minoanization or Mycenaeanization carry tacit indications of cultural contact, tentatively balancing between a *description of how* and an *explanation of why* changes in the material record are driven by hierarchical interactions between different socio-cultural groups (Figure 2.2). The technological dimension of such interactions is a field ripe for further study, and the potter's wheel in particular provides common ground for investigating millennia of prehistoric interactions in the Aegean.[50]

The traditional narrative for the innovation of the potter's wheel is a simple, linear explanation that correlates the emergence of the wheel-throwing forming technique to growing populations and social complexity that demanded routinized, increased production of ceramic products. However, recent work supports the first use of the potter's wheel in the Near East and Aegean regions as a finishing stage for hand-built, pre-formed roughouts, rather than the emergence of wheel-throwing a vessel from a lump of clay.[51] These roughouts are made by stacking hand-rolled coils of clay. Hand-built coiled pots are then finished without the use of rotative kinetic energy (RKE), in contrast to wheel-coiling, where the potter's wheel is used to thin,

50 Berg, 'Meaning in the Making'; Choleva, 'The First Wheelmade Pottery'; Lis et al., 'Mobility'; Jeffra, *The Archaeological Study*; Jeffra, 'A Re-Examination'; Kiriatzi, '"Minoanising"; Gauss and Kiriatzi, *Pottery Production*; Hilditch, *Reconstructing Technical Choice* and Hilditch, 'Analyzing Technological Standardization'.

51 Roux, 'Wheel Fashioned Ceramic Production'; Jeffra, 'A Re-Examination' and Choleva, 'The First Wheelmade Pottery'.

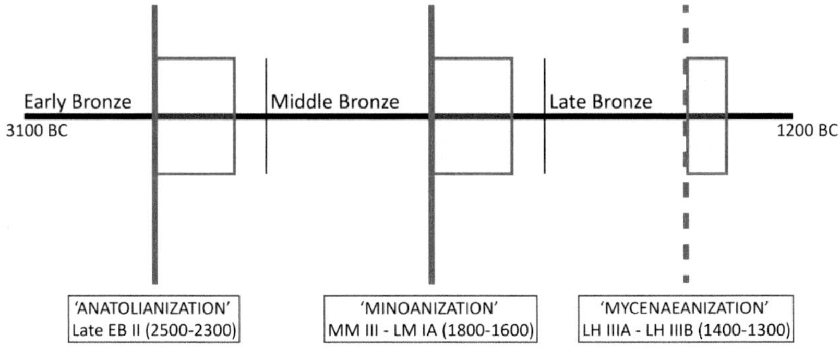


Figure 2.2: Overview of the Bronze Age Aegean'-ization' phenomena. Image prepared by Jill Hilditch and Loes Opgenhaffen, used with permission.

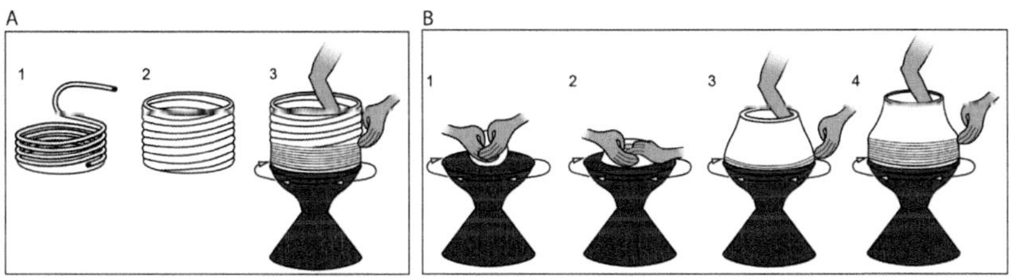


Figure 2.3: Comparison of the wheel-coiling and wheel-throwing methods of vessel construction. Image prepared by Caroline Jeffra, used with permission.

shape and finish the coiled walls, actions that require increasing amounts of RKE (Figure 2.3 for a comparison of the methods).[52] This means that we should see wheel-coiling as an intermediate or combination potting strategy, effectively bridging hand-building and the use of RKE techniques.[53] This distinction is important when considering how the potter's wheel spread throughout the potting communities of the Bronze Age Aegean.

In the Bronze Age Aegean, the appearance of the wheel was considered as two potentially independent events, connected to two different horizons of intensifying cultural contact.[54] The earliest identified horizon of wheel use appears during the later Early Bronze (EB) II period (*c.* 2500-2100BC), known as the Lefkandi I/Kastri phase, and is widely interpreted as the material result of increased trade in metals between social groups in the Aegean and

52 Roux and Courty, 'Identification'.

53 Jeffra, 'A Re-Examination', p. 31.

54 Knappett, 'Tradition and Innovation'; Choleva, 'The First Wheelmade Pottery' and Rutter, '"Minding the Gap"'.

(predominantly) western Anatolia, where the potter's wheel is known to have already been in use.[55] The vessels of this phase are accordingly known as 'Anatolianizing' shapes, reflecting close parallels to existing EB vessels from western Anatolian sites such as Liman Tepe, Çesme, and Troy.[56] In the later transition from the Middle Bronze (MB) to the Late Bronze (LB) period (1800-1600BC), the potter's wheel is considered as a technology of the Minoan culture of Crete. The potter's wheel then spread beyond the shores of Crete as part of a package of technologies that attest to growing Minoan power and influence, traditionally termed 'Minoanization', within the southern and central Aegean region.[57]

Diagnostic traces of wheel use within Cretan and non-Cretan but 'Minoanized' ceramic assemblages of the MB III to LB I have recently been reassessed and show that wheel-coiling techniques continue to dominate the ceramic record.[58] This would indicate that wheel-throwing was not an innovation linked to this transitional phase.[59] Instead, wheel-thrown pottery is necessarily a later technological development within the Aegean. This transition from wheel-coiling to wheel-throwing is poorly understood in Aegean archaeology and beyond. This is exacerbated by the fact that few ceramic assemblages offer the stratigraphic continuity to trace the trajectory of the potter's wheel as technological development. Stratigraphic continuity is essential for identifying and mapping modification of the *chaîne opératoire* from hand-building techniques to the incorporation of RKE. One Bronze Age Aegean site in particular, Akrotiri, known as 'the Pompeii of the Aegean', has afforded a closer look at the dynamics of technological change within ceramic manufacture.

## The case of Akrotiri on Thera

Akrotiri is an exceptional Bronze Age site situated on the island of Thera (modern day Santorini) among the Cyclades of the southern Aegean Sea. The site was buried under thick layers of pumice during a catastrophic volcanic

55 Renfrew, *The Emergence* and Alram-Stern and Horeijs, *Pottery Technologies*.

56 Şahoğlu, 'Interregional Contacts'.

57 Along with distinctive architectural features (lustral basins, switchback staircases), Linear A script and the vertical warp-weighted loom. See Hägg and Marinatos, *The Minoan Thalassocracy* and Macdonald et al., *The Minoans*. See Evely, 'The Potter's Wheel' and Evely, *Minoan Crafts* for an overview of the evidence for the potter's wheel on Crete.

58 Jeffra, 'A Re-Examination'; Jeffra, 'Forming Technology'; Gorogianni, Abell and Hilditch, 'Reconsidering Technological Transmission'.

59 Jeffra, 'A Re-Examination'.

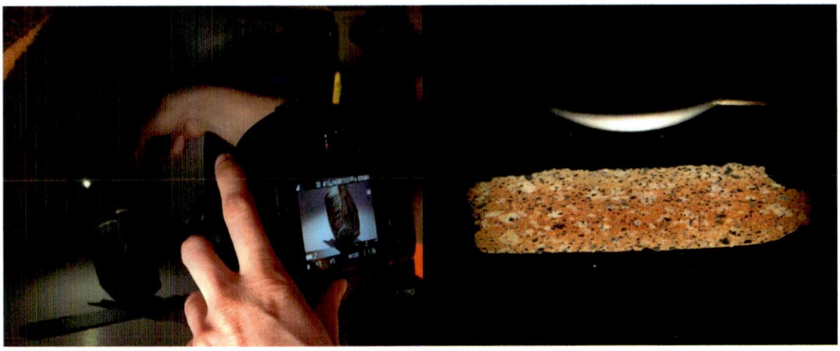

Plate 2.1: *Left*: Macrotrace identification with controlled light; *Right*: ceramic fabric analysis using petrographic analysis of thin sections. Photo: author.

eruption towards the end of the seventeenth century BC. Ceramic material recovered from excavations at Akrotiri has revealed a complex local ceramic tradition in the millennia preceding the eruption. The diachronic nature of the Akrotiri ceramic dataset provides an excellent platform from which to investigate processes of Minoanization. By focusing on the mid- to later phases of the Middle Bronze Age sequence (approx. 1950-1700BC), distinguished by the arrival of Minoan stylistic and technological features (wheel-coiled vessels) within the local ceramic assemblage, we can reconstruct a detailed understanding of what is inherently 'local' versus 'Minoanized' at Akrotiri. Compositional and textural analyses of archaeological vessels and geological sediments were carried out in tandem with surface macrotrace analysis to characterize the fabrics of the assemblage (Plate 2.1). Using this information, the *chaîne opératoire* for each of the wares was identified and then compared across a range of local Theran and non-local or Minoanized shapes to reconstruct the practices of locally based potters. Through this integrated analysis programme, we were able to assess whether shared practices as a community at Akrotiri differed over time in response to changing interactions with the Minoan world.

For the wheel finished 'Minoanized' vessels analysed at Akrotiri, such as the ledge-rim bowl, there seems to have been little attempt to refine the clay paste in imitation of the fine-grained fabrics of contemporary Minoan imports. Instead, these bowls were manufactured using locally compatible raw materials that showed little to no paste processing, effectively presenting the same production sequence as the local handmade pots, indicating the same technical decisions of the local clay recipe. It would seem, therefore, that the first use of wheel technology at Akrotiri was firmly integrated within the local tradition, rooted from the earliest stages of the production sequence. From this knowledge we can interpret that local potters at Akrotiri had sustained contact with Minoan potters, allowing them to learn the wheel technique, which they then incorporated into their own local potting traditions. This does not suggest political dominance by Crete, nor a strong itinerant or semi-permanent Minoan potting presence at Akrotiri. Instead, potters at Akrotiri, and therefore the people they were making pots for, actively chose to participate in certain Minoan ritual practices that required specific shapes, such as the ledge-rim bowl, for which deliberate formation techniques were used to enhance the intrinsic value of these novel artefacts.[60] A strong, local Cycladic tradition of pottery production was maintained, and with it, a well-defined 'non-Cretan' community identity within the settlement of Akrotiri.

In tandem with the scientific analysis of ceramics, a systematic programme of experiments was carried out to explore the wider material evidence, both products and tools, for wheel-coiling and wheel-throwing in the Bronze Age Aegean. A ceramic type set, designed to act as an analogue for the wide range of vessels corresponding to the Aegean Bronze Age, was created in order to improve our ability to differentiate between specific wheel use strategies through their macrotraces, facilitating deeper investigation of the relationship between these techniques.[61] Three primary variables were considered: vessel shape, forming technique, and clay type (degree of coarseness). The macrotrace results of the experimental type set were then applied to the study of archaeological material, consisting of key diachronic assemblages recovered from across the Aegean, including the later MB Akrotiri assemblage. The sequence of study described here is crucial; the experimental type set allows us to first establish the range of diagnostic traces associated with specific wheel-forming techniques. Following this, the traces on the Aegean material can be recognized as relating to one kind of wheel-forming or another.

The investigation at MB Akrotiri illustrates how we can assemble increasingly precise information about the practices and interconnections between communities of craftspeople. For example, small vessels were found to correspond with one of two different forming techniques: hand-building or wheel-coiling.[62] Conical cups, straight-sided cups, and hemispherical cups – shapes associated with Crete – were all wheel-coiled. Ledge-rim

60 See also Knappett and Hilditch, 'Colonial Cups?'.

61 The type set was based upon an expansion of earlier experimental work by Jeffra, *The Archaeological Study*, Jeffra, 'Experimenting Wheel-Coiling Methods' and Jeffra, 'Crafter'.

62 Jeffra, 'Forming Technology'.

bowls, also associated with Crete, were mostly wheel-coiled but also made by hand atop a woven mat. Lastly, locally-derived Cycladic cups were almost exclusively hand-made, with just two examples which were wheel-coiled. Emerging from this is a picture of technical negotiation, where shapes typical of Crete as well as shapes typical of Akrotiri and her neighbours were formed using wheel-coiling, alongside some cases of hand-building as well. In this way, the experimentally-derived data supports the presence of a network of communication between craftspeople on Akrotiri (forming Cycladic cups) and Crete (forming the other shapes) through which the transmission of potter's wheel technique took place.

So far, the technical details that suggest shared craft networks between Crete and the Cycladic community of Akrotiri are only visible, and therefore accessible, to archaeological specialists with the privilege of working at the site. Even communicating this dataset to other regional archaeological specialists is a challenge, but one which the visualization workflow is explicitly tackling. The visualization component starts with 3D scanning of the experimental type set to record, annotate and create 3D models of sherds or fragmentary vessels that display diagnostic macrotraces of specific wheel use techniques (Plate 2.2). This requires close collaboration with the experiment component, as recognizing which traces should be recorded and to what resolution is crucial. The next phase then involves 3D scanning of archaeological vessels to support the identification of wheel use techniques using traditional high resolution photography, and importantly to record these traces in a more accessible and easily manipulatable 3D format.[63] The 3D models of the experimental type set are digitally archived and made accessible through open access, online forums such as SketchFab and the Tracing the Potter's Wheel project website. Allowing specialists and non-specialist users to interact with the 3D models, to zoom in, out, or rotate as needed, aids future identification of forming traces on other archaeological assemblages. A technical manual has been devized to record the workflow of the visualization process and explicitly allow other scholars to replicate the 3D scanning process. In effect, through the production of annotated high resolution 3D models that highlight diagnostic macrotrace features,

63 Publication of Greek archaeological material in any format is subject to permission by the Greek Ministry of Culture and application for study and publication permits is carried out in collaboration with colleagues in the Greek Archaeological Service. The members of the TPW project are deeply indebted to Dr. Irene Nikolakopoulou for her generosity in inviting us to participate in the study of the MB Akrotiri ceramic assemblage, and to the site director, Prof. Christos Doumas, for allowing us to access the material at this incredible site. We hope there will be a future opportunity to 3D scan the ceramics from Akrotiri, once the relevant permissions are secured.

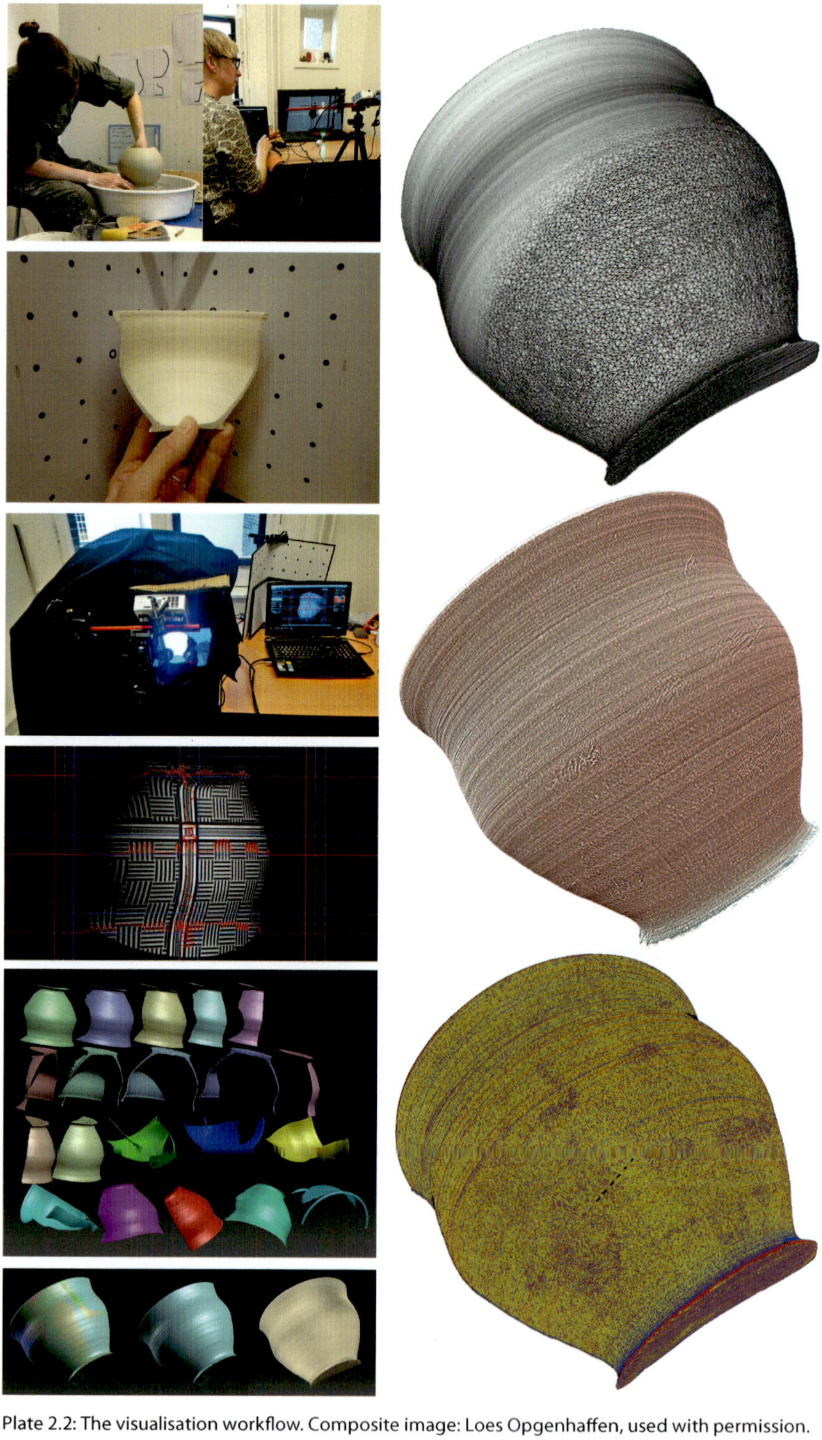

Plate 2.2: The visualisation workflow. Composite image: Loes Opgenhaffen, used with permission.

the analysis, experiment and visualization components combine to actively promote the reconstruction of technological practices among the sherd tables and within the minds of specialists and non-specialists alike. In this way, performative methods and material engagement in archaeological approaches can show us how understanding the past can inform and shape the future.

## Conclusion

Performative methods in archaeology provide a valuable heuristic tool for investigating the many behaviours and interactions of both producers and consumers of material culture. This chapter has outlined an integrated approach to material engagement across three arenas of archaeological action – experiment, analysis, and visualization – connected by an explicit engagement with the *chaîne opératoire* approach.

Technological choices were made by ancient potters, from raw material exploitation and preparation, to forming technique and firing strategy. These choices can be investigated through the study of physical traces on the vessel and composition, allowing archaeologists to reconstruct *chaînes opératoires* for specific vessel types. By identifying these sequences within archaeological assemblages, and tracing their use spatially and chronologically, archaeologists are able to consider the communities of practice in which potters were learning and performing their craft. Recognition of these craft communities is crucial for investigating new technologies such as the potter's wheel, where the adoption and transmission of such technologies are materialized through the social dynamics of craft community membership and interaction. The strength of archaeological approaches to material culture resides precisely in this ability to move from tangible objects to intangible behaviours and interactions of people in the past, as has been demonstrated for the ceramic assemblage from Akrotiri on Thera.

The case of Akrotiri encapsulates the depth and scope of research that is necessary at every site in order to use the *chaîne opératoire* approach to its full potential. This realization has led to the development of an innovative toolkit for the wider archaeological community so that they may replicate integrated methods and practices and generate larger numbers of compatible datasets that can be brought into a collective interpretation of this technological phenomenon. By building up from the foundational strengths of the *chaîne opératoire*, and explicitly sharing workflow practices, a common lexicon of terms for wheel use within ceramic production, and its study, is established. The benefits are twofold: interested non-specialists are able to acquire expertise and directly access the tools to gain competence, and active specialists are able to contribute their data and findings seamlessly. Given that ancient technologies pose large-scale regional and diachronic questions, fostering participation with peers and the integration of data from different scholars and research teams is critical. Just as the wheel was introduced into the existing *chaîne opératoire* of ceramic production, primarily as a bridging technique between hand-building and wheel-throwing techniques, the 3D scanner has been introduced to the archaeologist's toolkit, effectively bridging the representations of hand-drawn images and the realism of photographic techniques. This idea of a *tradition in transition* forms a common thread for the analysis of technological innovation in past and present practice.[64]

## Bibliography

Alram-Stern, Eva and Barbara Horejs, *Pottery Technologies and Sociocultural Connections between the Aegean and Anatolia During the 3rd Millennium BC*, OREA 10 (Wien: Austrian Academy of Sciences, 2018).

Berg, Ina, 'Meaning in the Making: The Potter's Wheel at Phylakopi, Melos (Greece)', *Journal of Anthropological Archaeology*, 26 (2007), pp. 234-252.

Bourdieu, Pierre, *The Logic of Practice* (Cambridge: Polity Press, 1990).

Broodbank, Cyprian, 'Minoanisation', *Proceedings of the Cambridge Philological Society*, 50 (2004), pp. 46-91.

Choleva, Maria, 'The First Wheelmade Pottery at Lerna: Wheel-Thrown or Wheel-Fashioned?', *Hesperia: The Journal of the American School of Classical Studies at Athens*, 81 (2012), pp. 343-381.

Dobres, Marcia-Ann, 'Of Great and Small Chaînes of Being: Toward Understanding the Sensual and Social Embodiment of Prehistoric Technology', in *Embedded Technologies: Reworking Technological Studies in Archaeology*, ed. by Brian Boyd and Bill Sillar (London: Routledge, 1999), pp. 1-19.

Dobres, Marcia-Ann, *Technology and Social Agency: Outlining a Practice Framework for Archaeology* (Oxford: Blackwell Press, 2000).

Dobres, Marcia-Ann, and Christopher R. Hoffman, 'Social Agency and the Dynamics of Prehistoric Technology', *Journal of Archaeological Method and Theory*, 1 (1994), pp. 211-258.

64 Opgenhaffen, *Visualising the Potter's Wheel.*

Dobres, Marcia-Ann, and John E. Robb, 'Doing Agency: Introductory Remarks on Methodology', *Journal of Archaeological Method and Theory*, 12 (2005), pp. 159-166.

Edmonds, Mark, 'Description, Understanding, and the Chaîne Opératoire', *Archaeological Review from Cambridge*, 9 (1990), pp. 55-70.

Evely, Doniert R. G., 'The Potter's Wheel in Minoan Crete', *Annual of the British School at Athens*, 83 (1988), pp. 83-126.

Evely, Doniert R. G., *Minoan Crafts: Tools and Techniques, Studies in Mediterranean Archaeology*, 92 (Göteborg: Paul Åström, 1993).

Gauss, Walter, and Evangelia Kiriatzi, *Pottery Production and Supply at Bronze Age Kolonna, Aegina. An Integrated Archaeological and Scientific Study of a Ceramic Landscape, Ägina-Kolonna, Forschungen and Ergebnisse* (Wien: Österreichische Akademie der Wissenschaften, 2011).

Gorogianni, Evi, Natalie Abell and Jill Hilditch, 'Reconsidering Technological Transmission: The Introduction of the Potter's Wheel at Ayia Irini, Kea, Greece', *American Journal of Archaeology*, 120 (2016), pp. 195-220.

Gosselain, Olivier, 'Materializing Identities: An African Perspective', *Journal of Archaeological Method and Theory*, 7 (2000), pp. 187-217.

Gosselain, Olivier, 'Social and Technical Identity in a Clay Crystal Ball', in *The Archaeology of Social Boundaries*, ed. by Miriam T. Stark (Washington D.C.: Smithsonian Institution Press, 1998), pp. 78-106.

Hägg, Robin, and Nanno Marinatos, *The Minoan Thalassocracy: Myth and Reality; Proceedings of the Third International Symposium at the Swedish Institute in Athens, 31 May-5 June 1982*, Vol. Skrifter Utgivna av Svenska Institutet i Athen 4 (Göteborg: Paul Åström's Förlag, 1984).

Hilditch, Jill, 'Analyzing Technological Standardization: Revisiting the Minoan Conical Cup', in *Understanding Standardization and Variation in Mediterranean Ceramics: Mid 2nd to Late 1st Millennium BC*, ed. by A. Kotsonas (Leuven: Peeters, 2014), pp. 21-33.

Hilditch, Jill, 'Ceramic Analysis in Greece', *Archaeological Reports*, 62 (2016), pp. 89-102.

Hilditch, Jill, 'Fabrics and Technology: Phases B and C', in *Akrotiri, Thera. Middle Bronze Age Pottery and Stratigraphy*, ed. by Irene Nikolakopoulou (Athens: Archaeological Society, 2019), pp. 377-470.

Hilditch, Jill, Duncan Pirrie, Carl J. Knappett, Nicoletta Momigliano, and Gavyn K. Rollinson, 'Taking the Rough with the Smooth: Using Automated SEM-EDS to Integrate Coarse and Fine Ceramic Assemblages in the Bronze Age Aegean', in *Insights from Innovation – New Light on Archaeological Ceramics: A Conference in Honour of David Peacock*, ed. by Emilie Sibbesson, Ben Jervis and Sarah Coxon (Southampton: The Highfield Press, 2016), pp. 74-96.

Hilditch, Jill, *Reconstructing Technical Choice, Social Practice and Networks of Exchange in the Middle Bronze Age of the Cyclades: The Ceramic Perspective*, Unpublished PhD thesis (Exeter: University of Exeter, 2008).

Hodder, Ian, 'Agency and Individuals in Long-Term Processes', in *Agency in Archaeology*, ed. by Marcia-Ann Dobres and John. E. Robb (London: Routledge, 2000), pp. 21-33.

Ingold, Tim, 'Materials against Materiality', *Archaeological Dialogues*, 14 (2007), pp. 1-16.

Jeffra, Caroline D., 'A Re-Examination of Early Wheel Potting in Crete', *Annual of the British School at Athens*, 108 (2013), pp. 31-49.

Jeffra, Caroline D., 'Experimental Approaches to Archaeological Ceramics: Unifying Disparate Methodologies with the Chaîne Opératoire', *Archaeological and Anthropological Sciences*, 7 (2015a), pp. 141-149.

Jeffra, Caroline D., 'Experimenting Wheel-Coiling Methods', *Arkeotek Journal* (2015b).

Jeffra, Caroline D., 'Forming Technology', in *Akrotiri, Thera. Middle Bronze Age Pottery and Stratigraphy*, ed. by Irene Nikolakopoulou (Athens: Archaeological Society, 2019a), pp. 471-477.

Jeffra, Caroline D., 'CRAFTER: Potting Techniques of the Bronze Age', *EXARC Journal* (2019b).

Jeffra, Caroline D., *The Archaeological Study of Innovation: An Experimental Approach to the Pottery Wheel in Bronze Age Crete and Cyprus*, Unpublished PhD thesis (Exeter: University of Exeter, 2011).

Jenkins, Richard, *Key Sociologists: Pierre Bourdieu* (London: Routledge, 1992).

Kiriatzi, Evangelia, '"Minoanising" Pottery Traditions in the South West Aegean during the Middle Bronze Age: Understanding the Social Context of Technological and Consumption Practices', in *Mesohelladika: The Greek Mainland in the Middle Bronze Age*, ed. by Anna Philippa-Touchais, Gilles Touchais, Sofia Voutsaki and James Wright, BCH Supplement 52 (Athens: École française d' Athènes, 2010), pp. 683-699.

Knappett, Carl J., 'Tradition and Innovation in Pottery Forming Technology: Wheel-Throwing at Middle Minoan Knossos', *The Annual of the British School at Athens*, 94 (1999), pp. 101-129.

Knappett, Carl J., and Jill Hilditch, 'Colonial Cups? The Minoan Plain Handleless Cup as Icon and Index', in *Plain and Simple. The Evolution and Significance of Plain Pottery Traditions in the Second Millennium BC Near East and East Mediterranean*, ed. by Claudia Glatz (Walnut Creek, CA: Left Coast Press, 2015), pp. 91-113.

Lave, Jean, and Etienne Wenger, *Situated Learning: Legitimate Peripheral Participation* (Cambridge: Cambridge University Press, 1991).

Leeuw, Sander E. van der, 'Giving the Potter a Choice: Conceptual Aspects of Pottery Techniques', in *Technological Choices: Transformation in Material Cultures since the Neolithic*, ed. by Pierre Lemonnier (London: Routledge, 1993), pp. 238-288.

Leeuw, Sander E. van der, Dick A. Papousek and Anick Coudart. 'Technical Traditions and Unquestioned Assumptions', *Techniques et Culture*, 18 (1992), pp. 145-173.

Lemonnier, Pierre, 'Introduction', in *Technological Choices: Transformation in Material Cultures since the Neolithic*, ed. by P. Lemonnier (London: Routledge, 1993), pp. 1-35.

Leroi-Gourhan, André, *Gesture and speech* (Cambridge: MIT Press, 1993).

Lis, Bartek, Maria Choleva and Štěpán Rückl, 'Mobility in the Bronze Age Aegean: The Case of Aeginetan Potters', in *The Transmission of Technical Knowledge in the Production of Ancient Mediterranean Pottery*, ed. by Walter Gauss, Gudrun Klebinder-Gauss and Constance von Rüden (Wien: Österreichisches Archaeologisches Institut, 2015), pp. 63-75.

Lock, Gary, and Brian L. Molyneaux, *Confronting Scale in Archaeology* (New York: Springer, 2006).

Macdonald, Colin F., Erik Hallager, and Wolf-Dietrich Niemeier, *The Minoans in the Central, Eastern and Northern Aegean: New Evidence, Monographs of the Danish Institute at Athens* (Athens: Danish Institute at Athens, 2009).

Mahias, Marie-Claude, 'Pottery Techniques in India: Technical Variants and Social Choice', in *Technological Choices: Transformation in Material Cultures since the Neolithic*, ed. by Pierre Lemonnier (London: Routledge, 1993), pp. 157-180.

Marsh, Erik J., and Jeffrey R. Ferguson, 'Introduction', in *Designing Experimental Research in Archaeology, Examining Technology through Production and Use*, ed. by Jeffrey R. Ferguson (Boulder, Colorado: University Press of Colorado, 2010), pp. 1-12.

Mauss, Marcel, *Sociologie et anthropologie* (Paris: Presses Universitaires de France, 1936).

Molloy, Barry P.C., 'Recent Advances in the Use of 3D Modelling for the Study of Archaeological Objects', *Open Archaeology*, 4 (2018).

Molloy, Barry P.C., and Marina Milić, 'Wonderful Things? A Consideration of 3D Modelling of Objects in Material Culture Research', *Open Archaeology*, 4 (2018), pp. 97-113.

Opgenhaffen, Loes, 'Visualising the Potter's Wheel: Developing Integrated 'Digital Science' Approaches for Identifying Wheel Use in Pottery Production', (forthcoming).

Opgenhaffen, Loes, Martina Revello Lami and Ivan Kisjes, 'Pottery Goes Public. Performing Archaeological Research Amid the Audience', *Open Archaeology*, 4 (2018), pp. 62-80.

Outram, Alan K., 'Introduction to Experimental Archaeology', *World Archaeology*, 40 (2008), pp. 1-6.

Peacock, David P.S., 'The Scientific Analysis of Ancient Ceramics: A Review', *World Archaeology*, 1 (1970), pp. 375-389.

Renfrew, A. Colin, *The Emergence of Civilisation: the Cyclades and the Aegean in the Third Millennium BC* (London: Methuen, 1972).

Revello Lami, Martina, Loes Opgenhaffen, and Ivan Kisjes, 'Pottery Goes Digital. 3D Laser Scanning Technology and the Study of Archaeological Ceramics', in *Proceedings of the 43rd Computer Applications and Quantitative Methods in Archaeology Conference 'Keep the Revolution Going' (CAA, 30 March – 3 April 2015), University of Siena*, ed. by Stefano Campana et al. (Oxford: Archaeopress, 2016), pp. 421-432.

Reynolds, Peter J., 'The Nature of Experiment in Archaeology', in *Experiment and Design: Archaeological Studies in Honour of John Coles*, ed. by Antony F. Harding (Oxford: Oxbow, 1999), pp. 156-162.

Roux, Valentine, 'Wheel Fashioned Ceramic Production during the Third Millennium BCE in the Southern Levant: A Perspective from Tel Yarmuth', in *Techniques and People: Anthropological Perspectives on Technology in the Archaeology of the Proto-Historic and Early Historic Periods in the Southern Levant*, ed. by Steven A. Rosen and Valentine Roux (Paris: Editions De Boccard, 2009), pp. 195-212.

Roux, Valentine, and Caroline D. Jeffra, 'The Spreading of the Potter's Wheel in the Ancient Mediterranean: A Social Context-Dependent Phenomenon', in *The Transmission of Technical Knowledge in the Production of Ancient Mediterranean Pottery*, ed. by W. Gauss et al. (Wien: Österreichisches Archäologisches Institut, 2015), pp. 165-182.

Roux, Valentine, and Marie-Agnès Courty, 'Identification of Wheel-Fashioning Methods: Technological Analysis of 4th–3rd Millennium BC Oriental Ceramics', *Journal of Archaeological Science*, 25 (1998), pp. 747-763.

Rutter Jeremy B., '"Minding the Gap": From Filling Archaeological Gaps to Accounting for Cultural Breaks. A 2013 Perspective on a Continuing Story', *American Journal of Archaeology*, 117 (2013), pp. 593-597.

Şahoğlu, Vasif, 'Interregional Contacts around the Aegean During the Early Bronze Age', *Anatolia*, 27 (2007), pp. 97-120.

Sassaman, Kenneth E., and Wictoria Rudolphi, 'Communities of Practice in the Early Pottery Traditions of the American Southeast', *Journal of Anthropological Research*, 57 (2001), pp. 407-425.

Schlanger, Nathan, 'Mindful Technology: Unleashing the Chaîne Opératoire for an Archaeology of Mind', in *The Ancient Mind: Elements of Cognitive Archaeology*, ed. by Colin Renfrew and Ezra B.W. Zubrow (Cambridge: Cambridge University Press, 1994), pp. 143-151.

Schlanger, Nathan, 'Techniques as Human Action: Two Perspectives', *Archaeological Review from Cambridge*, 9 (1990), pp. 18-26.
Shepard, Anna O., *Ceramics for the Archaeologist, Publication 609* (Washington DC: Carnegie Institute of Washington, 1968).
Sillar, Bill, and Michael S. Tite, 'The Challenge of 'Technological Choices' for Materials Science Approaches in Archaeology', *Archaeometry*, 42 (2000), pp. 2-20.
Stark, Miriam T., *The Archaeology of Social Boundaries* (Washington DC: Smithsonian Institution Press, 1998).
Wenger, Etienne, *Communities of Practice: Learning, Meaning and Identity* (Cambridge: Cambridge University Press, 1998).

## About the Author

**Jill Hilditch** is Associate Professor at Amsterdam Centre for Ancient Studies and Archaeology, University of Amsterdam. She specializes in the Bronze Age Aegean, ceramic analysis, and the reconstruction of production processes. She is PI of the Vidi project 'Tracing the potter's wheel' (2016-2021). This project uses the potter's wheel as a prism through which to investigate the transmission of craft knowledge amongst potting communities in the Aegean Bronze Age.

# 3. Making Musicians Think: The Problem with Organs

*Hans Fidom*

**Abstract**

Discussions regarding reconstruction, replication, or re-enactment in music can be fruitful only if musicians are experts in understanding the quality of the material with which they make their music: sounds. Such expertise can be acquired by analyzing and adapting the ways organists and organ builders deal with organ sounds, as each organ is an individual to a far greater extent than any other musical instrument. Organ builders discern how the thousands of pipes in an organ should sound and cooperate; organists have to able to understand the frames thus set. Generally speaking, it follows that composers' intentions are subordinate to musicians' and listeners' ones: music is something that sounds.

**Keywords:** organs, sound, situationality, voicing, music-making, listening, intentionality

However interesting discussions regarding reconstruction, replication, or re-enactment may be, musicians have no choice but to accept that the quality of the material they make their music with – sound – sets the frame for any decision.[1] So how can a musician assess sound, knowing that it exists in time only and never appears exactly the same way twice? What are its characteristics? How does it behave? Musicians playing portable instruments may not particularly feel the urge to let themselves be bothered by such questions, yet even they will need to check at least how the acoustics their

1 This text offers a first acquaintance with a larger project researching listening as a tool to develop new perspectives on conceiving, perceiving, and understanding music. The Orgelpark at Amsterdam will serve as a laboratory, inviting musicians, organ experts, and scholars to work together. The project is an initiative of the Chair Organ Studies at VU University Amsterdam.

Dupré, S., A. Harris, J. Kursell, P. Lulof, M. Stols-Witlox (eds.), *Reconstruction, Replication and Re-enactment in the Humanities and Social Sciences*. Taylor & Francis Group
DOI 10.5117/9789463728003_CH03

music is to be made in effects their sound palette. In this essay I propose dealing with what I like to call 'the problem with organs' as an effective way to practice sound quality assessment considering the listening strategies of organists, organ voicers, and organ consultants.

It is the highly inspiring attention scholars pay to listeners and listening that encouraged me to do so. Following the examples of philosophers like Theodor Adorno and Hans-Georg Gadamer,[2] who were among the first to position listening as a pivotal activity in musical situations, Gadamer even explicitly regarding organ music,[3] and of artists like Pauline Oliveros, who introduced the concept of 'deep listening',[4] new understandings of listening to music are being developed 'as we speak' by George Ellis Benson, Marcel Cobussen, Alicja Gescinska, Henkjan Honing, Jacob Lekkerkerker, Franziska Schroeder, Salomé Voegelin, and many others.[5]

The problem with organs meant here is that organs differ from one another significantly more than other instruments, to such a degree even that it seems best to say that *the* organ does not exist: each organ confronts with yet another sound concept. How do experts deal with that?

## Listening as an organist

The consequence of the fundamental diversity of organs is that striving to make convincing music on them leaves no choice but to listen carefully to the instrument to be played: what can it do, what not? How does the action 'feel', and how does it affect the sounds? In what ways do the acoustics of the space interfere, knowing that this specific organ has been made to sound optimally in this specific space? It may very well be that this preliminary

2 I would like to mention especially pivotal texts by Adorno and Gadamer here: Adorno, *Philosophie der neuen Musik*; Gadamer; *Die Aktualität des Schönen*.

3 Gadamer; *Die Aktualität des Schönen*, pp. 40-41.

4 Oliveros explains the concept here: http://ciufo.org/classes/ae_sp14/reading/deep_listening_intro.pdf. Checked on 4 February 2019. Interestingly, a comparable type of listening 'Het Grote Luisteren', i.e. 'The Great Listening', was introduced in 2005 by Elmer Schönberger; see his book *Het gebroken oor*.

5 Of high importance are Benson, *The Improvisation*; Cobussen, *The Field*; Gescinska, *Thuis in muziek*; Lekkerkerker, 'To be a Dancer'; Schroeder, *Soundweaving*; Voegelin, *Listening to Noise*. The new *Oxford Handbook of Music Listening in the 19th and 20th Centuries* necessarily focuses on 'circumstantial evidence' as it is impossible to ask listeners who are not present anymore how they listened. Regarding the approach of listening explored here, especially Wolfgang Gratzer's contribution to the handbook ('Is Listening to Music an Art in Itself – or Not?') is of interest: Gratzer thinks 'that "the art of listening" stands for creating meaning autonomously'.

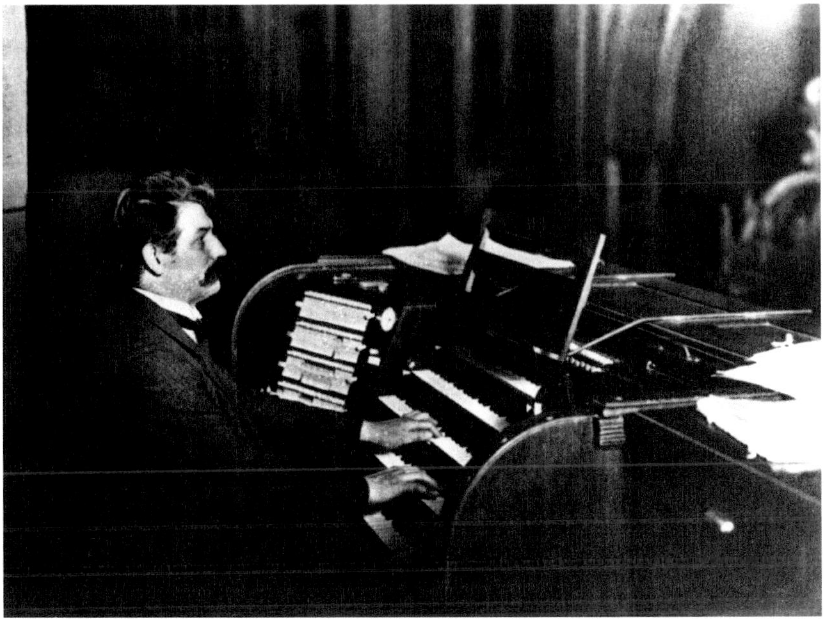

Figure 3.1: Albert Schweitzer, photographed on November 11th 1928, playing the Walcker organ at the Reinoldikirche in Dortmund. Schweitzer told Reinoldi-organist Gerard Bunk that he admired the organ. Yet it was in almost no part similar to what he called 'the true organ'.
Photo: Archive Gerard Bunk Gesellschaft, Dortmund

listening will affect major aspects of the music to be made, including timing and tempo, articulation and phrasing, and thus that one and the same score leads to very different 'musics'.[6]

In 1906, Albert Schweitzer, who was not only a famous doctor in Africa but a highly influential organ expert and organist as well, coined the term '*Die wahre Orgel*', 'the true organ'. It 'just' needed to be identified and henceforward be taken as a reference for any new organ to be built.[7]

In hindsight, that appears strange, since Schweitzer loved historical organs. The huge collection of historical organs[8] in Europe, spanning over five centuries, documents that every region 'produced' multiple sound

6 The word 'performances' is problematic here, as it implies that there must be something in place to be performed. This in fact betrays tacit idealism, as that something to be performed may be considered an ideal object. Instead, each music, in the sense of a thing that comes into existence through sound, is itself an artwork, not a more or less perfect 'iteration *of* an artwork'.
7 Schweitzer, *Deutsche und Französische Orgelbaukunst*; Cf. Fidom, *Diversity in Unity*.
8 Organs are most durable, as they provide separate sound generators per sound colour per pitch, i.e. pipes, which do not need to be exposed to human breath or direct touch in order to sound, and are made of sustainable materials such as tin, lead, and wood.

concepts over time. If one of these is to be named 'true',[9] what about the others? And: could any of these concepts ever fit the music from any composer perfectly? Or provide conclusive clues as to how to perform it?

It does not require elaborate listening sessions to detect answers to questions like these. Listen, for example via YouTube, to a few of the many so-called 'Bach organs' that have been built over the past few decades.[10] I would be the last one to criticize their sound concepts, yet the differences between them demonstrate that organ experts and organ builders, e.g. the ones that conceive and realize the sounds, do not agree on how the ideal Bach organ should sound, despite the fact that Bach's life, works and views have been researched more intensively than those of any other musician of the past. Or, listen to the many Arp Schnitger organs that have been restored over the past decades. Although Schnitger's significance in the field of organ building is comparable to that of Stradivarius in the field of violin building, restorers and other experts involved in attempts to represent, recreate, replicate Schnitger's sound concept restored Schnitger organs' sound in different ways, leaving 'us' unable to claim that 'we' agree on how Schnitger expected his organs to sound.

So, despite the dreams of organists like Schweitzer, and they were many in the 20th century, organs are not 'true' to any type of music. Instead, they are firmly rooted in everyday musical practices, where they invite, inspire and challenge musicians by providing them with different, often much more sound material than they look for. Musicians that are open-minded

9 Musicians and music scholars might be inclined to prefer the word 'authentic' here, yet, that has become rather contaminated by its discussion since the 1990s. A good introduction to this discussion is Cook, *Very Short Introduction*.

10 For example, the ones in Leipzig, Thomaskirche (2000); Dordrecht, Grote Kerk (2007); Vienna, Hofburg/Augustinerkirche (1985) – each of them is named 'Bach-organ' explicitly. YouTube offers ample recordings of these instruments. An interesting older 'Bach-organ' is the one in Bremen Cathedral (1966); a Bach-organ yet to be completed is being built in Leuven (the plan was to complete it in 2019, and then move it from the Predikherenkerk to the St.-Michielskerk). Compare the first four organs mentioned by checking these YouTube videos (I give some personal listening experiences to focus your attention [all videos checked on 16 February 2019]):

- Leipzig: https://youtu.be/sYD3nfNZVZ8
  silvery sound, gentle highs, middle voices not very clear
- Dordrecht: https://youtu.be/H-cuFpG7F3Q
  overall wooly sound quality, flexible wind pressure, middle voices audible well
- Vienna https://youtu.be/MBpohDd_v1w
  nasal sound, high frequencies powerful, middle voices a bit muffled
- Bremen https://youtu.be/GiQHv8z-p1w [4'20", especially 7'30"]
  high frequencies penetrating and clean, transparent, middle voices clearly audible

and willing to learn from the instrument to be played allow themselves the chance to reconsider their performance, and even question composers' intentions once in a while.

This reminds us of what Benson, Cobussen, and Schroeder propose, namely that music always contains a certain amount of improvization in the sense that musicians activate all kinds of 'traces' of past experiences, convictions, preferences, etc., when making music. If that music is made on an organ, these 'internal' traces may very well be activated by the 'external' ones the organ provides by presenting sounds the musician never considered before. Unsurprisingly, it is precisely this aspect of the diversity of organs, forcing us to attentively listen to them and, through that, revealing the joy and inspiration such listening can provide, that kept organists improvising through the centuries.

## Listening as a voicer

Organ builders build organs, but it is the voicers among them that finally determine the sound colours of the instruments. In practice, voicing is about slightly changing the geometry of pipes in such ways that they sound the way they are planned to. To be able to do that, the voicer has to acquire an 'image' of how each pipe, each stop and, finally, the organ as a whole should sound.[11]

How complex a task voicing is becomes clear when considering the structure of organs. Each organ consists of stops; each stop, controllable with the 'stop knobs' next to the organ keyboards and providing a specific sound colour, consists of as many pipes as the organ keyboard has keys. Assuming that the organ keyboards have at least 50 keys, even a small organ with only four stops already contains 4 x 50 = 200 pipes. The mid-size organs Johann Sebastian Bach used to play had circa 2,500 pipes, large organs 'normally' have about twice as many.[12] Now in order to make a stop produce a specific sound colour, the pipes of that stop need to resemble one another; every stop shows specific pipe forms and sizes.[13] There are two main 'families' among

11 This is the way one of the most influential voicers today, Munetaka Yokota (Tokyo), has phrased his task time and again.

12 One of the largest fully functioning organs in the world, at the Wanamaker Store in Philadelphia, has 28,750 pipes: https://wanamakerorgan.com/stoplist.php. Checked on 2 February 2019.

13 This is why measurements and scales of pipes are of decisive importance. To complicate matters a bit more, the sound colour 'inside' one stop may vary as well; for example, the lowest pipes of a Trumpet stop may have been made in such a way as to make them sound a bit muffled,

organ stops: flue stops (the way their pipes 'work' is comparable to that of recorders; obviously without finger holes) and reed stops (comparable to clarinets; no finger holes either).

The voicer determines the sounds of pipes – their colour, their power, their development – to a significant extent, despite the fact that the changes applied to their geometry are very small, sometimes almost undetectable. Even the effects of the material pipes are made of on their sounds[14] can be manipulated by the voicer.[15]

Adjusting the quality of organ pipe sounds is an artistic act of construction. A voicer will only be able to judge to what extent the sound of a pipe correlates to what is required if the reference 'image' is in place beforehand; in the process, that very image irrevocably undergoes changes, and so forth. As this process of interplay can be followed rather precisely by simply visiting and following a voicer, if only for a day (or two), it offers musicians a deeper way of understanding sound than 'just' listening to an organ as an organist, who is always already minding the music to be made. Voicers listen to the sound of every single pipe, or of a small set of pipes, as if it were a 'field' where all kinds of activity take place: there is rumble, jitter, there are all kinds of harmonics, some stronger, some weaker, and that not only in the 'stable sound', but also, and significantly differently, in the first instance, when the sound starts to exist, as well as in the last instance, when it stops living. How voicers respect the 'images' they work with and put them into perspective at the same time, deciding upon all these factors in a sound, may inspire musicians to recognize that *their* sound assessment skills can be enhanced in comparable interplay between, again, internal

whereas the pipes of that same Trumpet stop in the treble may have a slightly different scaling, as to make them sound, for example, a bit brighter.

14 As sound colour results from the proportions between the amplitudes of the harmonics of any sound, it is obvious that pipes with a relatively rough- or open-structured inner wall, which makes it hard for the less powerful upper harmonics to survive, will sound less bright.

15 This explains for a considerable part why Bach-organs and restored Schnitger organs do sound so diverse. Today's organ building practices do not always acknowledge this nonetheless rather physical fact. Quite often, organ builders tend to downsize the influence of voicers, claiming that a good pipe already should sound well immediately if made well, it being a doubtful sign if the voicer has to spend a lot of time on it. An often-used argument is that organ builders like Schnitger do not seem to have spent much time on voicing their instruments. Yet we have to keep in mind that Schnitger lived in the age of the guilds, in which developments in the arts proceeded slowly. His sound concept – and of course he most probably would not have been aware of even having one – was not challenged the way concepts today are; he 'just' had to apply and at times perhaps refine his own procedures. Today, however, organ builders have to decide what style a new organ should have, or how to revive the sound of an organ to be restored – which implies an enormous increase in responsibilities and skills on the part of the voicer.

and external influences, now, however, knowing the power and richness of the latter better.

## Voicing and research

Most organ building books do mention the topic of voicing, but keep it practical, 'just' explaining what effect a certain change of its geometry has on a pipe's sound. Yet, two research projects resulted in publications relevant here, as they go a bit further: one by acoustician Vincent Rioux (France) and one by organist João Segurado (Portugal). Rioux departed from a scientific perspective, Segurado, being a musician, developed and applied artistic research strategies. Both projects took place in Sweden, at the University of Gothenburg; since its start in 1995, the University's related Göteborg Organ Art Center GOArt has inspired many researchers to focus on organ-related topics.

Vincent Rioux published his thesis *Sound Quality of Flue Organ Pipes* in 2001. He collected listeners' perceptions and compiled 'classification lists' of 'descriptors' listeners used to describe sounds and sound aspects. Furthermore, he undertook 'a quantitative analysis' of this 'initial "qualitative" categorization'. Rioux mentions that organ pipes are not perfectly stable regarding the sound they produce. An adjustable 'hybrid pipe' was constructed, in order to 'follow in a rational way the geometrical adjustments occurring during a "real" voicing session on a "real" pipe'. As might be expected, such unavoidable reduction of aspects to take into account unavoidably leads to results that provide only the beginning of an understanding of how voicing works.[16]

João Segurado documented the way the organ builders Gerald Woehl and Claudius May-Woehl voiced the huge organ in Studio Acusticum in Piteå, North-Sweden. Part of the instrument was inaugurated in 2012; it still awaits completion. Linking the voicers' work to his own ways of playing the instrument during different voicing stages, Segurado decided to take the verb 'to voice' literally, as in 'giving the organ its voice', the 'ultimate

16 Rioux, *Sound Quality*, pp. 120, 131. Surprisingly, Rioux, otherwise rather meticulous, did not address the problematic use of loudspeakers as tools to assess sounds; loudspeakers, however, *do* change conception and perception of sounds – especially organ sounds. Audio expert Robert Harley opens chapter 6 in his *Complete Guide to High-End Audio* by stating that 'of all the components in your audio system, the loudspeaker's job is by far the most difficult.' A few lines later, he phrases positively yet clearly: 'Although even the best loudspeakers can't convince us that we're hearing live music, they nonetheless are miraculous in what they can do'.

question' being whether one could 'actually know something about the voicer's experience of listening'.[17] Thus, taking Woehl very seriously, who 'never really mentioned time as a parameter of voice or a major concern of voicing', Segurado states, referring to Sarah Voegelin's thinking on sound art, that 'we may [...] exclude the possibility of voice being an event or performance of sound-art [...]'. This illustrates to what complexity trying to think about sound in practice can lead, as opposed to the reduction Rioux felt he had to apply.

A useful middle way might be to keep track of the voicer's practice: as soon as voice is used to make music, it manifests itself in sound during a specific time in a specific place, making it in that specific situation[18] accessible to research, discussion, and assessment. Of course, the appearance of a sound can and will differ from situation to situation and from ear to ear, so its voice may be considered to be time- and spaceless, as Segurado supposes. Yet, that doesn't necessarily complicate assessing sound; it is 'simply' a matter of interconnecting the multiple appearances of a given sound, subsuming them to the string of situationally identifiable singularities they together form, and thus ascribing it identity.[19]

## Intermezzo: Loudspeakers and sound

An important aspect to be mentioned here is that listening skills today are different from the ones Schnitger or Bach had at their disposal. Not just because today's societies and cultures are fundamentally different from former ones, but especially because we listen to loudspeakers most of the time. Think of it: in the kitchen it is the radio or TV that sounds, in our living room we enjoy hi-fi or perhaps even high-end systems, in shops muzak drips from ceiling speakers, during concerts we experience stage systems, and in between, we block the sound of the outside world by putting in our smartphone earbuds. As a result, our ears have become so used to the physical inability of loudspeakers to reproduce sounds 'lossless' – a term borrowed here from the realm of high-end audio reproduction –, i.e. with the colour, power, timing and spaciousness the music had when it

17 Segurado, *Never Heard Before*, pp. 136-139.

18 Since this is so obvious, yet often neglected in everyday musical practice, I introduced the term 'situationality' as a tool to rethink music-making: Fidom, *Music as Installation Art*.

19 Situationality understood this way thus opposes idealism, or may be said to lead to 'inverted idealism': it negates effectively the idea that a score includes an ideal performance, or that a sound has an ideal form – both to be strived for by musicians and voicers.

was recorded, that we may be in danger of taking loudspeaker sound as a reference instead of literally 'immediate' – as in *not* mediated by any loudspeaker system – sound.[20]

## Listening as an organ consultant

During an organ consultant conference organized by the Dutch Cultural Heritage Agency on 29 January 2016, dedicated to listening and sound, the general opinion unsurprisingly was that sound would be too volatile to be discussed in depth. Furthermore, improving sound assessment skills by using recording gear was judged suboptimal; the quality of sound recording devices was doubted.

As a result of this attitude, and in fact ironically enough, it is exactly that very volatile thing, i.e. sound, that is the only reflection of the intuitions, craftsmanship, and artistry the respective restorers, voicers, and their consultants apply. Texts and/or recordings relating sounds before and after restoration, do not exist. Even dr. h.c. C.H. Edskes, internationally considered one of the most influential organ consultants of the 20th century, did not publicly share his sound assessment strategies, notwithstanding the fact that he led restorations of world-famous organs such as the ones in the Nieuwe Kerk at Amsterdam (restored in 1981) and the Martinikerk at Groningen (1984). What if he would actually have published his most probably valuable insights about sounds and about how to apply listening as a tool to judge them? Both instruments sound impressive. They even seduce, at least me, to consider that they produce sounds that might very well resemble the ones they made when they got their present appearance – respectively in 1651 and 1740. But by lack of accounts, I cannot be sure. In the case of the Martinikerk organ, the sounds may as well carry our ears towards 1450.

20 An interesting source on this topic is Harley, *The Complete Guide*. It is available in e-pub format, so it has no page numbers. Harley says in the first paragraphs of his chapter on loudspeakers: 'Although even the best loudspeakers can't convince us that we're hearing live music, they nonetheless are miraculous in what they can do. Think about this: a pair of loudspeakers converts two two-dimensional electrical signals into a three-dimensional "soundspace" spread out before the listener. Instruments seem to exist as objects in space: we hear the violin here, the brass over there, and the percussion behind the other instruments. A vocalist appears as a palpable, tangible image precisely between the two speakers. The front of the listening room seems to disappear, replaced by the music. It's so easy to close your eyes and be transported into the musical event. To achieve this experience in your home, however, you must carefully choose the best loudspeakers from among the literally thousands of models on the market. As we'll see, choosing loudspeakers is a challenging job'.

Or 1542, or 1628, or 1855, for those are (just some of) the years in which the instrument was changed as well.

Two recent organ sound related discussions may serve to give an impression of the ways organ consultants listen: one regarding a listening test organized in the course of the preparation of the concept of the so-called Utopa Baroque Organ in the Orgelpark at Amsterdam, the other regarding the arguments applied to substantiate the plans for the restoration of the Schnitger organ at Zwolle.

## The listening test preparing the Utopa Baroque Organ

The Utopa Baroque Organ was inaugurated on 21 March 2018. Its concept, facade, technology, and interfaces were developed by the Orgelpark itself.[21] The organ was built according to the principle of 'process reconstruction': it was tried to imagine how organ builder Zacharias Hildebrandt, who lived and worked in the same time and region as Johann Sebastian Bach, would have built an organ in a hall like the Orgelpark; to follow the resulting lead by actually building such an organ; and, finally, to make it sound as if it had already lived for over two-and-a-half centuries. The aim was not to build yet another 'Bach-organ' but to conceive a sound world that would fit historically informed performances of Bach's music as well as inspire making new music. That the Utopa Baroque Organ meets both demands was the wish of the Orgelpark.

The preparations started in 2013. The main parts to be reconsidered were the so-called 'wind-chests', the large wooden boxes (often meters long and over a meter deep, and about 20-30cm high), on which the pipes stand, according to the stops they belong to.[22] Bellows are used to pump air ('wind') into the chests and valves in the chests prevent the wind entering the pipes and making them sound ('speak'). The organist

21 The organ was built in cooperation with four organ builders. Elbertse Orgelmakers (Soest, The Netherlands), built the case, the bellows, the windchests, and part of the wooden pipes; Hermann Eule Orgelbau (Bautzen, Germany) built the other part of the wooden pipes and all metal pipes; Sinua (Düsseldorf, Germany) designed the hardware and software for the digital interface of the organ; Munetaka Yokota (Tokio, Japan) was responsible for the voicing.

22 The Utopa Baroque Organ has six wind chests: two for the so-called 'Hauptwerk', located in the middle of the organ, and playable on the lower keyboard; two for the 'Oberwerk', located over the Hauptwerk, and playable on the upper keyboard; and finally two for the 'Pedal', playable with the pedal (foot) keyboard. The Hauptwerk has 13 stops, the Oberwerk 13 stops as well, and the Pedal 7 stops. The facade of the organ shows this set-up explicitly.

controls these valves with the keys and the stop knobs. How that works is best understood by comparing wind-chests to chessboards, as they have a similar structure. Having eight rows along its 'x-axis' (named a-b-c-d-e-f-g-h) and eight rows along the y-axis (named 1-2-3-4-5-6-7-8), a chessboard resembles a wind-chest for an organ with eight keys and eight stops. Now let's suppose that the stop on row 3 has a trumpet-like sound (i.e. eight pipes with that sound, each on a different pitch), and stop 5 a fluty sound. Assuming that we like the sound they make together, we need to open stops 3 and 5 first, by pulling the respective stop knobs; that way, we open all valves under all pipes of stops 3 and 5. Playing key c would open the valve under row c and make pipes c3 and c5 sound together. Playing the keys c, e, and g, would make pipes c3, c5, e3, e5, g3, and g5 sound together, etc. The possibilities are great, but what's not possible is assigning any pipe or set of pipes to any key, so that, for example, pipes c3, e6, g1 and h7 would sound on key d.

The double function of the Utopa Baroque Organ required that this grid remained in place in order to enable playing early music in historically informed ways. On the other hand, the organ should allow overriding the grid as well in order to enable musicians to compile new sounds for new music. This meant that every pipe had to have its own valve, which in turn should be controllable by a dedicated electromagnet. All this implied that the organ would need to have two sets of keyboards and stop knobs: a traditional one controlling the traditional key and stop valves, and a twenty-first century one with a computer aiding the organist to control the electromagnets.

Interestingly, the so-called 'spring chest' (a specific type of wind chest) in the Huss/Schnitger organ at Stade, Germany (1675) provided the perfect model for the wind-chests the Utopa Baroque Organ needed, as it provided a valve for each pipe. Yet one issue irritated: it appeared that conventional organ electromagnets, needed to control this many valves, might disturb the wind flow to the pipes in case the organ would be played on its traditional keyboards, thus compromising its baroque sound concept.

One of the main sessions of the International Orgelpark Symposium in 2014 aimed at finding solutions to this problem. Around forty organ experts and organ builders from all over the world were invited to take part in a listening test in the Orgelpark, as that obviously needed to be the place to assess the sounds to be tested (the acoustics an organ has to function in determine major choices regarding its sound). A little test organ was built and positioned on the balcony were the Utopa Baroque Organ would eventually stand. It was built in such a way that the same pipes could be heard on two systems, one with objects in the wind flow – electromagnets – and one without. Would the experts detect any difference by ear?

The listening experiment was based on having the same pipe(s) sound twice in short motifs, using both systems; the listeners, each equipped with a response form, had the 'simple' task of indicating whether they detected a difference, and if so, how they would describe it. The organ had two stops: a 'Principal' (having a bright, stringy sound) and a 'Gedackt' (a darker, softer sound).

The test took place on 6 and 7 June. The results were remarkable. A considerable number of respondents detected differences, even when the motifs sounded twice in the same sound colours on the same system.[23] Nonetheless, there were many respondents as well that did not detect differences. The Orgelpark concluded that there were insufficient grounds to consider mounting electromagnets in the wind flow unproblematic.

The reason to be so cautious was that the respondents detecting differences found it hard to describe what they heard. Being experts, they apparently required themselves to be able to detect differences between the systems by ear, if only because conventional wisdom in nineteenth and twentieth-century organ building literature has it that wind chest design has a huge influence on an organ's sound quality. To prevent biasing future listeners this way, the Orgelpark decided to not include any electromagnets in the wind chests. After quite a bit of research, a different, much smaller magnet, originally developed for industrial purposes, appeared appropriate, as it could be mounted outside the wind flow.

It would be nonsensical to conclude from the test that organ experts are not excellent listeners. On the contrary, they are able to identify many sounds of many pipes, recognize the sound concept they belong to, and sometimes even the specific organ they're in. Apparently, it is possible to compile a mental databank of sound by listening over and over attentively to pipes, stops, and complete organs. Yet such sound memory systems can and will interfere with the information the ear is transferring to one's brain and influence its perception of the aural input. Which is to say that the way organ consultants listen extends the lesson to be learned from organists and voicers: unbiased listening is impossible, even if one desires it.

23 Published in *Orgelpark Research Report* #5/2, pp. 30-31.

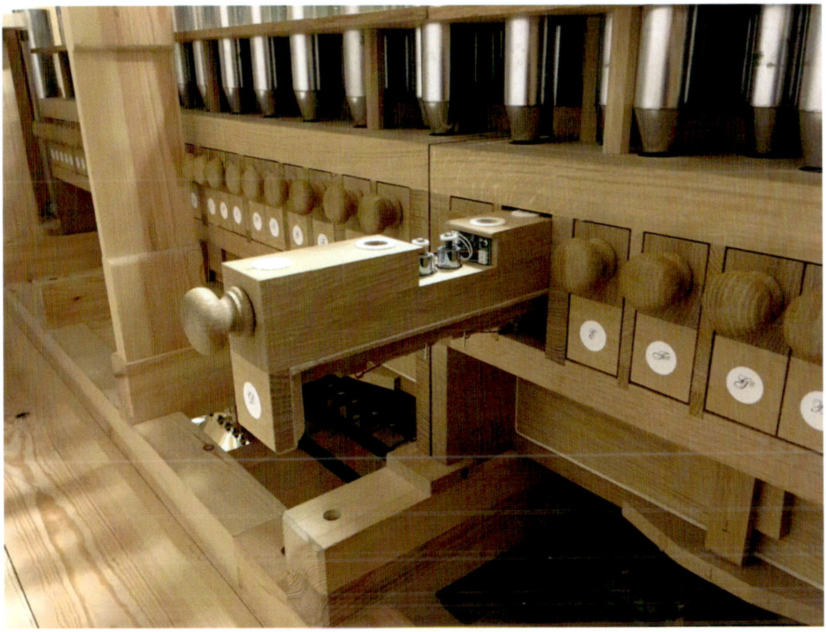

Plate 3.1: Close-up view of a few of the over 1,600 magnets in the wind chests of the Utopa Baroque Organ. Photo: author.

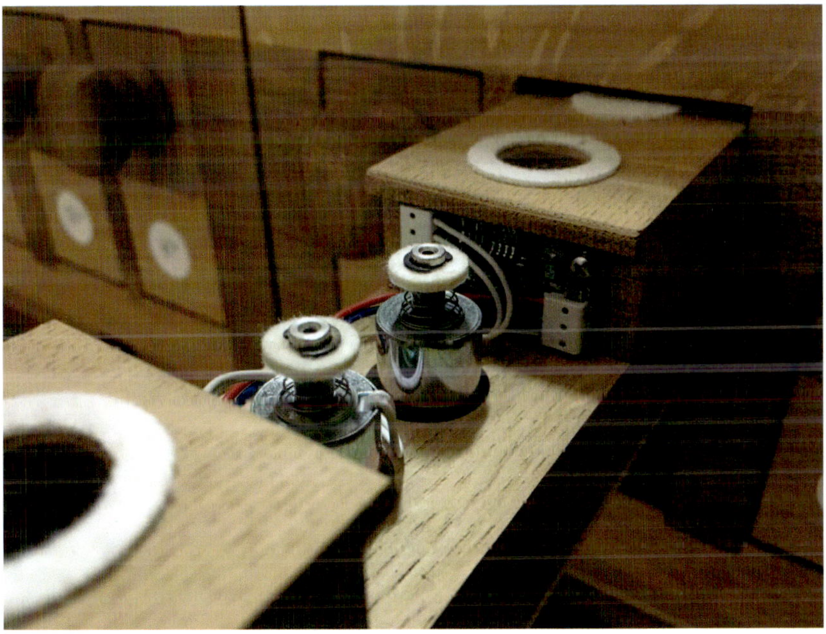

Plate 3.2: Close-up view of a few of the over 1,600 magnets in the wind chests of the Utopa Baroque Organ. Photo: author.

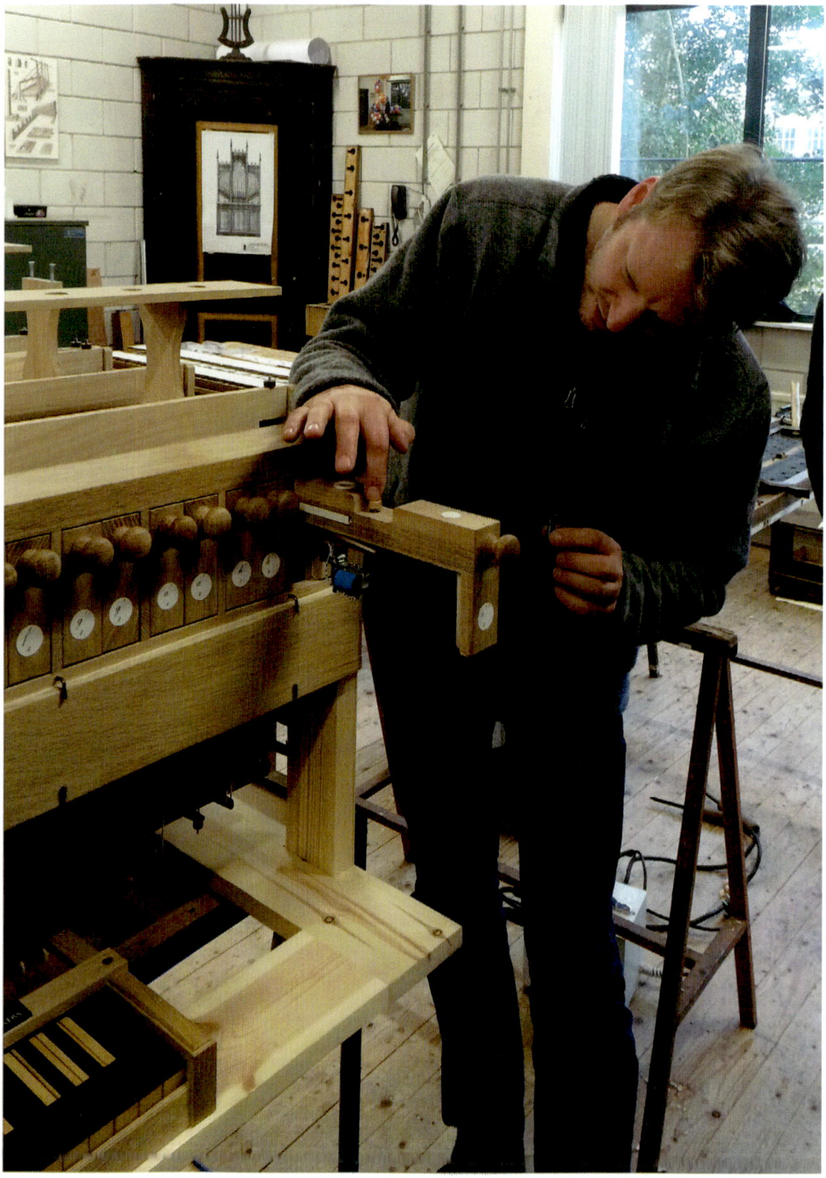

Plate 3.3: The test organ used for the listening test in 2014 in the Elbertse workshop, Soest. An organ builder has pulled out one of the beams on which the valves are mounted; a classic electromagnet (with the blue body) is clearly visible. The organ builder pushes the respective valve open with his finger. Photo: author.

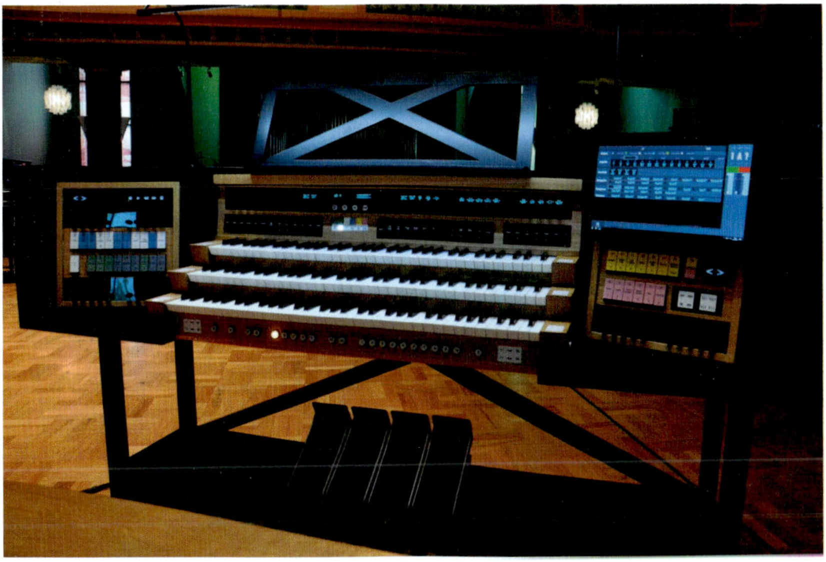

Plate 3.4: The freestanding console with which the Utopa Baroque Organ can be played as well has touch sensitive keyboards, allowing controlling the valves in the wind chests very precisely. Compare Plate 3.5. Photo: Orgelpark / Joyce Vanderfeesten.

## The Schnitger organ in Zwolle

The preparation for the upcoming restoration of the organ in the Grote Kerk in Zwolle documents how opinions about a sound concept can rather quickly change over only a few decades.

The Zwolle organ has been internationally renowned all its life: On its completion in 1721, it was the largest in the Netherlands. It was built by Arp Schnitger; after his death in 1719, his sons Franz Caspar and Johann Jürgen completed the instrument. Following the natural course of organs' lives, the instrument was changed a few times, especially in the nineteenth century. In 1950, it was decided that organ builder Dirk Andries Flentrop would restore it. It turned out that enough Schnitger material had survived in order to try and reconstruct the state the organ had been in in 1721. In the late 1980s, the organ began to need attention. The reeds were 'revoiced', and in 2000/2001 the flue stops were 'optimized as well as possible'.[24] Subsequently, it was decided to restore the organ in phases. The first phase was taken care of in 2008: Restoration of the action (the connection between keys/stops and

24 https://www.flentrop.nl/restauratie/zwolle_grot.html. Checked on 2 February 2019. See as well the impressive 1000+ page archival research report consultant Cees van der Poel compiled in 2014 (only available as pdf), *Verleden en huidige staat van het orgel.*

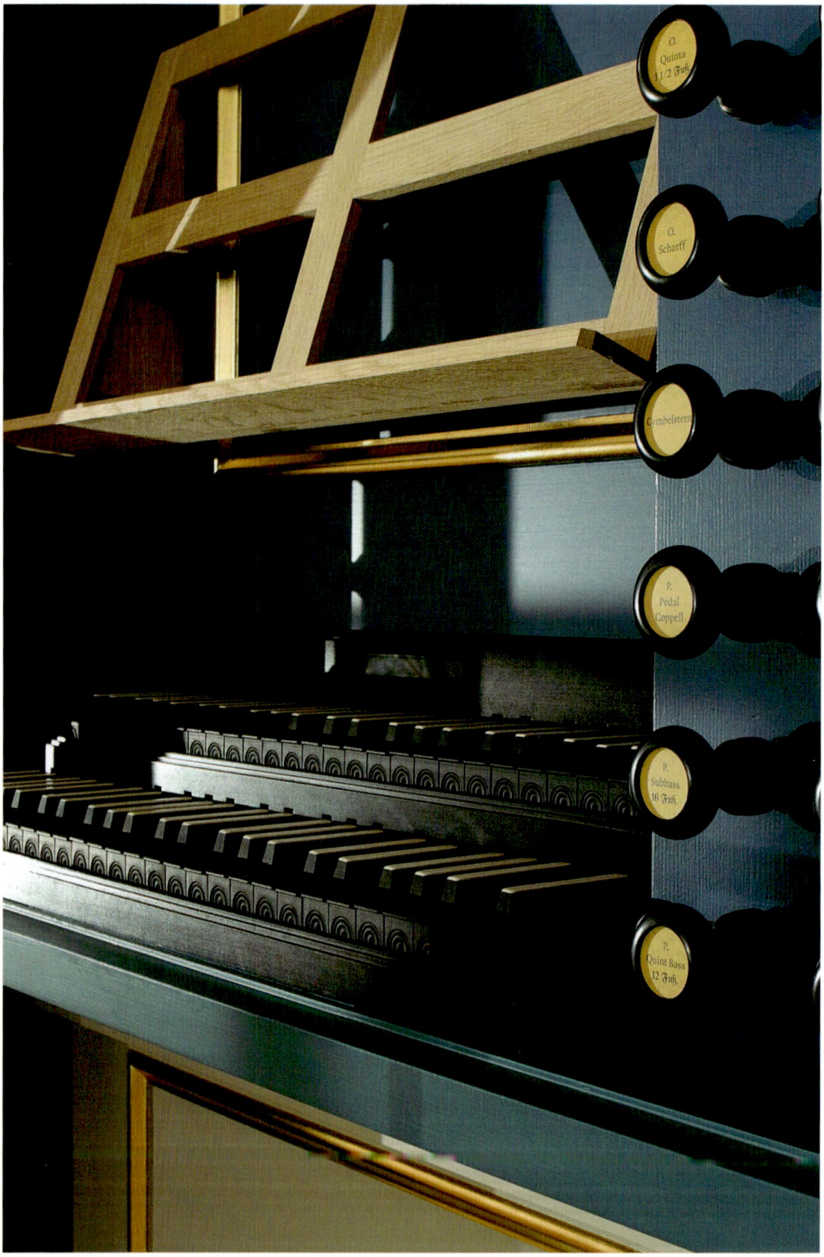


Plate 3.5: The built-in keyboards of the Utopa Baroque Organ are mechanically connected to the valves in the wind chests. Photo: Orgelpark / Joyce Vanderfeesten.

the wind chests). In 2013, the wind supply system was restored. Whereas other historical organs have lost their original bellows, the unique 12-bellow Schnitger system survived at Zwolle. The next step will focus, among many other things, on the sound of the organ. It has been decided to strive for the same goal as in 1950, and to complete the task Dirk Andries Flentrop had set himself. Having realized that he could not be entirely successful, he decided to keep many original parts of the organ safely close by, humbly hoping that future generations might be able to continue and finish his work. Many of these parts may now be reinstalled.

The other side of the story of the Zwolle organ is that it became rather famous between 1955 and the late 1980s. Charles de Wolff, a conductor/organist who managed to enthuse audiences for organ music that otherwise would have neglected it, saw the Zwolle organ as one of the finest available.[25] He considered its sound fit not only for baroque music (such as the works by Johann Sebastian Bach), but as well for highly complex nineteenth-century organ music (such as the huge choral phantasies by Max Reger), and mid-twentieth-century compositions (including György Ligeti's *Volumina*, requiring the organist to make all pipes sound at the same time). De Wolff recorded no less than 18 LP's in Zwolle. A quote from an interview in 1996: 'My connection with this organ has always been tight. It keeps fascinating with its tremendous richness. Describing the sound? One has to experience it. I can combine stops in most remarkable ways, yet everything just sounds good.' And 'I never felt the urge to temper the strict sound. […] No, maintaining the current state is my preference.'[26]

Another organist who found the Zwolle organ right for Bach was the French organist Michel Chapuis, who, in 1968, recorded part of his LP series with Bach's complete works at Zwolle. Listening to these today shows a remarkable taste for refined stop combinations, i.e. sound colours. It is important to remember here that Chapuis loved the bright, clean and transparent sound new organs were given after World War II, as did, by the way, the even more famous Marie-Claire Alain in her first integral Bach organ works recordings (not in Zwolle). Apparently, Chapuis saw in the Zwolle organ the historic evidence that the post-war organ sound aesthetics were the right ones, at least for Bach's music.

A third indication that the sound Flentrop had given the Zwolle organ was considered musically very relevant, was the foundation of the Stichting Schnitgerprijs Zwolle in 1963. The goal of the foundation was to 'raise

25 De Looze, 'Een Schnitger'.
26 De Looze, 'Een Schnitger'.

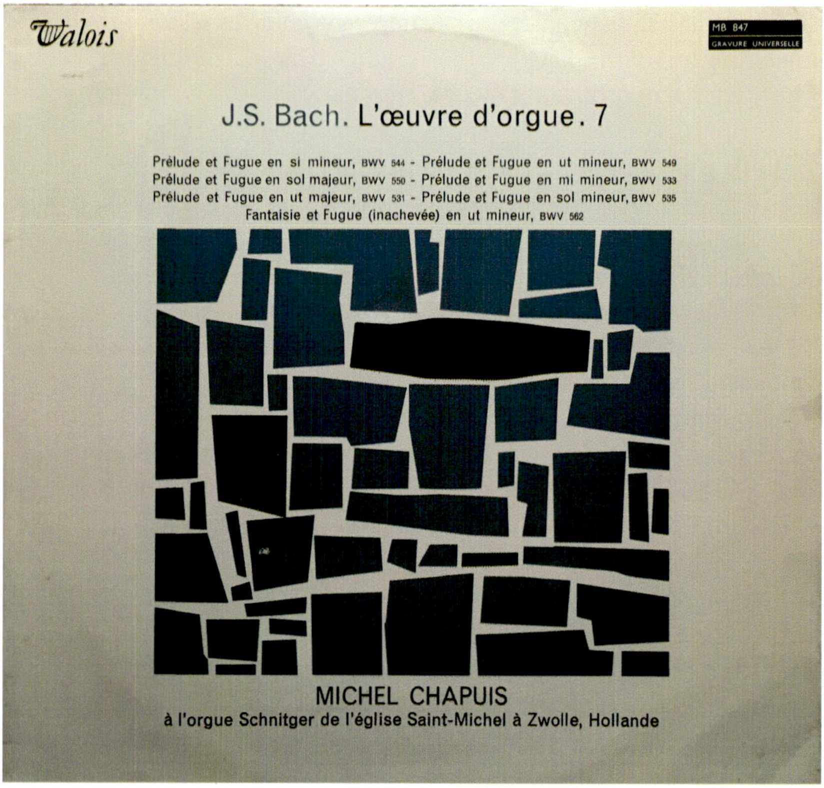


Plate 3.6: Cover of one of the two Bach-LP's Michel Chapuis recorded at Zwolle in 1968. Photo: author.

attention' for the organ, as well as 'making this organ continuously serve the enrichment of contemporary organ music.' It organized a bi-annual international composition competition that lasted until 1976 and assigned composers to compose new music in the years in between. In 1967 it started a cooperation with the International Organ Festival at St Albans, England; compositions that had won the Zwolle competition would be performed there as well. Archives document that the competition was internationally welcomed; institutions and persons from 27 countries corresponded with the foundation.[27]

Since the early 1990s, the sound new and restored organs were given in the 1950s and 1960s has been increasingly considered to need 'correction'.

27 The archives of the foundation are listed here (checked on 20 February 2019): https://www.archieven.nl/nl/zoeken?mivast=0&mizig=210&miadt=141&micode=1517&milang=nl&mizk_alle=schnitgerprijs&miview=inv2#inv3t1. The website includes a short history of the foundation as well.

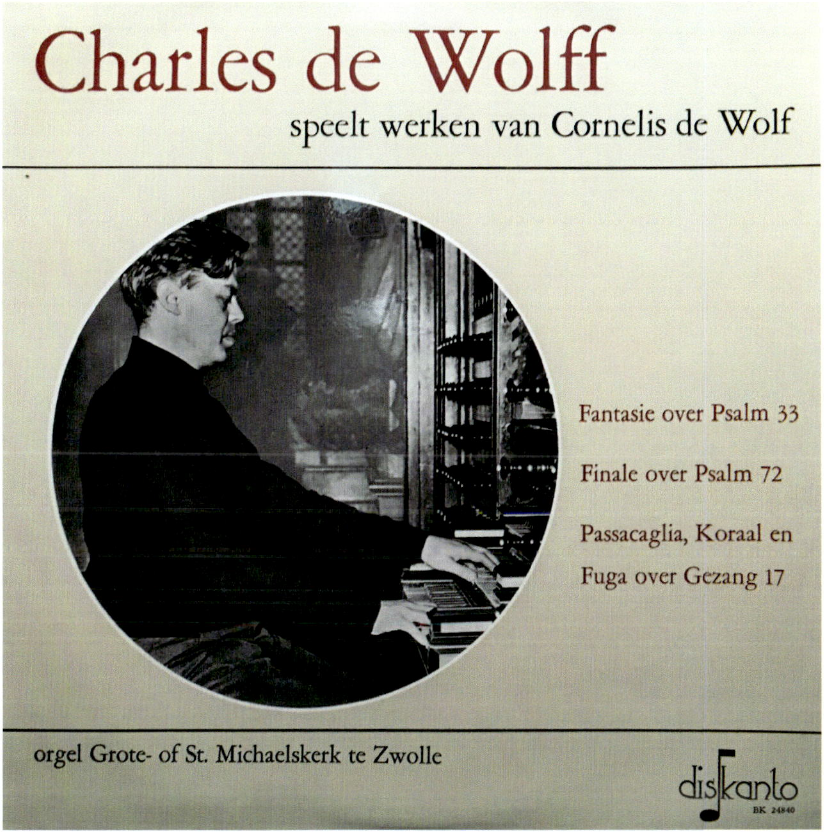


Plate 3.7: One of the earlier LP's Charles de Wolff recorded at Zwolle. Without date [1966]. Photo: author.

In particular, the energy of high frequency pipes has been reduced in quite a few organs; one of the first to get corrected, in 1990, was the organ at the Kloosterkerk in The Hague, built by the Danish firm Marcussen in 1966. Flentrop's website mentions that the activities that took place in Zwolle in 2001 actually concerned an 'intonatiecorrectie' as well.[28]

28 https://www.flentrop.nl/restauratie/zwolle_grot.html. Checked on 5 February 2019. Typically, these corrections include making toeholes (the holes through which the pipes are blown) of flue pipes larger, and consequently widening the windways (which is to say that the entrance of the wind to the pipe, the toehole, and the exit of the wind, the windway, are changed). The main argument to do so is that whereas this manipulation of a pipe's geometry is relatively non invasive, it sounds significantly easier and less tense as a result. It often indeed seems as if the pipes begged for this. Yet, claiming that the changes could be considered non-invasive, being ever so small, betrays a vision on sound that prefers the visual over the auditory: it may be the case that the eye can (almost) not detect changes in pipe geometry, but if the ear can easily detect them, they need to be qualified as invasive, all the more as there is no going back.

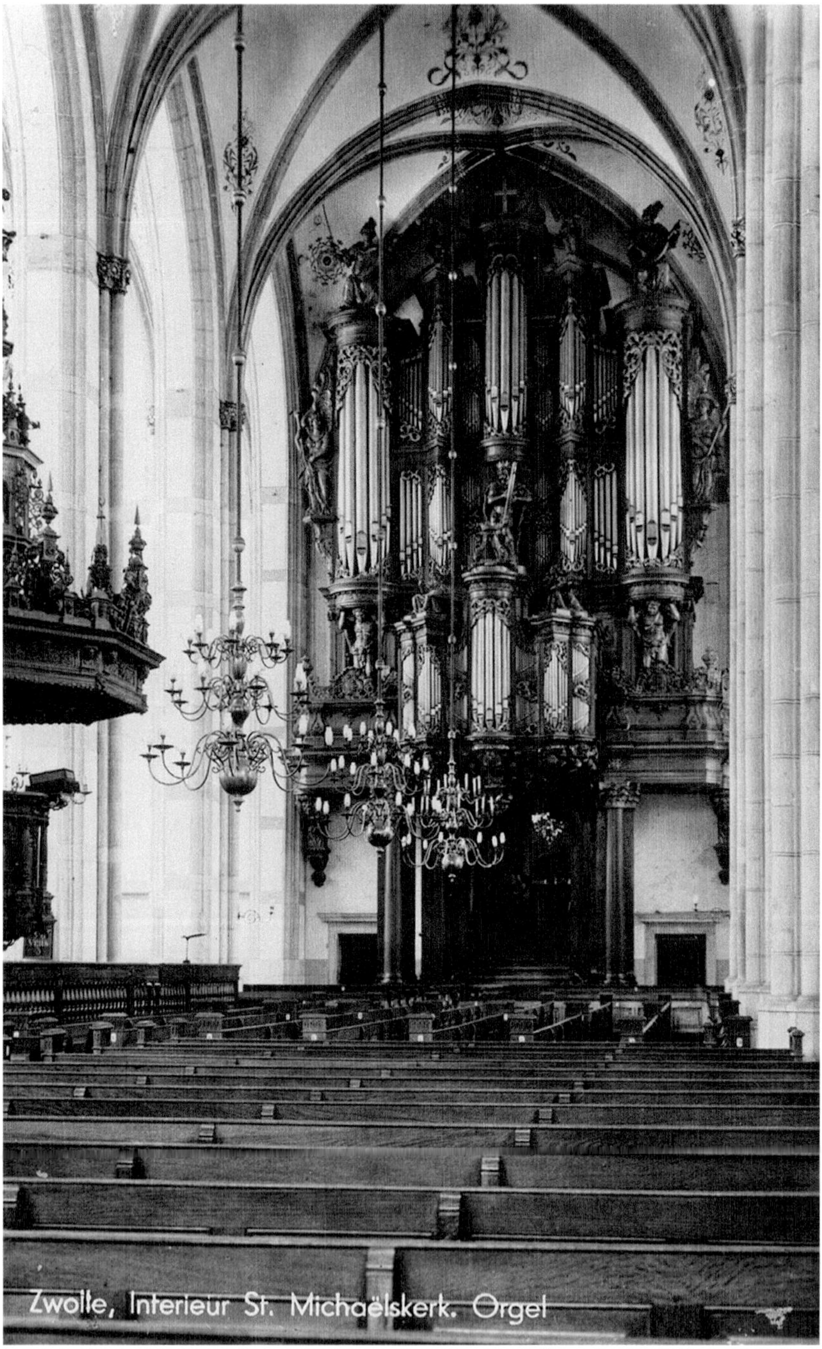


Figure 3.2: The Schnitger organ at Zwolle, Grote Kerk, on a photo card from the 1970s. Meanwhile, the benches have partly been removed. Photo: author.

The question is: Did organ builders in the 1950s really not know better when they gave these instruments their typical sound? Or did they actually strive for clear, high frequencies and a certain tension? It may very well turn out that 'we' today will appear to have done exactly what is despised in the majority of publications on organ history, namely pretending to know better than our predecessors, including, in Zwolle, Charles de Wolff, Michel Chapuis, and, to add yet another famous organist, André Jolivet, who in 1969 composed *Mandala* for the Zwolle organ.[29]

Five Sound Heritage students from VU University Amsterdam researched the 'Zwolle case' in 2018. They concluded that 'the consequence of "going back" [to 1721] is the loss of evolved sound'. Asking the experts involved why Schnitger's sound concept would need to be considered superior over the later one, it surprised the students to be confronted at times with answers betraying a 'highly romanticized view of Schnitger and "authenticity"'. They noticed as well that the interviewees agreed that it should be considered 'impossible to fully return to the original', as the 1955 sound concept had already been changed in the process of the past decades. The students' conclusion: 'A balance is needed: whether proposing a simple repair, conservation or restoration, some historical information is always going to be stripped away – not only impacting the artefact but also limiting the opportunities of future generations.'

## Conclusion

Musicians work with a most volatile material: sound. The ways organists, organ voicers, and organ consultants deal with the diversity of organ sound may inspire them to try and understand it better, as it is the basis for any decision regarding how to make music.

Organists listen to the instrument to be played first, as they need to get acquainted with the sound material it provides. Forcing the sounds to fit preconceived idea(l)s about how a piece should sound, for example inspired by the wish to express composers' intentions, or making it resemble any 'true' organ, is normally not what organists do; the sounds are given, hence it is the presentation of the music that needs to be adapted. It may even be the case that unexpected sounds inspire new musics.

29 Other relevant compositions include *Concerto per organo, due Trombe, due Corni en due Tromboni* by Albert de Klerk (1967), and *Music for Organ* by Wim de Ruiter (1972), and *Hymn* by Jo van den Booren (1985).

Organs not only challenge to listen attentively, they enable doing so in a very gentle way as well, as their sounds can be alive as long as necessary.[30] This opens the opportunity for musicians to not only study how organists apply and develop listening skills but check other organ experts' (organ voicers and organ consultants) listening strategies as well.. Voicers are the ones giving organs, which can have thousands of pipes, their multi-faceted sound; consultants decide 'by ear' about sound and apply that ability when developing organ building or restoration strategies.

What voicers do is best understood as an act of construction, rather than reconstruction; the sound qualities they literally 'have in mind' are both referred to and changed during the voicing process. By doing so, voicers in fact suggest strategies towards making convincing music; musicians may apply *their* skills in a comparable way, i.e. by activating a reciprocality between the concept of what they wish their music to be and their situational understanding of the sounds to be worked with – one 'simply' influences the other. Furthermore, following a voicer for a day or two reveals how complex even single sounds can be.

Organists' and voicers' practices thus suggest that it does not have to be considered a problem that performances of a given score will sound (more or less) different time and again; the identity of 'musics' based on that score can be conceived by understanding them as singularities to be subsumed to a set; the set being identifiable without requiring considering its diversity as a problem.

Following organ consultants when listening professionally adds an extra lesson, namely that it is not possible to focus exclusively on the information the ear sends to our brain, even if one really wants to. Especially when asked to assess sounds, expert listeners will want to prove they are experts indeed and safeguard their reputation: Do I detect what I should detect in this sound? Being an expert seems to be troublesome; by building up a sound memory system, that very system is bound to interfere with what the ear detects.

It does not surprise that the organ field itself has trouble researching sound in depth; even organ sound remains volatile, hence not easy to assess, which leaves room to foster strong intuitions. A major one is that recovering or replicating a sound concept from, say, the eighteenth century, is possible despite evidence – i.e. restored organs – proving that the opposite is true. The case of the Zwolle organ shows that this intuition can be so powerful that the sound that made the organ famous in the second half of the twentieth century most probably will have to disappear. It betrays a late trace of typical

30 In Halberstadt, Germany, a 639 year (or longer) lasting performance of John Cage's Organ2/ ASLSP started in 2000. Website: https://www.aslsp.org/de/ (checked on 5 February 2019).

nineteenth-century belief in progress: it may be the case that organists loved the Zwolle organ between 1955 and the late 1980s for good reasons, but 'we' know better now.

In short, the not so tacit suppositions that start resonating immediately when words like replication, reproduction, and re-enactment are used, should be suspended when music is at stake. The sound, being the material the music is to be made with, has to be assessed first. The way experts deal with the diversity of organs suggests how that may be done. It leads to a state of mind that accepts that music is about paying attention to a time and space-dependent succession of sounds. Given that frame, it becomes possible to determine to what extent replication, reproduction, and re-enactment are conceivable at all.

## Bibliography

Adorno, Theodor, *Philosophie der neuen Musik* (Tübingen: Mohr, 1949).

Attali, Jacques, 'Noise: The Political Economy of Music', in *The Sound Studies Reader*, ed. by Jonathan Sterne (London: Routledge, 2012), pp. 29-39.

Benson, Bruce Ellis, *The Improvisation of Musical Dialogue* (Cambridge: Cambridge University Press, 2003).

Carlsson, Tobias, *On Dynamic Behaviour of Wind Systems for Pipe Organs*, (Gothenburg: Chalmers University of Technology, 2002).

Cobussen, Marcel, *The Field of Musical Improvisation* (Leiden: LUP, 2017).

Cook, Nicholas, *Very Short Introduction to Music* (Oxford: Oxford University Press, 1998).

Cressman, Daryl, *Building Musical Culture in Ninetheenth-Century Amsterdam* (Amsterdam: Amsterdam University Press, 2016).

Fidom, Hans, *Diversity in Unity. Discussions about Organ Building between 1880 and 1918* (Dieren: KNOV, 2002).

Fidom, Hans, 'Music as Installation Art', in *Orgelpark Research Report #2*, ed. by Hans Fidom (Amsterdam: VU University Press, 2012), pp. 15-39.

Fidom, Hans, 'The Utopa Baroque Organ at the Orgelpark', in *Orgelpark Research Report #5/2*, ed. by Hans Fidom (Amsterdam: VU University Press, 2017), pp. 15-72.

Gadamer, Hans-Georg, *Die Aktualität des Schönen. Kunst als Spiel, Symbol und Fest* (Stuttgart: Reclam, 1977).

Gescinska, Alicja, *Thuis in muziek. Een oefening in menselijkheid* (Amsterdam: De Bezige Bij, 2018).

Harley, Robert, *The Complete Guide to High-End Audio* by Robert Harley (Carlsbad, California: hifibooks.com, 2015).

Honing, Henkjan, *Iedereen is muzikaal: wat we weten over het luisteren naar muziek* (Amsterdam: Nieuw Amsterdam Uitgevers, 2009).
Lekkerkerker, Jacob, 'To be a Dancer on the Organ', in *Improvisation. Orgelpark Research Report #3*, ed. by Hans Fidom (Amsterdam: VU University Press/Orgelpark, 2017/2), pp. 587-620.
Looze, Gert de, '"Een Schnitger die niet verveelt". Al 275 jaar rollen forte-klanken uit tegen de gewelven van de Zwolse Grote Kerk', in *Reformatorisch Dagblad* (16 August 1996).
Pirsig, Robert, *Zen and the Art of Motor Maintenance* (New York: Morrow and Co, 1974).
Poel, Cees van der, *Verleden en huidige staat van het orgel in de Grote Kerk van Zwolle* (Zwolle: Stichting Schnitgerorgel Zwolle, 2014).
Rioux, Victor, *Sound Quality of Flue Organ Pipes. An Interdisciplinary Study on the Art of Voicing* (Gothenburg: Chalmers University of Technology, 2001).
Schönberger, Elmer, *Het gebroken oor* (Amsterdam: Meulenhoff, 2005).
Schroeder, Franziska, *Soundweaving: Writings on Improvisation* (Cambridge: Cambridge Scholars Publishing, 2014).
Schweitzer, Albert, *Deutsche und Französische Orgelbaukunst und Orgelkunst* (Leipzig: Breitkopf & Härtel, 1906).
Segurado, Joao, *Never Heard Before. A Musical Exploration of Organ Voicing* (Gothenburg: University of Gothenburg, 2015).
*The Oxford Handbook of Music Listening in the 19th and 20th Centuries*, ed. by Christian Thorau and Hansjakob Ziemer (Oxford: Oxford University Press, 2019).
Voegelin, Salomé, *Listening to Noise and Silence. Towards a Philosophy of Sound Art* (New York: Continuum, 2010).

## About the Author

**Hans Fidom** is Chair of Organ Studies at Vrije Universiteit, Amsterdam, and a leader at Orgelpark Research Amsterdam. Questioning the ways music and musical instruments act in practice, thus trying to understand the interrelations of artistic and academic research, is central to Fidom's work. He supports his research by being active in the field himself as an organist and an organ expert. Hans Fidom's PhD was a socio-musicological investigation into the way in which discussions in periodicals and other written records influenced the development of German organs around 1900. From 1996 to 2006, Hans Fidom was editor of the magazine *The Organ*. He was active as secretary of the Foundation International Organ Competition Haarlem and as advisor to the Arts Council of Groningen. He is examinator of the Dutch school for organ expertise.

# 4. Making Sound Present: Re-enactment and Reconstruction in Historical Organ Building Practices

*Julia Kursell and Peter Peters*

**Abstract**

Pipe organs have always reflected the artisanal knowledge and skills of their makers. Since the 1990s, replicating historical instruments has opened new ways of studying practices of knowledge and making that revolve around historical organs. In this chapter, we argue that the idea of how a pipe should sound always involves a certain degree of reenactment of the building process. We follow the history of explicating concepts of sound, from one of the most prominent sources for organ builders and reconstructors written by the 18th-century monk Bédos de Celles via an explication of why sound was not centrally discussed in sources before the mid-19th century up to the historically informed pipe making and voicing by Japanese organ builder Munetaka Yokota.

**Keywords:** pipe organ reconstruction, Dom Bédos de Celles, concepts of sound, organ pipe voicing, history of sound concepts

## Introduction

In his treatise *L'art du facteur d'orgues* ('The Art of the Organ-Builder'), French Benedictine monk and organ builder Dom François Bédos de Celles (1709-1779) described what is nowadays seen as the conundrum of organ sound: 'The entire instrument can be very well built, the pipes perfectly well manufactured and conditioned, and still this can be a very bad organ'.[1]

1 Bédos de Celles *L'Art du facteur d'orgues*, p. 425. 'Tout l'Instrument peut être très-bien construit, les Tuyaux parfaitement bien faits & bien conditionnés, & cependant faire un fort mauvais Orgue.' (our translation)

Dupré, S., A. Harris, J. Kursell, P. Lulof, M. Stols-Witlox (eds.), *Reconstruction, Replication and Re-enactment in the Humanities and Social Sciences*. Taylor & Francis Group
DOI 10.5117/9789463728003_CH04

Making the pipes 'speak', as Bédos continues to explain, requires careful attention: 'this instrument being made only to be heard, it is essential to give it a good and pleasing harmony'.[2] Many delicate operations, called 'voicing'[3] are necessary to make sure that the wind passes optimally through the sound-producing mechanism. Only after having voiced all the pipes can the organ builder begin to tune them, i.e. to set them to the appropriate pitches.

Today, voicing is held to be the moment at which the organ builder conveys the signature sound quality to the instrument. The tweaking of pipes that are almost ready to play, but do not sound properly yet, seems to conceal the secret of the good organ builder. The particular timbre of the instrument will depend on the skill of the voicers, and it seems that Bédos was the first to utter this. What listeners eventually appreciate in the instrument – whether consciously or not – is in the hands and ears of these experts. When entering contemporary workshops, one often finds the wall decorated with plates from historical treatises such as Bédos's. Especially organ builders with an interest in the long history of organ building – actually a regular feature in the craft – seem to hold him in high esteem. Even though organ builders have adopted new practices, including machinery that was unavailable when Bédos wrote his treatise in the second half of the eighteenth century, there seems to be a sense of connecting back to this historical moment of making the art of organ building accessible.

The sound of the organ, as the two authors of this chapter contend, has always been the central problem in organ building. Yet, sound means something different to Bédos than to today's organ builders. This problem becomes particularly relevant when today's organ builders try to approximate the historical sound of the instruments. This entails the necessity to question not only materials and operations, but also ways of hearing and listening. Sound's fleeting nature has not changed since the time of Bédos, but the ways in which sound can be reproduced have changed dramatically. This has affected the notion of sound itself. The intent to understand what organs sounded like in the past necessarily implies addressing sound according to one's present notion of it.

How can we then make sense of this additional element – the notion of sound –, if we want to understand practices of re-enactment in organ building? Obviously, recreation, re-enactment and reconstruction will always be guided by today's conditions for learning to hear. This article

2 Bédos de Celles *L'Art du facteur d'orgues*, p. 425. 'Cet Instrument n'étant fait que pour être entendu, il est essentiel de lui donner une bonne & agréable harmonie.' (our translation)

3 See on this also the contribution by Hans Fidom in this volume.

proposes a framework that involves three chronologically arranged case studies that allow us to reconstruct three approaches to sound's fleeting nature. The first is that of the French monk and organ builder Bédos; the second concerns a shift in the understanding of sound that began during the second half of the nineteenth century; the third focuses on one expert in the reconstruction of historical organs, namely Munetaka Yokota, who was responsible for the casting and voicing of the pipes in several projects of reconstructing historical instruments.

Historical pipe organs are artefacts with a biography. Built for spaces with their own functions and acoustics, they are unique. Their biographies, sometimes spanning hundreds of years, make pipe organs interesting objects of musicological and organological study.[4] As 'mirrors of their time', pipe organs not only reflect the musical and artistic ideals and practices of a certain period, but also the artisanal knowledge and skills of their makers. As such, they offer an important contribution to an epistemic history of musical artefacts that focuses on the persistence of practical knowledge through technologies and materials.

Thus, organ building, in its richness of information, is a fruitful object in the study of re-enactment practices. For disclosing this knowledge, we will introduce a twofold approach to methodologies of reconstruction, re-enactment and replication. We discuss, on the one hand, the replication and reconstruction of the material object, and on the other hand the way in which historically distant concepts of sound and hearing get in the way of the community of organ builders and researchers who also need to make the result accessible to the present concepts of sound and hearing. We follow a chronological order to tease out the differences between the historical settings from which we reconstruct the concepts of sound and their relations to manual and aural practices.

The three parts of this chapter instantiate a different logical order of these two approaches: Bédos naturally begins with casting the metal sheets from which the pipes are made to then take the organ builder through the construction of a pipe and its voicing. Casting is central in this case study, as it epitomizes the extent to which the organ builder depended on carrying out the necessary operations in one go. Even though the good sound of the organ is the goal during this moment, it is at the same time very distant from this operation. By contrast, nineteenth-century scientists and organ builders found themselves confronted with a new malleability both of their basic materials and of sound, and accordingly also attempted to free the discourse

4 In musicology, 'organology' refers to the study of musical instruments more generally.

from the accidental features inherent to craft. We argue that this allegedly independent concept of sound is a necessary precondition for today's organ builders' practices. Today's reconstructing craftspeople don't simply attempt to reverse the logical order of building and listening, but they are in fact very conscious of the very reversal, as we will seek to instantiate using the example of Yokota. Hence our working hypothesis: Organ building has always relied to a certain degree on practices of re-enactment. What has changed is the degree to which the idea of malleability of sound guides the re-enactment process.

## Organ building in the eighteenth century – Casting the pipe in one go

### Dom François Bédos de Celles

Dom Bédos was a Benedictine monk who became widely known as an organ adviser and organ builder. Born in the Hérault in France, he entered the congregation of St. Maur in the region of Toulouse at the age of seventeen. He later moved to the abbey of Sainte-Croix in Bordeaux, where he became the abbey's secretary. Three years after moving he finished his first organ, which was built for the monastery church. How he had come to learn the craft is not fully known. Very likely he had been apprentice to the Toulouse based Jean-François Lépine, as the congregation was interested in having its own organ builder.[5] Two of the organs Bédos built are still in use. The one for Sainte-Croix was later transferred to another church in Bordeaux, the second is located at the basilica Notre-Dame des Tables in Montpellier.

Already before 1760, Bédos had made himself a name as a savant. He was elected to the *Académie des Sciences* of Bordeaux, and in 1760 went to Paris to finish a treatise on sundials and gnomology. Starting from 1763, he lived in Paris permanently, where he died at the abbey of Saint Denis in 1779. Bédos is best known for the treatise *L'Art du facteur d'orgues*. Published between 1766 and 1778, its four volumes were part of the *Déscriptions des arts et métiers* that were commissioned by *the Académie Royale des Sciences*. Its 113 volumes provided a detailed account of the state of eighteenth-century manufacturing, craftsmanship, and trades. *L'Art du facteur d'orgues* is the most detailed historical document on organ building in the eighteenth century, and it is still used by organ builders as a reference or source of inspiration.

5 On the biography of Dom François Bédos de Celles see Steinhaus, *Wege zu Dom Bedos*; Steinhaus, 'Bédos de Celles', https://www.mgg-online.com/mgg/stable/47819, checked on 2 December 2018.

Bédos's treatise is in line with the overarching project of a representation of craftsmanship. The four volumes addressed the insider and, in addition, also attempted to embed organ building in a broader context. The first volume of the treatise set the stage by introducing general mechanical principles, an overview of tools, and an introduction to the various divisions in an organ. The second volume started off from distributing competences. Here, Bédos addressed the commissioners rather than the organ builders themselves, reminding them of what had to be considered once the decision to build a new instrument had been taken. Most often, Bédos insinuated, commissioners were unable to anticipate what the envisaged space would allow them to have. It was therefore crucial to involve the organ builder from the very beginning. In the ensuing chapters, this second volume expounded in great detail the actual building and construction of the various parts of an organ. The third volume addressed the organists, instructing them how to maintain their instrument. The fourth and final volume added descriptions of concert organs and small organs, as well as cylinder organs.

As is characteristic of the volumes of the *Descriptions des arts et métiers*, Bédos's treatise is accompanied by rich illustrations of excellent quality. The plates added to each of its volumes show detailed illustrations of organ parts, tools, and working procedures. In this combination of text and images, the publication shared the same stylistic characteristics of the *Encyclopédie*. While the *Déscriptions* did not have the same reach as the *Encyclopédie* seen on the longer historical perspective, illustrations such as those in the organ treatise can be seen to adhere to the same policy of publicizing knowledge. As Charles Kostelnick has argued, visualizing knowledge stood at the centre of the *Encyclopédie* project.[6] Existing technical drawing conventions were used, such as perspective, cutaways and exploded views to see the inside of objects, as well as letters and figures as labels for linking image and text.[7]

In addition, the craftsmen were represented as interacting with technology and performing practical tasks in a physical and cultural context such as the manufacturing shop.[8] This points to a shift from knowing entirely through the mind to knowing through experience and sensory perception. As Kostelnick explains, this shift was characteristic of Enlightenment epistemology.[9] Another illustrative convention in the *Encyclopédie* was the use of images to narrate a process: 'Rhetorically, each plate integrates the temporal elements of its story into a coherent narrative over space and time and is often

6 Kostelnick, 'Visualizing Technology'.

7 Kostelnick, 'Visualizing Technology', p. 444.

8 Kostelnick, 'Visualizing Technology', p. 447.

9 Kostelnick, 'Visualizing Technology', p. 447f.

accompanied by numerous drawings of tools and equipment visualized in the narrative.'[10] John R. Pannabecker has examined how the resulting method of uniform representation by Denis Diderot and Louis-Jacques Goussier, highlighting smooth process and lack of obstacles, eventually omits the tacit knowledge, experience and problem-solving that go with the inherent difficulties and failures of actual work.[11] He analyses how contemporary writers like Claude-Henri Watelet and Benoît-Louis Prévost attempted to represent some of the more intuitive, tacit aspects of the arts and reflected more on matters of personal judgment than did either Diderot or Goussier.

## Casting pipes

To give a better idea of how Bédos used text and images to transmit the art of organ building, the process of casting pipe metal serves as a good example. The chapter on this subject has three sections, each divided into numbered paragraphs. Plates are referred to in the text and sometimes also in the margins. Three plates relate to this chapter (see our Figures 4.1, 4.2 and 4.3). The first plate presents a view of the organ builders' workshop with the melting pot on the left, a slanted casting bench and several objects and tools. Two workers operate a sliding box, into which one of them pours the molten alloy. Some of the tools are shown in more detail below, among which, again, some function in a way that resembles an exploded view.

The first, relatively short, section of the chapter explains which tin to use and how to test it: English tin is the best, tin from Hamburg and the Netherlands should not be used. Testing can be done by using a cupped test stone and watching the texture of the surface of the melt – the object in question can be seen in Figure 4, plate LXIV (our Figure 4.1) lying on the floor and labelled 'G'.[12] The second section explains how the melting pot is constructed; it should be placed in such a way that it does not hinder the workers. This refers to Figures 1 to 3 on the same plate, to end up in the overview of the workshop in Figure 4. Next, the casting bench is described, as well as the remaining tools to be seen on this plate.

A lengthy description ensues of how to construct the casting bench in such a way that it will not warp as a result of the heat of the molten metal. There are two ways to position the casting table: Slanted, which is most

10 Kostelnick, 'Visualizing Technology', p. 448.

11 Pannabecker, 'Representing Mechanical Arts'.

12 It does not seem logical that the test stone lies on the ground, which means that rather than representing an actual situation, the image is stylized to give room to the various objects mentioned in the text.

common according to Bédos and shown in the overview of the workshop as well as in the less detailed entry of the *Encyclopédie* (our Figure 4.4), and horizontal, which is shown on a different plate and is the one that Bédos himself prefers. That plate (LXVI, our Figure 4.2) also shows the construction of the table. After explaining that the table can be covered with layers of cloth, molton or linen, Bédos describes the construction of the casting box and the rails that guide it along the edge of the table.

The third section is devoted to the actual casting. To establish the temperature of the molten metal, a piece of white paper should be used: If it stays white, the metal should be heated more. This and other details suggest that Bédos knew from his own experience, or from experienced workers, how to cast. The illustration is now referred to, allocating letters to the two collaborating craftsmen. When casting the molten alloy in the casting box, worker C should press the box tightly to the table, whereas worker D should keep the casting spoon low to prevent the metal from flowing over the hands of worker C. A skimmer is used to clean the molten metal, and a stick to stir it. The right moment to start sliding is when the surface becomes 'grainy' or 'sandy'. While walking along the bench, the box has to be pressed tightly, and at the end of the table one should walk faster because the temperature has dropped already, which causes the metal to solidify. Surplus tin is collected in the trough at the end of the table. Bédos makes clear that the goal of the correct casting procedure is to get metal plates that have the right thickness. Plate LXXII shows the design of a double casting box in Figure 4.3. Such a box enables the casters to even more precisely control the thickness of the metal sheet.

Overall, the tools are described and visualized in great detail, including human figures and even objects that seem mundane, such as the broom to clean up spilled tin in the workshop illustration. They present situations – e.g. the casting bench and the box in their relation to the melting furnace – that are assembled for a specific purpose, such as controlling the thickness of the metal plate. The use of paper to establish the temperature or the position of the casting spoon to prevent accidents can be seen as instances of verbalizing and visualizing tacit knowledge or embodied skills. Yet, the most important purpose of the description is to convey the rationale of this step in manufacturing the pipes. The missing element of this rationale is the sound of the pipes. Any of the steps can eventually result in, and be controlled by, listening to the pipes during voicing. For the organ builder this loudly tacit aspect could not possibly be addressed. The goal – good sound – was always implied, yet without any means to differentiate within this concept according to the required procedure.

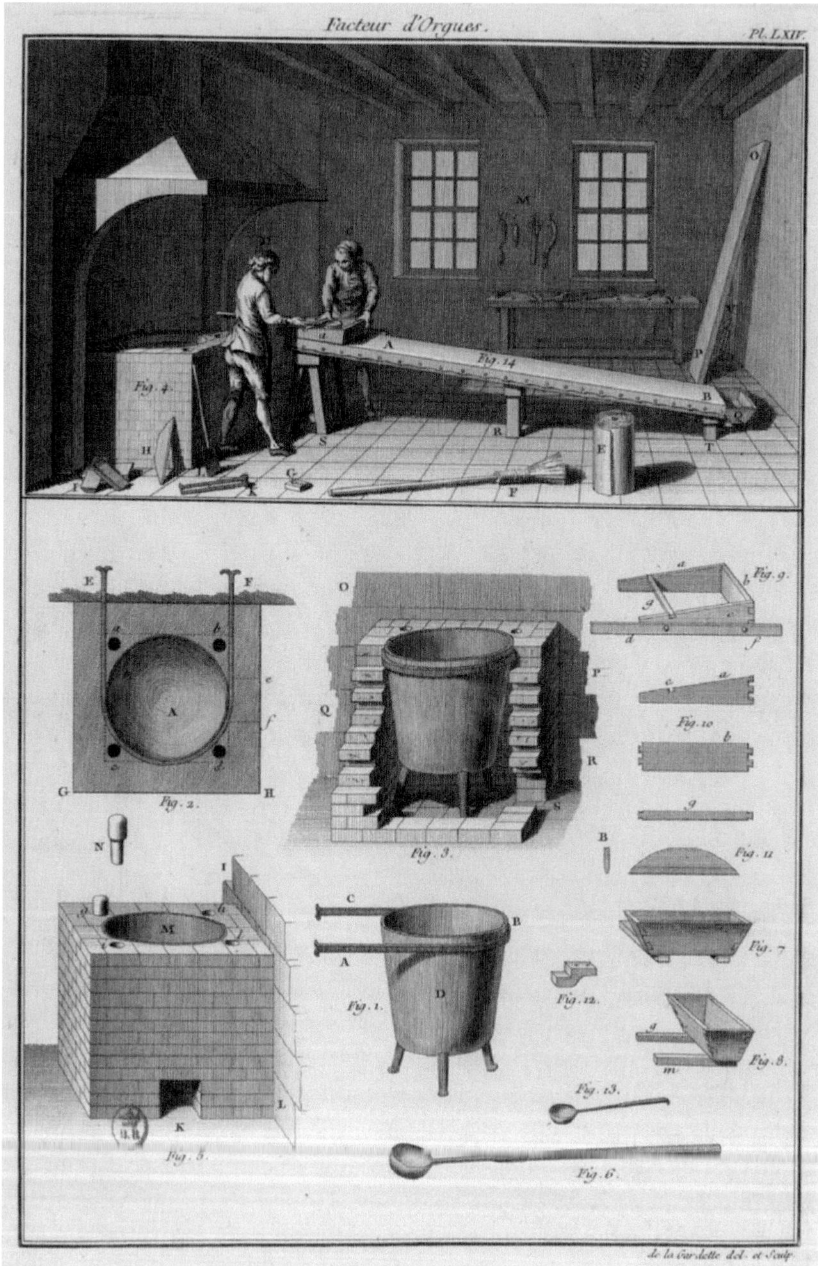


Figure 4.1: Bédos, *Facteur d'Orgues*, Plate LXIV. Photo: IMSLP – Petrucci Music Library. IMSLP – Petrucci Music Library, https://imslp.org/wiki/.

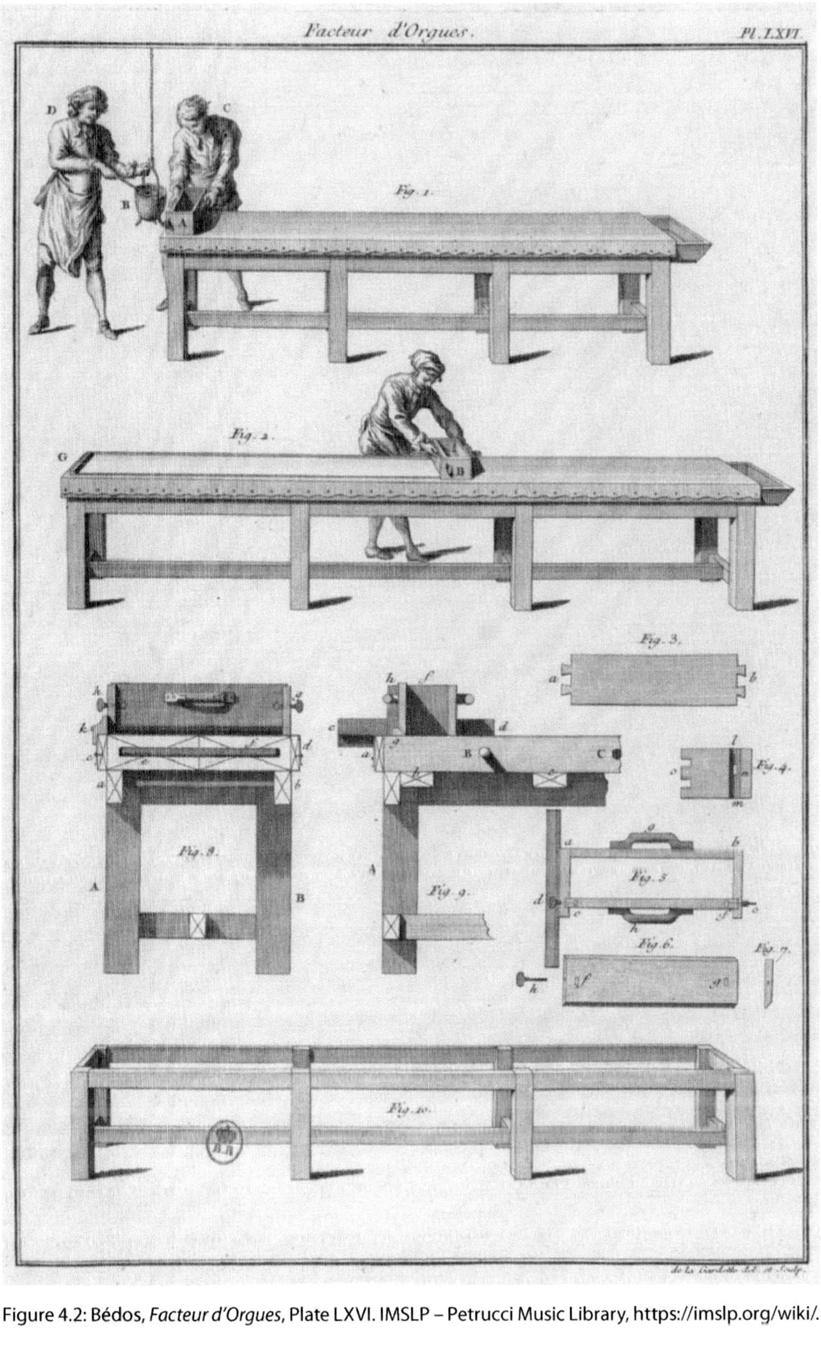


Figure 4.2: Bédos, *Facteur d'Orgues*, Plate LXVI. IMSLP – Petrucci Music Library, https://imslp.org/wiki/.

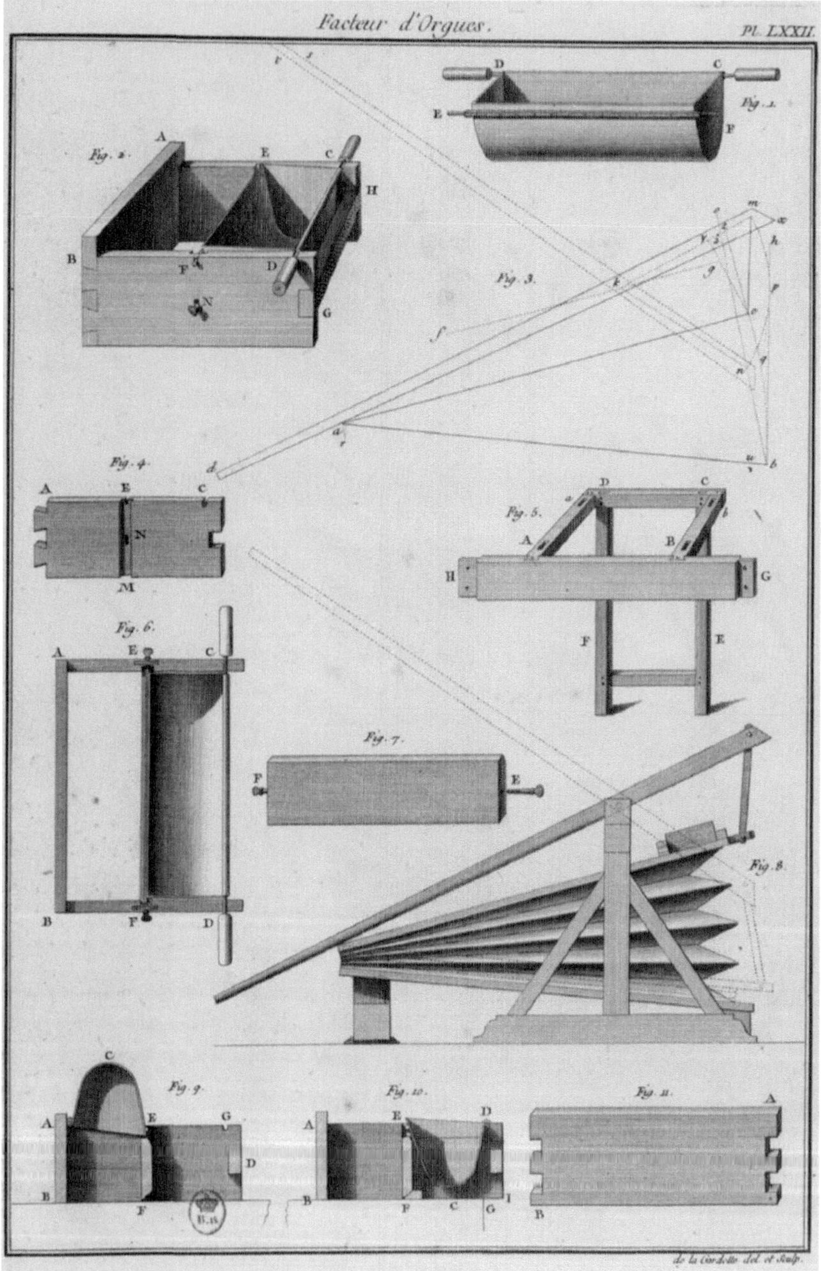


Figure 4.3: Bédos, *Facteur d'Orgues*, Plate LXXII. IMSLP – Petrucci Music Library, https://imslp.org/wiki/.

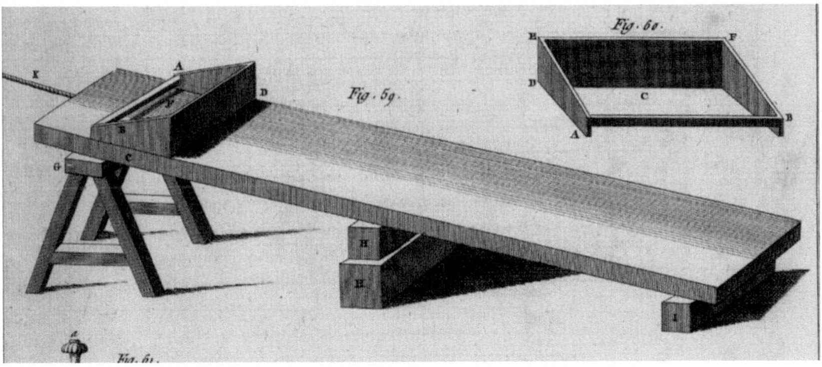


Figure 4.4: *Encyclopédie*, Plates, vol. 5, Plate X, excerpt. IMSLP – Petrucci Music Library, https://imslp.org/wiki/.

## The discussion of sound in *L'Art du facteur d'orgues*

The description of casting metal is perhaps closest to the general aims of the encyclopaedic projects such as the *Descriptions des arts et metiers*. The aspect of sound, however, makes this craft peculiar, as becomes clear from the passage cited at the beginning. Any gesture or operation that the organ builders carry out is intended to contribute to the sound of the musical instrument. This starts with the very first subjects discussed in Bédos's treatise. The advice that the place for the organ should be chosen carefully and the dimensions of the instrument considered with respect to the architecture and the financial means that are available, in the event, are meant to secure its functioning as a musical instrument. The organ could be too large or placed in the wrong spot, not to speak of architectural problems, such as a structure that would not hold the heavy weight of the pipes or start vibrating with them; any unreasonable decision can eventually spoil the organ. This aim of the craft – to give the instrument a good and pleasing harmony – is explicated in the treatise only when it comes to voicing, yet its implicit presence differs to some extent from notions of tacit knowledge that result in less ephemeral effects than sound.

To understand the peculiar nature of sound as an element of practical knowledge, two aspects in the overarching function of sound must be distinguished. One is conceptual, the other pragmatic. A concept of sound accompanies all the steps in making the organ. Yet, although the experiential knowledge of former generations of organ builders flows into them and is explicated to a great extent in Bédos's treatise, the effect of the operations cannot be controlled directly. Bédos also addresses such a moment of direct control when he discusses voicing. In this operation, actual sound must be heard. Here, sound acts as the effect of the tweaking of the pipes so as to enable them very concretely to function properly – and that means to produce good sound.

Bédos's description of voicing is telling in this respect; it contains the only moment when the sound becomes the subject of the discussion. That is to say, sound is being discussed when it occurs in direct interaction with the material object. By changing the position of the upper or lower lip, for instance, or by raising or lowering the languid, the metal plate that separates the pipe body from the pipe foot and creates the windway.[13] He used terms such as 'soft' or 'sharp' to describe the sound of the pipe in question. Such qualifying descriptions always go hand in hand with concrete actions. If the builder used the right materials, followed the correct scaling of the pipes and the cutups, and assured a good wind pressure from the very beginning, a pipe would speak almost automatically. Voicing thus was understood to be the finishing moment in a process, which in one single proper way prepared harmonious sound.

To find out what an organ sounds like, the organ builder of Bédos's times had to go to an organ and play it. The immensely complex procedure that was needed for materializing this instrument, of course, relied on experience and, by the same token, on the transmission of tacit knowledge. Most of the steps of the procedure had to be done without being able to carry out an immediate test. The casting of the metal, which is crucial for a pipe that is thick and thin, soft and hard at the ideal spots, resulted in a characteristic spectrum of overtones, as we would say in a more contemporary technology. Yet, in the process of making a pipe, the moment of casting was far from the moment when the pipe first resounded. This comes to the fore in the moment of voicing. If the casting went wrong, voicing would not remedy the problem.

## Organ building in the nineteenth century – Malleable sound

In 1887, Swiss organist Carl Locher published a booklet in which he describes the organ stops. He also included an entry about voicing. Although badly paid, voicing should be seen as 'the actual *art* in organ building,' he asserted.[14] He considered it one of the most important operations in organ building, since the instrument's sound depends on it. As did Bédos, Locher insisted

13 The windway is the narrow opening between the lower lip and the languid.

14 Locher, *Erklärung der Orgel-Register*, p. 26.

that an excellent instrument may turn out to sound bad, if it is not properly voiced. He went on to provide criteria for successful voicing. Thus, the result should be 1). a proper definition of the pipes' sound with regard to high and low tones; 2). it should enable the pipes to 'speak' easily and quickly, 3). convey them an appropriate sound quality (*Klangfarbe*), and 4). adjust the volume to the space in which the instrument is located. Finally, Locher included 5). temperament, that is to say the adjustment of tuning to the varied purposes of polyphonic music, in his list.[15]

In this list, the appearance of the term *Klangfarbe* is telling, since it emerges in the German language only during the nineteenth century, most often as the translation of the French *timbre*.[16] Locher provided a separate entry for the term, arguing that it was critical in defining the topic of his publication: the qualities that distinguish one stop from another must be referred to as their *Klangfarbe* or tone colour. The term itself, as he went on to explain, was introduced by Hermann von Helmholtz in his 'classic' book *On the Sensations of Tone as a Physiological Basis for the Theory of Music*.[17] Locher dedicated his brochure to Helmholtz, who, in return provided a friendly note that Locher listed in the beginning together with acclamations from other celebrities in the organ world of his times, such as the Munich-based composer Josef Rheinberger or the Weimar cantor Alexander Wilhelm Gottschalg.[18]

Most importantly, *Klangfarbe* also emerged as an aim in and of itself for the organist, because the very art of using the stops could be seen as directly intervening in sound quality and creating all kinds of shades from the means at hand. Mixture stops combining several high pitched pipes on one note provided prefixed combinations of overtones, a famous example being the organs of Gottfried Silbermann. But in general, the organist was responsible for finding the best combinations or *Registermischungen*. For this art, the organist had to study the *Klangfarbe* of each stop, so as to be rewarded by a beautiful sound in their combination.[19]

The background for this eulogy to tone colour was Helmholtz's research into the frequency components of musical sound. He was first to generate almost pure tones with just one frequency. Combining these into various

15 Locher, *Erklärung der Orgel-Register*, p. 27.

16 Locher, *Erklärung der Orgel-Register*, p. 28. On this see Kursell, *Epistemologie des Hörens* and Dolan, *The Orchestral Revolution*.

17 Locher, *Erklärung der Orgel-Register*, p. 29.

18 Locher, *Erklärung der Orgel-Register*, p. IX.

19 Locher, *Erklärung der Orgel-Register*, p. 34ff. (on Mixture stops), p. 36 (on Silbermann) and p. 50ff (on Registermischungen).

patterns, he was able to verify whether the ear could distinguish patterns of such components that differed in nothing but the intensity of the frequencies involved. This eventually made the theorem of Jean-Batiste Joseph Fourier about periodic waves being analysable into sinusoidal components accessible for an unambiguous formal description of sounds. As a result, musical intervals changed their status. They could no longer be said to represent mathematical proportions straightforwardly, as had been assumed since the legendary Pythagorean experiments, but they had to be acknowledged as consisting of complex combinations of frequencies that differed according to the concrete composition of each tone involved. In return, physiology was now able to explain how the ear could cope with its complex task. The ear had to be seen as an analogy to a mathematician using the theorem of Fourier, and each distinguishable tone as an equivalent to the solution of an equation in Fourier's terms.[20]

Helmholtz's findings boosted not so much the symbolic description of sounds – which was left to technicians – but they entered the imagination. As media theorist Friedrich Kittler has argued, the moment in which sense data were disconnected from their sources in the world outside of our minds, media history took off.[21] One could now imagine a friend's voice speaking through the telephone and music sounding from the radio, because the electric energy was now understood to pass through the wire, and not the voice or the sound themselves. By the end of the nineteenth century, acousticians went so far as thinking of sound in terms of circuit diagrams.[22] Meanwhile, sound was represented more and more as being malleable in and of itself.

For organ builders, this malleability had a very different and no less important aspect. Parallel to these conceptual reflections in scientific research, the process of organ making changed. At a time when the use of steam power radically altered manufacturing and transportation practices, the organ workshop, 'once the epitome of cutting-edge mechanical expertise, was rapidly becoming an anachronism'.[23] By the second half of the nineteenth century, organ building was dominated by what Barbara Owen has called 'the organ factory'.[24] Phases in the organ building process that were previously dependent on embodied skills, such as the casting and planing

20 See Kursell, *Epistemologie des Hörens*.
21 Kittler, *Discourse Networks 1800/1900*.
22 Wittje, *The Age of Electroacustics*.
23 Owen, 'Technology and the Organ', p. 213.
24 Owen, 'Technology and the Organ', p. 221.

of the metal sheets for pipes, were now adapted to the ease of machinery. Bédos would advise organ builders to cast the sheets as closely as possible to the final thickness that was required for the type of pipe in question. This would save the organ builder the time and effort of scraping the metal manually until it had the required thickness and it would prevent material loss. Scraping machines changed the economy. Since the thickness of the sheet no longer mattered because the machine would do the work, organ builders gave up on casting sheets close to the final thickness. Under the new circumstances this proved less efficient.

The question of how to adapt the organ to new scientific insights and technological developments as well as new aesthetic ideals was taken up by organ builders and organ theorists. One of them was Johann Gottlob Töpfer (1791-1870) whose book *Die Orgelbau-Kunst nach einer neuen Theorie dargestellt und auf mathematische und physikalische Grundsätze gestützt*, published in Weimar in 1833, had a strong influence on organ building practices in the following decades.[25] From their theoretical debates and practical experiments, a new organ aesthetic emerged that can be summarized in three rules, as Hans Fidom puts it in his doctoral dissertation on German organ building between 1880 and 1918.[26] These rules aimed at a unity of pitch, timbre and dynamics, thus subjecting organs to a parametrized concept of sound. All individual pipes should sound the same in terms of their potentially independent manipulation according to one of these parameters. This was realized, on the one hand, by referring back to the formula Töpfer had suggested for standardizing the scaling of pipes.[27] On the other hand, Helmholtz's insights into the nature of hearing were set to work in a concrete instrument. For instance, his research indicated adding many high overtones to a fundamental frequency would result in a sharper, brighter sound, and vice versa. This insight led to the strategy to have many 8' stops in relation to higher stops. They were seen as representing the reference to the parametrized sound or, in other words, a zero-level for all three parameters.

The practice of voicing also shifted over the course of these developments. From its function as the final stroke to a process that already contained the essence of what was to be achieved, it turned into the actual secret of the

25 Töpfer, *Die Orgelbau-Kunst*.

26 Fidom, *Diversity in Unity*.

27 Töpfer's 'Normalmensur' is a rank of pipes based on an internal diameter of 155.5mm (6.12in) at 8' C (the lowest note of the modern organ compass) and a mouth width which is one-quarter of the circumference of such a pipe. The scale provides for a reduction in diameter of the pipes by half at every succeeding seventeenth pipe.

craft. Like the casting itself, the finishing of the sound was now considered an act of manipulation, guided by some ideal of sound. Voicers developed new techniques to adjust the character of the 'speech' and the sound of flue pipes. An example is making nicks, i.e. small cuts in the languid bevel that enable the voicer to reduce unwanted noise in the sound. Making nicks to smoothen the pipe sound was not common before the eighteenth century but became more or less a standard procedure in late eighteenth and nineteenth century organ building.

As can be seen from Locher's brochure, the organ building craft was becoming aware of these dramatic changes. The sound of the pipe was now handed over to the voicer, and the conundrum of good sound within a parametrized sound space was placed into their hands, ears, and imagination.

## Reconstructing organs

In recent decades, replica's and style copies have become part and parcel in the investigation of historical organ culture. Against the background of historically informed performance practices that emerged in the 1960s, there was a renewed interest in understanding how seventeenth- and eighteenth-century organ builders made their instruments. This led to detailed documentations, comprising organ cases and specifications, wind chests and wind systems, and the characteristics of the pipes. These characteristics include the pipes' dimensions and measurements, their metallurgical composition and the shape of the pipe mouths. Such documentation was essential in carrying out restoration projects aimed at bringing organs back to their original state as much as possible. As the German organist and scholar Harald Vogel has pointed out from the point of view of restoration ethics, however, later additions to the organ cannot simply be taken away. Also, the historical and use values have to be balanced.[28] This put new emphasis on replica building rather than restoring original instruments. Vogel resumes that '[w]e have actually come to the point now where replicas come closer to the original sound than the surviving originals themselves. Building new instruments in historical styles is the path that we must take in the future.'[29]

The actual building of these replicas or style copies not only required a close reading and interpretation of sources, both archival and material, but also a relearning of historical organ building skills. In the remainder of the

28 Watson, *Artifacts in Use*.

29 Vogel, 'Organ Restoration', p. 345.

chapter, we show that it was crucial in replica projects to understand how the sound quality of historical pipes was related to their production and anatomy. To actually make replicas of pipes, the researcher-builders relied on a variety of written sources, such as church contracts, organ inspection reports, and treatises about organ building and relevant practices, such as metal casting. In addition, they carried out extensive experiments to learn how to make pipes according to the historical information. As we will observe, the ultimate aim of contemporary builders is to recreate the sound of historical instruments, which will eventually bring us back to the question of whether and how sound was made present in these historical sources.

## Relearning pipe making skills

Many of the insights and ideals from the paradigm of historicizing organ building culminated in the 1990s in the milestone North German Organ Research Project of the Göteborg Organ Art Center (GOArt) at the University of Gothenburg in Sweden.[30, 31] The aim of this project was to construct an organ the way it might have been built by the famous organ builder Arp Schnitger in late seventeenth century Northern Germany.[32] The builders took the case of the 1699 Schnitger organ in the Lübeck Dom, of which only photographs remain after it was destroyed during a bombing raid in 1942, as a reference. The famous Schnitger organ in the Hamburg St. Jacobi church was also destroyed during the Second World War, but because the pipes had been stored, the surviving pipework of the organ could be used as a reference.[33]

A central assumption in the North German Organ Research Project was that recovering the sound of historical organs required relearning artisanal building techniques that were displaced by modern building techniques

30 Speerstra, *The North German Organ Research Project.*

31 Other organ replication projects have been carried out in Germany, the Netherlands, Korea, and the US.

32 It was built by the Göteborg Organ Art Centre (GOArt) of the University of Gothenburg, and financed by grants from, among others, the Swedish government and the European Union.

33 The North German Baroque organ was inaugurated in 2000. In the following years, members of the team were involved in two other projects to build replicas of historical organs. One was carried out at the Eastman School of Music in Rochester, designed after specifications of an instrument built by Adam Gottlob Casparini in 1776 in Vilnius, Lithuania. The other project, a new organ for the Anabel Taylor Chapel of Cornell University in the United States, aimed at reconstructing the tonal design of an instrument that Arp Schnitger had built for the Charlottenburg-Schlosskapelle in Berlin in the first decade of the eighteenth century.

over a century ago. Hans Davidsson, the initiator of the project and back then affiliated to the School of Music and Musicology of at the University of Gothenburg as organ teacher and musicologist, argued that from an artistic point of view, replicating antique organs became desirable as a reaction against modern organs produced by industrial methods. When organs became subjected to industrial production techniques, he wrote in 1993, 'the main aim was no longer to attain the highest quality possible; instead factors such as capacity and profit became predominant [...] piece by piece, the accumulated experience of the skilled craftsmen disappeared. Thus the end result came to be determined more by the production process itself than by aesthetical or stylistic aims'.[34]

How was the understanding of pipe making related to the sound quality of the pipes in the GOArt-project? And what did the researchers learn about these sounds through actually making the pipes? To answer these questions, we follow the work of Munetaka Yokota who was responsible for the pipe making and voicing for the Örgryte church organ.[35] Yokota developed his idea to work according to seventeenth- and eighteenth-century principles when he studied to become an organ builder.[36] Historically, Yokota argues, craftsmen worked with what was at hand.[37] This was true for the materials they used, but also for how they processed them. Instead of using machines to make a thick piece of wood thinner and throw away a lot of the material, they would look at the required strength and the qualities of the wood that was available. Using local materials thus required adaptation to these local circumstances.

This situated focus on how materials were handled and shaped can help to understand why well-made old pipes sound good, Yokota argues:

34 Davidsson, 'The North German Organ Research Project', p. 9.

35 Yokota was also involved in the New Baroque Organ project at the Orgelpark in Amsterdam. Here, a new organ was built after the style of the German organ builder Zacharias Hildebrandt (1688-1757). Peter Peters has done extensive ethnographic research on this project. Some of the interviews with Yokota that were done in the context of this research project were used in this chapter.

36 The sound of historical organs had fascinated him ever since he was a fourteen-year-old school boy. He bought an LP recording made by the German organist Fritz Heitmann in 1938 on the Schnitger organ in the chapel of the Charlottenburg Castle in Berlin. Because the organ was destroyed in the Second World War, the recording is the only way we can still know how this instrument sounded. In addition to this document there are photographs and measurements of the pipes. Having listened many times to this and other recordings of historical European organs, Yokota concluded that these 'old organs sounded better than new ones'. See Yearsley, 'The Organ-Building'.

37 Interview with Munetaka Yokota, 16 May, 2014, Orgelpark in Amsterdam.

> These old pipes have a beautiful balance between the fundamental pitch and overtones, a balance between 'musical' sound and 'noise,' as well as a good sense of balance between the strength, length, and character of the speech and the sustaining tone. (…) Was the old sound partly a product of the aging of the materials, or could we reach this level of quality again in a modern instrument? Essentially I define 'good sound' as sound that has a sense of life.[38]

To give the pipes a good, lively sound, the researchers and builders, including Yokota, had to learn the 'pattern language' of Arp Schnitger, a term they took from Christopher Alexander.[39] Schnitger and his fellow organ builders knew what an organ looked like and sounded like, because they knew the patterns of the language that made up all of the things that it would have to do. 'Reading' historical organs in combination with other primary sources, the researchers at GOArt tried to define and understand what Schnitger's pattern language was. 'Then the ways of working had to be understood, copied, tried out, and finally performed, in a continuous dialectic interplay between theory and practice.'[40] This process reconstruction did not aim at imitating but at learning. Imitating, Yokota argues, would be senseless precisely because one would only copy site-specific solutions that were developed under unique circumstances with local materials and tools.[41] Learning why historical builders developed certain solutions is thus key to making new pipes that sound as good and lively as the surviving old pipes.

One of the processes that was reconstructed at the Göteborg Organ Art Center (GOArt) in the 1990s was the casting of the metal for the pipes. As explained above, sheets were cast by pouring a molten lead-tin alloy into a wooden casting box, which has a narrow opening at the bottom, and then sliding this box over a casting bench. The melt would flow out of the box and spread into a thin layer covering the bench, cool down and solidify into a metal sheet. Important parameters in the historical casting process were the speed with which the metal cools down and the thickness of the sheet. Yokota used what he calls a tap testing method to analyse the material from which a historical pipe was made, its scaling, and wall thickness. Tapping on the metal with his fingernail gave him two tones: The metal tone produced by the pipe body and the air tone produced by the air column inside the pipe.

38 Yokota and Ruiter-Feenstra, 'Historical Pipe Metal Casting', p. 165.
39 Alexander, *The Timeless Way*; Alexander et al., *A Pattern Language*.
40 Speerstra, 'An Introduction', p. 19.
41 Interview with Munetaka Yokota, 16 May 2014, Orgelpark in Amsterdam.

A higher metal tone indicates a harder metal. It turned out that even though the chemical composition of a modern alloy would be almost identical to its historical counterpart, it would be much softer than in the old pipes. Combining this diagnosis with the fact that old pipes are usually thinner, Yokota concluded that it was necessary to know more about how the metal was cast and what the relationships were between the materials, tools and construction method.

Historical documents, such as organ contracts, organ building treatises and secondary sources were studied to gain information about the composition of historical pipe metal.[42] Metal samples of remaining organ pipes were analysed by using spectrometrical methods and the alloys of the pipe metal for the new organ were established.[43] The next step was to actually reconstruct a seventeenth-century practice of casting the metal sheets. In modern organ building traditions, the casting bench is made of stone or wood and covered with a fire-retardant cloth. In historical traditions, however, the casting bench was made of stone and covered with a linen cloth or a layer of sand. An important question was whether the material properties of the pipe metal were dependent on using sand or cloth. Yokota and his team established that casting on sand gives the metal a completely different quality than when cast in modern ways. Due to the sand bed beneath the molten metal, the metal cools quickly, causing the pipe metal to become harder than modern pipe metal. The right type of sand, the right proportion of impurities or trace materials in the metal, and the right casting temperature all proved to be of vital importance to the end result.

What role did descriptions and images in historical treatises play in the reconstruction of the casting process? In our interview with Yokota, he underlined that written sources such as Bédos's treatise could provide only part of the information he needed to make and voice the pipes.[44] The voicing through which a pipe gets its final sound character was hardly mentioned at all in the treatises, he said, and the pipe's sound therefore could only be

42 Examples of historical organ building treatises that were used are Johann Phillipp Bendeler, *Organopoeia* (1609); Salomon de Caus, *Les raisons des forces mouvantes* (1615), Diderot and d'Alembert, *L'Encyclopédie ou dictionnaire raisonné des sciences, des arts et des métiers*; and Dom Bedos de Celles, *L'art du facteur d'orgues* (1770). See below, footnote 44.

43 Yokota, 'Historical Pipe Metal Casting', pp. 165-167.

44 Interview with Munetaka Yokota, 27 August 2018, Orgelpark in Amsterdam. Elsewhere, he indicates five sources: Salomon de Caus's *Les raisons des forces mouvantes* (Frankfurt: 1615), Johann Philipp Bendeler's *Organopoeia* (Merseburg: 1690), Dom Bedos de Celles's *L'Art du facteur d'orgues* (Paris: 1770), Adlung's *Musica Mechanica Organoedi* (Berlin: 1768), and Diderot and d'Alembert's *L'Encyclopédie* (Paris: 1780). See Yokota and Ruiter-Feenstra, 'Historical Pipe Metal Casting', p. 167.

investigated by carefully observing historical pipes. Overall, the treatise *L'Art du facteur d'orgues* by Bédos de Celles proved to be the most systematically written and detailed, he added. Yet, it should be critically assessed because it is not clear if Bédos was a craftsman himself and his book represents French practices from the mid-eighteenth century only.

**An inbuilt history of hearing?**

For Yokota and his team at Gothenburg University, relearning the skills of Arp Schnitger and the workers in his workshop proved to be arduous. Hundreds of casting experiments were carried out before sheets could be cast that closely resembled the metal from Schnitger's pipes. During the casting experiments that were carried out between 1995 and 1997, practical questions arose: Did Schnitger cast on sand or cloth? If he cast on sand, what sand layer thickness did he use? What type of sand did he use? Did he mix any oil or water into the sand and if so, what type of liquid and how much did he use? Finding the answers required patience; Yokota writes:

> Many times, it is tempting to take modern shortcuts or to find a series of immediate solutions to specific problems; such solutions, however, often overlap and create a domino effect of new problems. Seventeenth-century builders must themselves have sought immediate solutions to problems, but they had many more technical limitations. Therefore they had to solve these problems in terms determined by their practical conditions. In addition, wood and metal have more complicated properties than modern materials such as plastic and steel. Our goal remained that of discovering the so-called simple techniques of the earlier organ builders, where each aspect of the process positively affected the others without creating unnecessary complications.[45]

Yokota's approach to pipe making was far more complex than simply re-enacting historical craftsmanship as it is handed down in treatises from sources such as Bédos. Re-enacting them would have meant to copy situations like the casting procedures that Bédos describes and shows. Instead, Yokota's approach started from the question of how to explain the lively sound of historical pipes, harder and thinner than modern pipes. His strategy was to understand the casting process from its historical context by doing practical experiments based on information that he gathered from

45 Yokota, 'Historical Pipe Metal Casting', p. 168.

a variety of sources, including treatises, but first and foremost the pipes' sound and shape. He aimed at understanding the 'grammar' of pipe making, a certain logic that is local and situated, yet at the same time can still be followed under the conditions of the present. This meant that much of the information that Bédos gives, even the hands-on practical information about determining the right temperature of the alloy or the actual construction of the casting bench, was not relevant. On the other hand, in solving the problem of casting on sand Bédos did not give enough information, but the article on casting lead in the *Encyclopédie* did.[46]

During our interview in August 2018, we discussed with Yokota the issue of sound. As explained above, Bédos does write about the pipe sound in his tenth chapter of the second volume, which focuses on the voicing of the pipes. In fact, Bédos himself calls it the most important chapter of his book, since it is about the sound of the organ. Yokota points out that Bédos's rendering of the voicing is at odds with nineteenth- and twentieth-century ideas about voicing as an 'art' in itself. 'There is no room for human judgment to get into it. It is only in the twentieth century that voicers get this artist status that they can destroy the organ sound or make it beautiful.'[47] For organ builders like Bédos, the sound of their instruments could not be imagined separately from the material artefact and artisanal knowledge. In his text, sound was built in rather then made explicit as a phenomenon that should or even could be manipulated in and of itself. The approach of Yokota and his team to recreate the sound of historical organs such as those made by Arp Schnitger and others mirrors this attitude in that it started from understanding the apparatus and logics that historical organ builders put in place. The historically informed reconstructions relied on experimenting and understanding rather than copying and re-enacting. Following concrete evidence from a variety of sources and putting preconceived ideas about baroque sound aside, they tried to learn the aesthetics of seventeenth- and eighteenth-century organ sound from what well-built pipes had to say.

## Conclusion

If you employ Munetaka Yokota and his team today – whether for reconstruction of a historical instrument or building a new one – you must be aware that his work as a voicer may take a long time – whether the pipes are cast

46 Yokota, 'Historical Pipe Metal Casting', p. 176.

47 Interview with Munetaka Yokota, 27 August 2018, Orgelpark in Amsterdam.

according to present day standards or historical written and material sources and re-learned casting skills. The reason is that you, as his employer, will expect the result to re-enact the sound only. You may not be willing to return to eighteenth-century restrictions of subjecting oneself to getting one chance only for obtaining a desired result. Nor will you necessarily be able to put it in the appropriate location – an eighteenth-century church that is to say –, and thus the very point of departure for Bédos. But this may not be necessary. Today's reconstruction practices in the world of sound nest the 'ancient sound' into today's habits of hearing – producing thereby something unheard of. And for this, we need a skilled voicer.

## Bibliography


Alexander, Christopher, Sara Ishikawa and Murray Silverstein, *A Pattern Language: Towns, Buildings, Construction*, Center for Environmental Structure Series (Oxford: Open University Press, 1977).

Alexander, Christopher, *The Timeless Way of Building* (Oxford: Oxford University Press, 1979).

Bédos de Celles, Dom François, *L'Art du facteur d'orgues*, 4 vols. (Paris: 1766-1778), Vol. 2 (1770).

Davidsson, Hans, 'The North German Organ Research Project at the School of Music and Musicology, University of Göteborg', *Svensk tidskrift för musikforskning*, 1 (1993), pp. 7-27.

Dolan, Emily I., *The Orchestral Revolution: Haydn and the Technologies of Timbre* (Cambridge, MA.: Cambridge University Press, 2013).

Fidom, Hans, *Diversity in Unity: Discussions in Organ Building in Germany between 1880 and 1918* (Amsterdam: Vrije Universiteit Amsterdam, 2002).

Kittler, Friedrich A., *Discoursive Networks 1800/1900*, trans. by Geoffrey Winthrop (Stanford, CA.: Stanford University Press, 1990).

Kostelnick, Charles, 'Visualizing Technology and Practical Knowledge in the Encyclopédie's Plates: Rhetoric, Drawing Conventions, and Enlightenment Values', *History and Technology*, 28 (2012), pp. 443-454.

Kursell, Julia, *Epistemologie des Hörens. Helmholtz' physiologische Grundlegung der Musiktheorie* (Paderborn: Fink, 2018).

Locher, Carl, *Erklärung der Orgel-Register mit Vorschlägen zu wirksamen Register-Mischungen* (Bern: Nydegger und Baumgart, 1887).

Owen, Barbara, 'Technology and the Organ in the Nineteenth Century', in *The Organ as a Mirror of Its Time: North European Reflections, 1610-2000*, ed. by Kerala Snyder (Oxford: Oxford University Press, 2002), pp. 213-229.

Pannabecker, John R., 'Representing Mechanical Arts in Diderot's "Encyclopédie"', *Technology and Culture*, 1 (1998), pp. 33-73.

Speerstra, Joel, 'An Introduction to the North German Organ Research Project', in *The North German Organ Research Project at Göteborg University*, ed. by Joel Speerstra (Gothenburg: Göteborg Organ Art Center / Göteborg University, 2003), pp. 15-20.

Steinhaus, Hans, 'Bédos de Celles, François, Dom', in *MGG Online*, ed. by Laurenz Lütteken (Kassel/Stuttgart/New York, 2016): https://www.mgg-online.com/mgg/stable/47819.

Steinhaus, Hans, *Wege zu Dom Bedos. Daten, Dokumente, Deutungsversuche* (Köln: Dohr, 2001).

Töpfer, Johann Gottlob, *Die Orgelbau-Kunst nach einer neuen Theorie dargestellt und auf mathematische und physikalische Grundsätze gestützt* (Weimar: Hoffmann, 1833).

Valeriani, Simona, 'Facts and Building Artefacts: What Travels in Material Objects?', in *How Well Do Facts Travel? The Dissemination of Reliable Knowledge*, ed. by Peter Howlett and Mary S. Morgan (Cambridge: Cambridge University Press, 2011), pp. 43-71.

Vogel, Harald, 'Organ Restoration in the Twentieth Century', in *The North German Organ Research Project at Göteborg University*, ed. by Joel Speerstra (Göteborg: Göteborg Organ Art Center, 2003), pp. 341-345.

Watson, John R., *Artifacts in Use: The Paradox of Restoration and the Conservation of Organs* (Richmond, VA: OHS Press, 2010).

Wittje, Roland, *The Age of Electroacustics: Transforming Science and Sound* (Cambridge, MA.: MIT Press, 2016).

Yearsley, David, 'The Organ-Building of Munetaka Yokota', *Counterpunch*, 3 December (2010).

Yokota, Munetaka, and Pamela Ruiter-Feenstra, 'Historical Pipe Metal Casting Techniques in Seventeenth Century North Germany', in *The North German Organ Research Project at Göteborg University*, ed. by Joel Speerstra (Gothenburg: Göteborg Organ Art Center / Göteborg University, 2003), pp. 165-186.

## About the Authors

**Julia Kursell** is Professor of Musicology at the University of Amsterdam. Before coming there, she was a research fellow that the Max Planck Institute for the History of Science and at Bauhaus University Weimar. She is one of the founding editors of the journal *History of Humanities* and one of the directors of the Vossius Center for the History of Humanities and Sciences at the University of Amsterdam. She is the author of two books: *Schallkunst. Eine Literaturgeschichte der Musik in der frühen russischen Avantgarde* (Sagner 2003) and *Epistemologie des Hörens: Helmholtz' physiologische Grundlegung der Musiktheorie* (Brill 2018).

**Peter Peters** is Professor at the Faculty of Arts & Social Sciences at Maastricht University, and director of the Maastricht Centre for the Innovation of Classical Music (MCICM). He has published on music, artistic research, innovation in classical music, and previously on time and travel in technological cultures. His current research topics focus on innovating classical music practices, especially symphonic music.

# 5. Reconstructions of Oil Painting Materials and Techniques: The HART Model for Approaching Historical Accuracy

*Leslie Carlyle*

**Abstract**

The relationship between artists' material choices, the preparation and application of materials, and the consequences – the appearance and material properties of the art-work – are studied using Historically Accurate Reconstructions Techniques (HART). This model provides a systematic means to evoke past practices. Reconstructions focus on a range of topics, from the flow properties of Van Gogh's paint to the colour of natural chalk used in artists' preparation layers. As the operator physically prepares and uses the materials, the steps towards the moment of painting are re-enacted. The products of the re-enactment, the documentation and the reconstruction itself, provide evidence – visual, physical and chemical – for comparison with historic oil paintings.

**Keywords:** oil paint, reconstructions, historical accuracy, database research, documentation

## Introduction to the HART Model

In the fields of art and of conservation-restoration, the term 'reconstruction' can have many meanings. A reproduction, or copy, can be made in order to understand how a specific painting was created and has a long and distinguished history in the classical training and education of artists.[1]

1 *European Paintings*, ed. by Hermens.

Dupré, S., A. Harris, J. Kursell, P. Lulof, M. Stols-Witlox (eds.), *Reconstruction, Replication and Re-enactment in the Humanities and Social Sciences*. Taylor & Francis Group
DOI 10.5117/9789463728003_CH05

In conservation training courses, undertaking a reproduction or copy is a successful strategy to help students understand and interpret an artist's technique prior to carrying out a conservation-restoration treatment on the original (see Stols-Witlox in this volume).

The term 'Historically Accurate Reconstruction Techniques' (HART) was coined in the early 2000s to describe a specific approach to the investigation of artists' material choices, including their preparation (using recipes) and application in the act of creating an oil painting. This research model, which was developed during the HART Project,[2] is a significant departure from the aims involved in making oil painting reproductions or facsimiles. It relies on the use of materials appropriate to the time of the recipe(s) with the aim of producing historical models at the material level, not only in terms of surface appearances.

## What is meant by historical accuracy?

As will be seen, the search for historically appropriate materials for replicating or reconstructing recipes leads directly into a practical cul-de-sac. It is obviously not possible to recreate the past. However, attempting to is in itself a highly edifying goal. Not only do we learn unexpected things by aspiring to historical accuracy, we also ask new questions of what has already been discovered.

For example, chemical analyses of historical paintings frequently identify the presence of calcite, commonly in the form of chalk, in the preparation layers (or grounds) used to coat canvases and panels prior to painting.[3] Reconstructions using unprocessed chalk, rather than modern cleaned and treated chalk, show that the unprocessed material mixed in oil is not the pure creamy-white we get with modern chalk, but is in fact dark beige (Plates 5.1a & b).[4] In this case the new questions posed of an old discovery (the presence of chalk) are: What colour was it when it was used? Were early oil-based chalk grounds dark beige? Does that explain artists' descriptions

2 The HART Project (Historically Accurate Reconstructions Techniques) 2002-2005, hosted by the Netherlands Institute for Cultural Heritage (ICN) in Amsterdam, was part of The De Mayerne Programme, funded by the Netherlands Organisation for Scientific Research (NWO), see Carlyle, 'Historically Accurate Reconstructions'; Carlyle, 'HART Report'; Carlyle, 'Representing Authentic Surfaces'.

3 Stols-Witlox, 'Historical Recipes'; Stols-Witlox, *A Perfect Ground*.

4 Carlyle, 'HART Report', pp.101-106.

of grounds that were tan or leather coloured?[5] When did chalk begin to be processed in such a way that it maintained its white colour in oil?[6]

In order to pursue this work, to experience the outcome of following old recipes (a form of re-enactment), and to create reference materials for future analysis and study, it is necessary to adopt a concept of historical accuracy which is entirely relative. As outlined elsewhere[7] and briefly summarized here, the scientific term for accuracy as defined in ISO Standard 5725-1, is useful for illustrating what 'accuracy' can mean for reconstructions using historical materials.[8] ISO 5725-1 illustrates accuracy in a bullseye diagram where it is shown as a measure of the relative distance from a goal, or in ISO terms, trueness. Plate 5.2 shows how aspirations for historical accuracy, such as choosing representative recipes from large collections of documentary sources, or sourcing materials at least somewhat appropriate to the time, takes researchers closer to the ISO bullseye of trueness than if we blindly reconstruct so-called historical oil paint using commercially available oil and modern pigments and fillers, most of which bear only a passing resemblance to those available in the past.

## The HART Model components

The HART Model evolved from research into historical documentary sources covering recipes and instructions for painting in oil. The intention was to understand what material choices artists were making and why, and how they proceeded to construct their paintings.[9] This work naturally led to the question of what written instructions actually meant in practice. Reconstructions were made to explore how the materials might behave during preparation and application, and to investigate whether they could be compared with what is found in paintings contemporary to the recipes.

5 Carlyle, *'The Artist's Assistant'*, p. 177.

6 The role of ground colour in Netherlandish paintings from the sixteenth- and seventeenth century is the topic of the research project *Down to the Ground, a Historical, Visual and Technical Analysis of Coloured Grounds in Netherlandish Painting, 1550-1650*, (University of Amsterdam/TU Delft). This five-year project connects art historical, art technological and scientific research to the question of the role of ground colour, using both material and digital reconstructions. https://www.nwo.nl/onderzoek-en-resultaten/onderzoeksprojecten/i/57/31757.html.

7 Carlyle, 'Towards Historical Accuracy'.

8 For a detailed discussion of scientific measurements and the ISO standards, see Heinicke and Heering, 'Discovering Randomness'.

9 Carlyle, 'A Critical Analysis'; Carlyle, *'The Artist's Assistant'*.

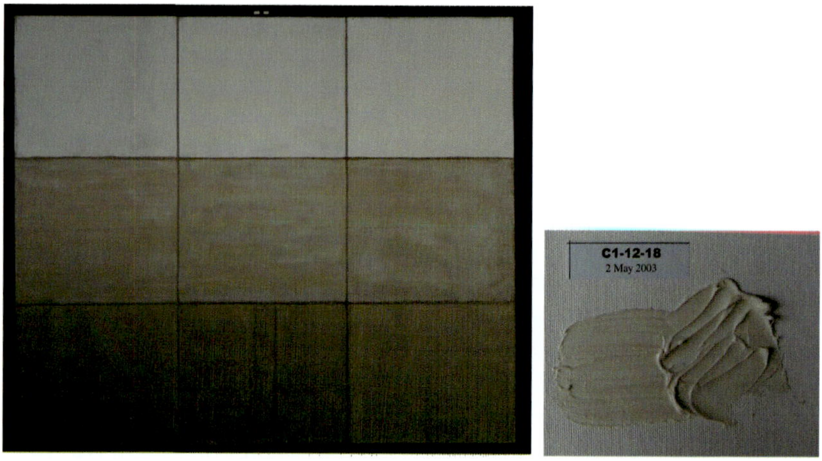


Plate 5.1 (a-b): a) HART ground reconstructions using unprocessed chalk in glue binder (top row), a glue-oil emulsion binder (middle row) and an oil binder (bottom row). b) modern chalk from an artist's supplier ground in oil, compare with bottom row of a). (Carlyle 'HART Report'). Photo: author.

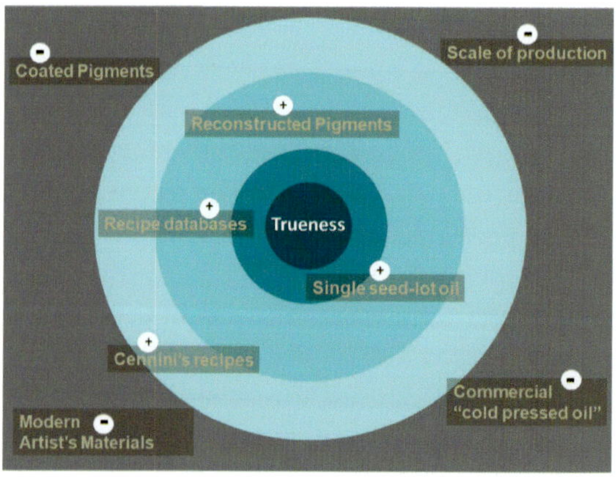


Plate 5.2: Adaptation of ISO 5725-1 bullseye definition of "accuracy" as applied to historical reconstructions, illustrating how the relative closeness to "trueness" can be evaluated. Note the position on the bullseye of the use of a single source recipe (e.g. from Cennini) compared with recipes selected from a large collection (recipe databases), and the use of commercially prepared "cold pressed linseed oil" relative to a linseed oil extracted from a single seed-lot. (Illustration first published in Carlyle, 'Towards Historical Accuracy'). Photo: author.

From the beginning, when undertaking reconstructions, an effort was made to find historically appropriate materials and methods. This step proved to be central; the very act of asking whether a given material was appropriate for a reconstruction uncovered an area of research that previously had not been systematically explored within a comprehensive research model. In addition, in reconstructing an instruction or recipe there were inevitably decisions to adopt one interpretation over another, with consequent compromises.

The assembly of materials and the physical samples from recipe reconstructions (varnishes, oil paints, preparation layers and composites) results in a unique reference collection for comparative visual and chemical analyses in the study of historical oil paintings. The thorough documentation of all aspects of the reconstruction and of the individual components employed results in 'highly characterized reconstructions' (HCRs). As will be seen, this greatly enhances their value as reference materials. What follows is a detailed look at the three interconnected parts of the HART research model.

## Part I. Reconstructions using representative recipes and historically appropriate ingredients

One of the cornerstones of the HART Model is that the practical instructions and recipes selected for reconstructions are representative. That is, they represent instructions which exist in a particular context since they were chosen from a substantial body of information[10] (see Part III describing documentary source research).[11]

### Historically appropriate materials and historical accuracy

Amongst the first findings in the HART Project was the difference in morphology and behaviour in oil paint of traditional stack-process lead white and twentieth-century, or modern lead whites (Figures 5.1a, b & c). Scanning

10 One of the goals in founding the ICOM-CC working group, Art Technology Source Research (ATSR), was to draw attention to a significant body of historical technical literature beyond popularized examples such as the fourteenth century manuscript by Cennino Cennini. See ATSR Conference post-prints by Archetype Publishers Ltd, London, and for medieval technical manuscripts see Mark Clarke (e.g. Clarke, *Tricks of the Medieval Trades*).

11 Although technically the first step, discussion of the use of a database as a research tool to identify and understand historical materials and representative recipes is reserved until Part III, as its value and importance become clearer after the discussions in Parts I & II.

Electron Microscope images (SEM-EDX backscatter) of cross-sections from traditional paintings consistently revealed a range of particle sizes, from significant lumps to finely divided particles. This behaviour in oil could only be achieved with historical reconstructions of stack-process lead white; modern sources of lead white exhibited small, uniform and finely dispersed particles.[12] Similarly, commercial sources of modern processed chalk did not display the clumping behaviour visible in the cross-section from the historical painting (Figure 5.1b). As Figure 5.1c illustrates, clumping could only be achieved by using unprocessed chalk since the product prepared for industry has undergone various treatments such as the addition of anti-clogging agents. As noted above (Plate 5.1), the differences in colour in oil between the raw material and the commercial product were also very significant.

There are many more examples of modern materials that do not reflect what was in use in the past. While individual pigments may be very similar to historical pigments in their chemical formulae, modern industrial manufacture means that they may exhibit significantly finer particle sizes as well as having received a range of surface treatments and coatings.[13] Pigment surface treatments (intended, for example, to stabilize, aid dispersion in the binder, or improve flow properties of coatings) may interrupt or inhibit pigment/binder interactions that would have occurred in historical oil paints.

Modern sources of artists' linseed oil are also unlikely to be representative of oil available in the past. The oil source itself is a good example for exposing the complexity of identifying appropriate material and for demonstrating the compromises involved. The oil yield and ratio of components in linseed oil is influenced by many factors, such as growing conditions (e.g. geographic location, temperature) and the plant cultivar.[14] Commercially available artists' linseed oil is reported to be a blend of linseed sources and has been pre-processed before it reaches the market (i.e. to remove the water-soluble component present in freshly extracted oil). Since it was not possible in the HART Project to establish a single linseed cultivar that would have been available for a particular time or place, or even to establish the level of purity that historical sources of linseed oil possessed,[15] linseeds were procured from a single biological grower and used consistently throughout. Because oil pressing on an industrial scale introduces

12 Carlyle et al., 'Historically Accurate Ground Reconstructions'.

13 Salis Gomes et al, 'Pigment Surface Treatments'.

14 Oomah and Kenaschuk, 'Cultivars'.

15 Carlyle, *'The Artist's Assistant'*, pp. 24-25.

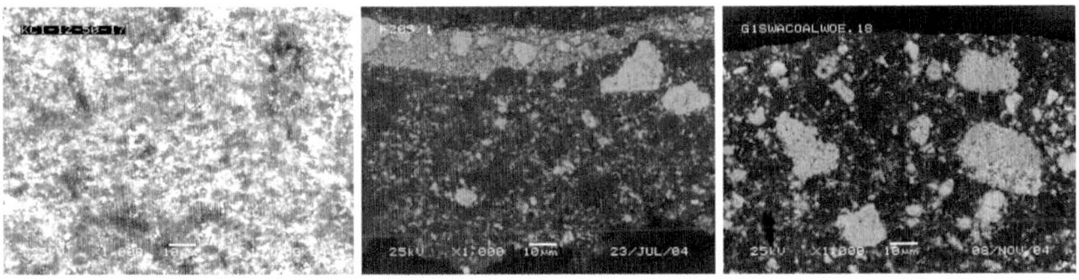

Figure 5.1 (a-c ): Scanning Electron Backscatter images (SEM-BSE) of cross-sections of reconstructions in comparison with one of Van Gogh's grounds: a) a reconstruction using a mixture of modern lead white with modern chalk; b) a sample from the Van Gogh ground, showing a bottom layer of lead white and chalk, with a top layer composed primarily of lead white; c) a reconstruction of the bottom layer in b) using traditionally prepared stack process lead white with unprocessed chalk (no anti-clogging agents). Note the large lumps of lead white pigment evident in b) and c) but not in a). Cross sections a) & c) prepared by Tatiana Ausema and Kathrin Pilz, HART Project, and b) by Muriel Geldof, Cultural Heritage Agency of the Netherlands (RCE). SEM-BSE by Kees Mensch Shell International Chemicals B.V., Amsterdam. Photo: author.

other variables (i.e. heat, pressure, and contamination with other oils), a custom-made stainless-steel oil press was used to control these variables. In the case of the oil source, the ideal of historical accuracy gave way to the compromise of producing reconstructions with a single consistent oil source, making them directly comparable chemically as well as being highly characterized.

Similar issues with historical accuracy involve problems of scale, both in amounts and in time; for example production records from the nineteenth century colourman Winsor & Newton (W&N) reveal that from twenty to seventy gallons (91 to 318 litres) of oil would be maintained at 212°F (100°C) for up to eight days in the preparation of drying oils for paint making, whereas laboratory reconstructions of drying oil typically involve much smaller amounts over a significantly shorter period. This presumably influences the chemical nature of the final product.[16]

## Representative recipe use: literal recipe reconstructions and model recipes

In some cases, a recipe from a single source can be shown to be representative in the context of an overall recipe collection. In such cases the recipe is then followed as closely as possible. In other cases, where recipes are complex and/or proportions are not reported, the reconstruction may be made using a simplified model recipe to reduce variables and for consistency throughout a project. In that case the proportion of ingredients can be established by comparing a significant number of records (see below) and determining

16 Carlyle et al., 'A Question of Scale'.

the proportions which appear most frequently. This working recipe is an editorial synthesis based on a number of documentary sources.

An illustration of a reconstruction which can be followed literally from a single historical recipe is the step-by-step instructions from Winsor & Newton's nineteenth century archive for an oil painting preparation layer (ground) dated 1871. To date, W&N's archive is the only source found which offers detailed information for nineteenth century commercially prepared grounds. The 1871 recipe, like the other production records from their archive, reveals that their commercial practices were far more complex in terms of materials and preparation steps than have been found in recipes published in artists' instruction manuals.[17]

An example of choosing the proportion to use for a representative recipe is the artists' paint addition (or Medium) 'Megilp', which was a popular nineteenth century product used to adjust oil paint flow characteristics and transparency. It consists of a mixture of lead treated linseed oil with concentrated mastic varnish. To determine the proportions to use in reconstructions, a large recipe collection was consulted. Analysis of 26 recipes or descriptions revealed that the most common proportion was 1:1 (fifteen of 26 recipes), with 2:1 being the second most frequently mentioned (eight of 26 recipes). After 1849, recipes were consistent in calling for the 1:1 ratio, making this proportion 'representative' for post 1849.[18]

An example of the creation of a model recipe is the choice of ingredients and production process for a standardized drying oil. After evaluating the many nineteenth century recipes for treating linseed oil prior to use, which frequently called for multiple drying agents, a model drying oil was produced based on recipes with only two ingredients (oil and lead (II) oxide). Fixed proportions and amounts were heated together to a consistent temperature (150°C), to limit experimental variables.[19]

*Reconstructions and samples from oil paintings, comparative analyses*

The evaluation of the visual and analytical evidence in paint micro-samples from historical oil paintings can be greatly improved by comparison with reconstructions (Plates 5.3a-e). This was taken to a new level in the HART Project as a consequence of the need to produce viable reconstructions of Van Gogh's preparation layers (or grounds) as a first step in investigating his

17 Carlyle et al., 'Historically Accurate Ground Reconstructions'; Carlyle et al, 'A Question of Scale'.

18 Carlyle, *'The Artist's Assistant'*, p. 103.

19 Carlyle, 'MOLART Fellowship', p. 39.

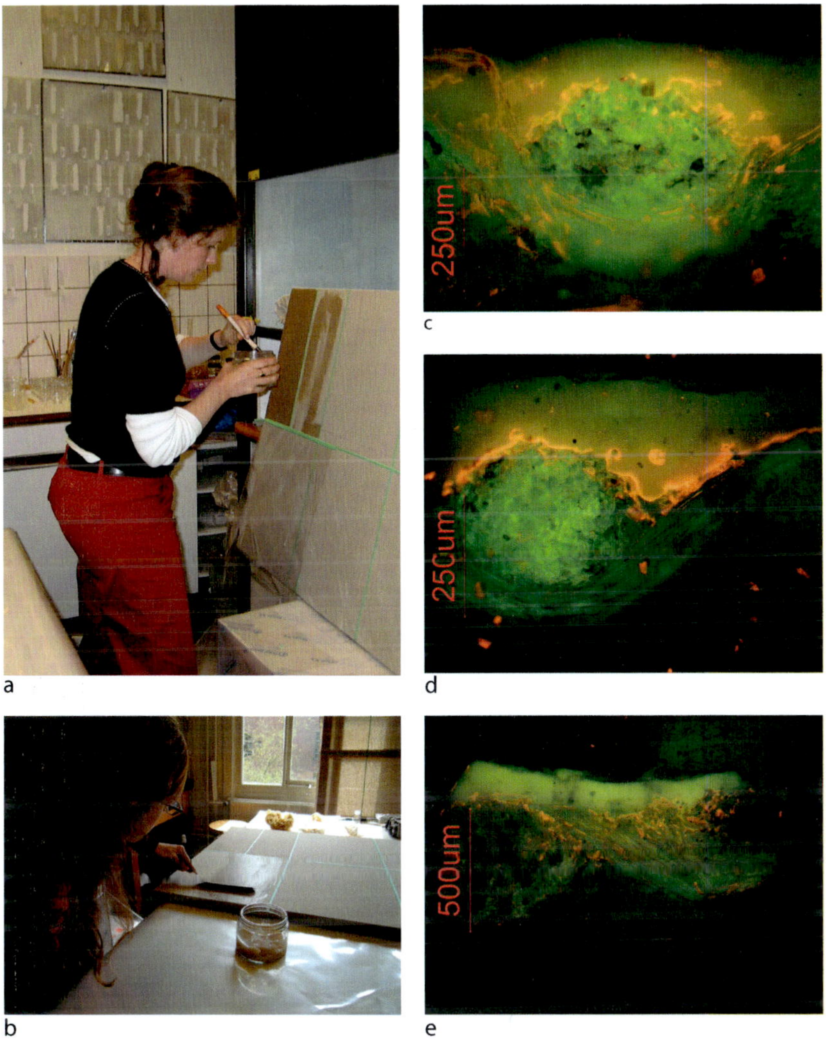


Plate 5.3 (a-e): The application of glue-size to a canvas (the first step in preparing the ground) during the HART Project. Historical records indicate that it could be applied as a fluid or gel. a) warm glue-size applied in a fluid state (image of Maartje Stols-Witlox); b) room temperature glue-size applied in a gel state (image of Katherin Pilz); c) cross section from canvas in a), a fluorescent stain shows the location of the glue size in orange; d) cross section from canvas in b) showing a continuous layer of glue when applied as a gel, e) cross section from an actual oil painting where the glue-size distribution appears most similar to c) suggesting it was applied in a fluid state. The state of the glue layer can play a role in the painting's response to moisture (cross section images by Diana Conde, see Carlyle & Conde, 2011).

paint and how it behaved upon application. While it had been customary to simply report on the presence of different materials, for example, that a given ground in a Van Gogh painting contained both lead white and barium sulphate, the desire to reconstruct that ground led to questioning the proportion of lead white and barium sulphate actually present.

In parallel to this question was another regarding the role of the barium sulphate. A popular additive to lead white in the nineteenth century, the influence of barium sulphate in terms of its behaviour in oil paint and its appearance was not clear. A series of lead white oil paints were therefore prepared with stepped proportions from five percent to 75 percent by weight barium sulphate. Cross-sections of the reconstructions were viewed by Scanning Electron Microscopy backscatter imaging (SEM) and Energy Dispersive X-ray analyses (EDX) software was used to confirm the proportions present. This revealed that the software was reporting percentages that were significantly inaccurate. The need to calibrate the instrument with known mixtures had not been previously appreciated. This was also the case with chalk/lead white ratios.[20]

Issues with detection limits and the ability of a given analytical instrument to identify the presence of a relatively minor ingredient, which could nevertheless have a profound influence on the behaviour of oil paint during and after application, have been explored using HART HCRs. For example, finding analytical evidence of copal oil varnish reported by artists to have been mixed into their oil paint (e.g. as a Medium to modify paint flow) has been challenging, even using state of the art analytical instruments.[21] Recently, a set of HCRs of hand-ground oil paint with stepped proportions of a reconstructed copal oil varnish[22] were used to test the capacity of different analytical protocols and instruments to successfully detect this material in the oil paint matrix.[23] An improved understanding of the composition of copal oil varnish through its preparation according to historical recipes, and of the detection limits for characteristic marker compounds, facilitates the study of this material in historical paintings.

20 Haswell et al., 'Van Gogh's Painting Grounds'; Haswell et al., 'Qualitative Determination'; Haswell et al., 'The Examination'; Carlyle et al., 'Historically Accurate Ground Reconstructions'.

21 Van den Berg et al., 'Recognition of Copals'.

22 Carlyle, 'Representing Authentic Surfaces'.

23 Langley et al., 'Scenes from the Life'.

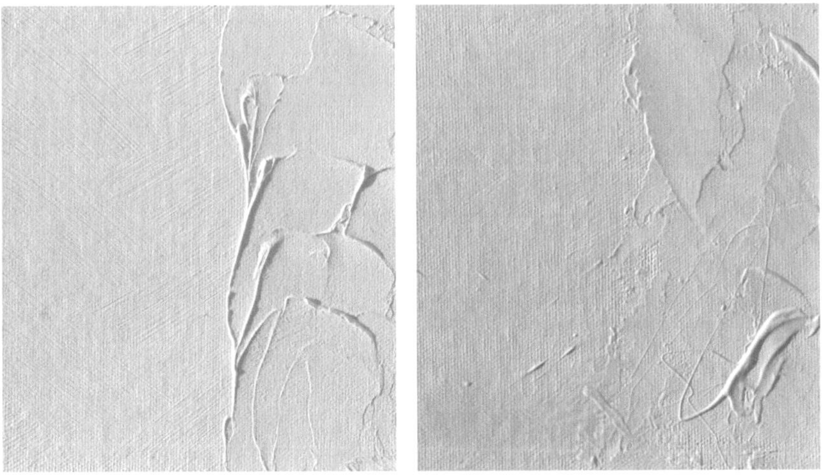

Plate 5.4 (a-b): Showing the difference in paint flow properties (rheology) a) when lead white paint is made with linseed oil heated to 150C versus b) when it has been heated to 300C. The same lead white pigment was used in both examples. (Carlyle, MOLART Fellowship). Photo: author.

*Re-enactment: experiential and visual evidence of oil paint flow properties*

The way that oil paint behaves under the brush or palette knife during application (its rheological properties), was rarely described in any detail in the British nineteenth century handbooks and manuals studied. However, an interest in paint flow qualities was implicit in the many recipes provided for artists' Mediums since their addition to the paint influenced flow and transparency. Given how crucial the flow properties of the paint are to the final image, whether fluid or paste-like or somewhere in between, it is surprising that this feature did not receive marked and consistent attention.

When a series of recipes for processing the oil prior to use in paint were explored using a single linseed oil source, the influence of oil processing was found to be surprisingly profound.[24] Untreated freshly expressed linseed oil, hand-ground with stack-process lead white, resulted in a paste-like paint, whereas the traditional oil processing method of washing the oil in water produced a lead white paint that was quite fluid. Similar effects were seen when the oil was heated either to 150°C or 300°C (see Plates 5.4a & b). Why this range of behaviours was not described in late eighteenth-century and nineteenth century British sources is likely explained by the wide availability of pre-processed oil from the artists' colourmen. Even when hand-grinding their own oil paint with powdered pigments, artists would have been selecting from a limited range of oil products which would in turn

24 Carlyle, 'MOLART Fellowship'

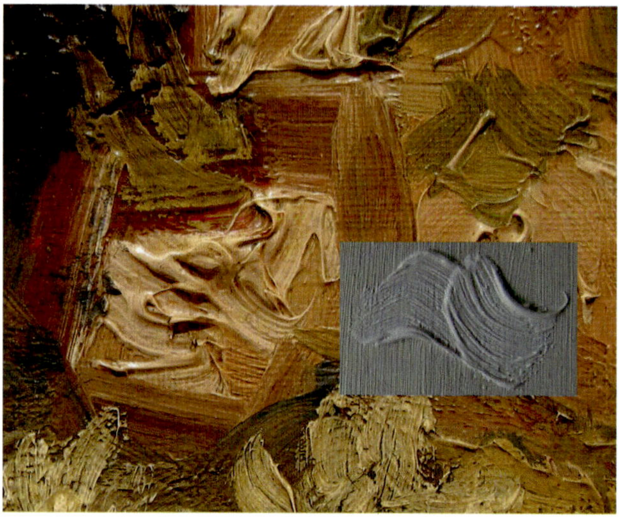

Plate 5.5: Detail from *Vase with Chinese Asters and Gladioli*, Vincent Van Gogh, August-September 1886, Van Gogh Museum, Amsterdam (Vincent van Gogh Foundation). Insert: HART Project paint formulation which achieved similar paint flow and brushwork characteristics (Carlyle, 2005a).

produce a limited range of behaviours (e.g. nineteenth century colourmen catalogues generally listed only the following: Purified Linseed Oil, Poppy Oil, Nut Oil, Pale Drying Oil, Strong Drying Oil and Fat Oil).[25]

Since the responsiveness of the lead white oil paint was so distinct, depending not only on the oil processing, but also on the effect of extenders such as chalk, barium sulphate, magnesium carbonate, beeswax, and aluminium stearate, it was necessary to define a vocabulary for describing how the oil paint handled during application. Interestingly, it was also found that the paint's behaviour was radically affected by the use of a brush versus a palette knife, and the use of the paint industry's tool: the fixed-distance draw-down bar. Some of the paints prepared could easily be brushed but would not form uniform draw-downs, whereas others which worked well with the palette knife were too stiff to brush out easily. The HART team members, preparing and painting out various lead white oil paint formulations in an effort to recreate the visual evidence of paint flow found in Van Gogh's paintings, worked together to describe the 'feel' of the different paints in an attempt to develop a common vocabulary. Words such as 'flexible', 'springy', 'dead', 'unresponsive', and descriptions like 'has strong/weak cohesive strength' were chosen.[26]

25 Carlyle, *'The Artist's Assistant'*, pp. 337-340.

26 Carlyle, 'HART Report', Chapter 4, Appendix IV, pp. 290-296.

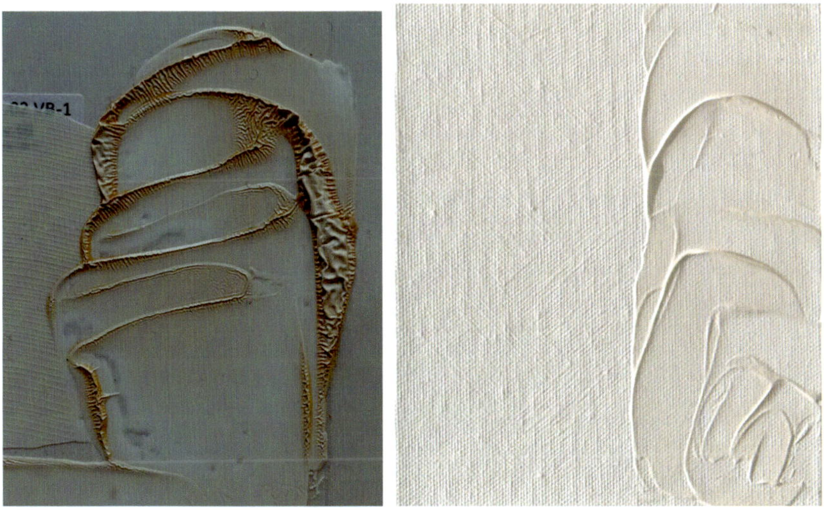

Plate 5.6 (a-b): Showing a) a lead white oil paint applied onto a polyester film in 1999, photographed in 2018. On the left side the thinly applied paint appears in good condition, however where applied in a thick layer on the left, there is extensive wrinkling and oil separation (the dark coloured material is oil at the surface of the paint); b) shows the same paint applied to a commercial Artist's Board. To date (2019) no visual changes have been observed since image b) was taken in 1999 (photograph by Carl Bigras, Canadian Conservation Institute). MOLART Fellowship sample BH8DM2: modern lead white hand-ground in linseed oil heated to 150°C with lead acetate, with the 19th century artists Medium, Megilp added (Carlyle, 2001b). Photo: author.

An unexpected outcome of this work was that this close and exacting attention to the paints' flow and appearance after application led to a heightened awareness of paint surfaces on historical paintings and a new visual literacy. The way the paint appeared on a Rembrandt versus a Van Gogh or one of his contemporaries could be 'read' in the now frozen features of a brush stroke; what the fresh paint must have been like in order to produce a fluid line or a thick impasto, its fundamental characteristics, became visible. The reconstructions demonstrated that the paint composition itself dictated the way the paint appeared in application; the material combinations gave results that were largely predictable and characteristic of that particular formulation (Plate 5.5).

## The importance of the substrate for oil paint studies

A further interesting development emerging from the HART experiments were the observations resulting from the way the oil paint behaved when applied to a variety of different substrates, ranging from reconstructions of painter's grounds on canvas, to modern artists' boards, ceramic tiles, and polyester film. That the paint is significantly influenced by the substrate was the subject of a great deal of commentary in the British eighteenth- and nineteenth century handbooks and manuals and was later identified as an area of focus in the much broader and deeper study of grounds carried out by Stols-Witlox.[27]

To achieve uniform thicknesses for study and analysis of the oil paints, draw-downs (paint applied with a metal fixed-distance applicator) were applied to transparent polyester film. This film was also used for brushing out any paint left over. While thin paint films on such a non-absorbent surface dry as expected, the same paint brushed out in thick applications behaves very differently, exhibiting defects such as wrinkling, shrinkage and oil separation (forming exudates of oil at the surface) (Plate 5.6a). Some of these effects are not visible for decades. The same paint applied thickly to artists' boards (with commercially prepared grounds over a canvas-like texture) or over ground reconstructions do not exhibit such problems (Plate 5.6b).

### The importance of storage conditions for oil paint reconstructions

As the reconstructions undergo decades of natural aging, the importance of keeping oil paint surfaces apart (not stacked together) and allowing surfaces to be exposed to the air becomes more and more evident. After sixteen years, thin polyester films used as dust covers over fully dried paint samples began to stick to the surfaces of some, necessitating horizontal storage with dust covers elevated above their surfaces. Some examples of oil paint applied in 2004 to glass slides, with polyester film dust covers applied after they had fully dried, were completely liquified by 2017.

## Part II. Highly characterized reconstructions as reference materials

### Documentation

What makes the reconstructions 'highly characterized' and therefore important as reference materials and for future research is that in the HART Model each material used, their combination and the results are all carefully documented. The individual materials used are labelled with the date of acquisition and the supplier and stored alongside the reconstructions.

27 Stols-Witlox, 'Historical Recipes'; Stols-Witlox, *A Perfect Ground*.

The materials in any given reconstruction can be identified using its unique and detailed code. An example of the exacting information for a given HART oil paint reconstruction is provided in Carlyle, 'Practical Considerations' p. 109. The sample label includes the date made, the project number, sample number and substrate and the combination of materials used. The materials' code then leads back to precise information on that ingredient. For example, in the code SWAB1-08-05: 'SWA' refers to reconstructed stack-process lead white pigment (S) Batch 2.6 produced by supplier Jeff Seynaeve during the summer of 2003. This was washed in water (WA) and dried prior to use. To this was added B1, an unprocessed barium sulphate from the Sachtleben Company (in this sample present at five percent by dry weight (05) with 95 percent SWA lead white). The code includes the oil used, (08): oil extracted from Electra Linseeds in 1999 and water washed in 2003.

A strip of the metallic lead (roofer's lead) used by Jeff Seynaeve[28] to reconstruct traditional stack process lead white pigment has been kept, as have the original flakes of untreated lead white pigment which are stored according to the batches received and their date. Samples of the original linseeds purchased in 1999 are kept along with the linseed oil extracted, including the date of extraction. Oils which were further processed (water washing, heating with lead (II) oxide, etc.) are also part of the MOLART and HART materials collection. All dry pigments and chemical compounds are preserved as well.

This level of detail in the documentation as well as the preservation of the starting materials has proved essential for recent work on radiocarbon ($^{14}C$) dating of lead white pigment alone and in oil paint. Dated reconstructions of stack-process lead white pigment and the same pigment then made into oil paint have been invaluable for confirming the accuracy and reliability of this new method for $^{14}C$ dating of micro-samples.[29]

## Part III. Historical technical documents: a database research methodology

This section details a research methodology developed over several different recipe database projects which have designed-in options for data extraction

28 Jeff Seynaeve, a Belgian artist and picture restorer with a strong interest in historical painting techniques, reconstructed stack-process lead white for the HART Project using the traditional method (Carlyle, 'HART Report'). We visited him to take notes and photograph the process he was using so that we could fully understand the conditions. The images are in the 'HART Report'.

29 Hendriks et al., 'Selective Dating' and Hendriks, 'The Art of Microscale Radiocarbon Dating'.

such that the database itself becomes an interactive research tool. Examples are described here as illustrations of the requirements for future online databases if they are to become viable as research tools for a variety of users and not just repositories of searchable information.

While collecting large amounts of data, for example in the form of descriptions of painters' materials and how to use them, has been made easier with the advent of the Web, the sheer volume of information is formidable.[30] A method to investigate the contents of these technical treatises is essential, since repetitions and variations are impossible to evaluate and understand without a system to track editions and query the data.

## Database design

The investigation of technical historical sources is not by its nature a linear process which simply follows the chronology of the records. Designing and populating a database and unearthing new documents proceed as parallel activities. At the simplest level, records and books are often found out-of-order in terms of their dates and editions with data entry completed on a second or third edition before the first is located. At another, more sophisticated level, the records themselves require repeated evaluation not only because of the acquisition of new sources, but also because of new interpretations. This means that the system of posing and even identifying queries is essentially recursive: a constant process of circling around and back to records entered previously.[31]

## Tagging/linking records

Interpretation involves grouping records through a system of 'tagging (annotating)' or 'linking'. For example, in recipes for 'Bitumen Tube Paint'

30 Examples include the Winsor & Newton Nineteenth-century Archive Database (Clarke and Carlyle, 'Page-Image Recipe Databases' (2005a&b)) which contains over 15,000 records, and a database currently under construction by Joana Devesa as part of her doctoral studies at Universidade NOVA de Lisboa on historical recipes for cleaning oil paintings. It contains information from 130 sources published between 1600-1900, and currently holds over 900 records (Devesa et al., 'Early Cleaning'). See also the Cologne Database for painting materials and reconstructions http://db.cics.th-koeln.de/start.fau?& (and Oltrogge, 'The Cologne Database'), as well as the Artechne Database: https://artechne.wp.hum.uu.nl/artechne-database/ developed under the direction of Sven Dupré.

31 In her doctoral work on the preservation of Performance Art, Hélia Marçal describes the process of documenting these works as 'recursive' in the same sense that it is being used here: the researcher must revisit what appears to be the same information over and over, bringing new insights and views each time, informed by what has been learned since the information was initially encountered (Marçal, 'From Intangibility to Materiality').

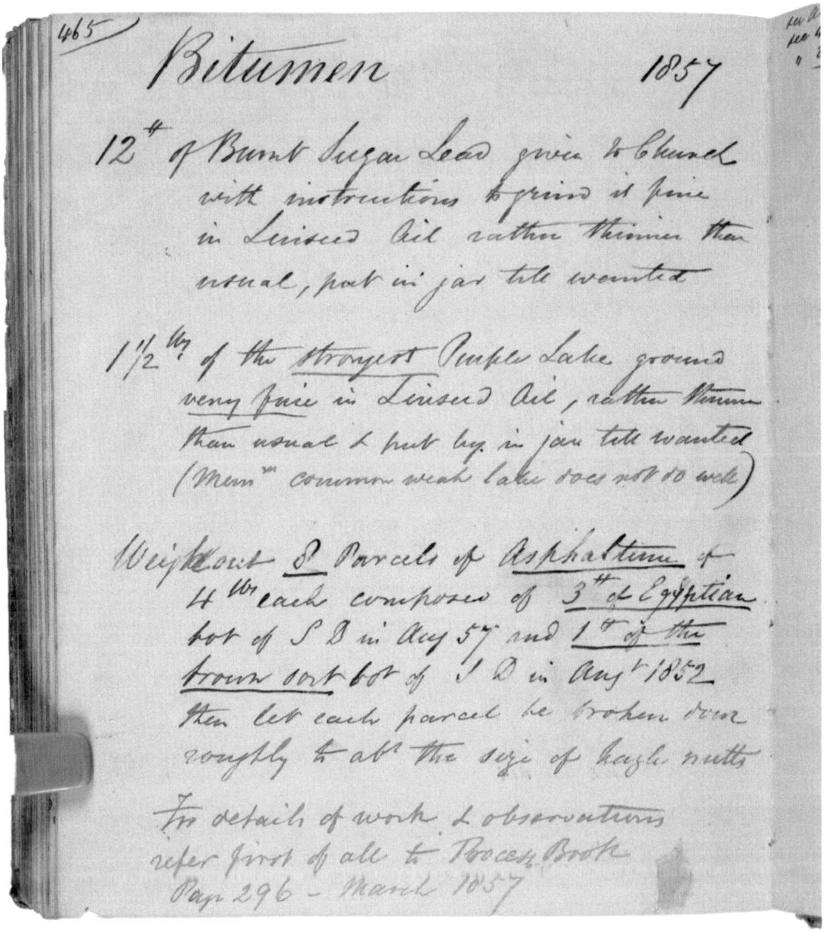
465
Bitumen 1857
12# of Burnt Sugar Lead given to Church
with instructions to grind it fine
in Linseed Oil rather thinner than
usual, put in jar till wanted

1½ lb of the strongest Purple Lake ground
very fine in Linseed Oil, rather thinner
than usual & put by in jar till wanted
(Mem^m common weak lake does not do well)

Weigh out 8 Parcels of Asphaltum of
4 lbs each composed of 3# of Egyptian
bot of S D in Aug 57 and 1# of the
brown sort bot of S D in Aug^t 1852
then let each parcel be broken down
roughly to ab^t the size of hazle nuts

For details of work & observations
refer first of all to Process Book
Pag 296 - March 1857

Plate 5.7: Page-image from the Winsor and Newton 19th century Archive Database (page code V2P465). The third line in the third paragraph contains the source of their asphaltum which was "Bot of S D in Aug 57" [bought from Sarah Druke August 1857] see endnotes 32 & 33. Photo: author.

in the Winsor & Newton Nineteenth Century Archive Database (W&N Database), one of the ingredients was 'Bitumen C'.[32] Initially we pictured a large bag or crate of raw bitumen labelled C'. This made sense in view of the company's habit of recording the source of the materials called for in a

32 The ingredients in W&N's 'Bitumen for Tube Paint' are of particular interest since a significant number of nineteenth century oil paintings suffer from severe paint defects commonly referred to as 'Bitumen Cracking' or 'Alligatoring' (due to the resemblance to alligator skin of the distorted dark surfaces they exhibit). Raquel Marques is currently engaged in doctoral studies of this phenomenon at Universidade NOVA de Lisboa, using the HART Model to combine technical source research, reconstructions and visual/chemical analyses of affected paintings.

given recipe: e.g. Egyptian Asphaltum bought off 'S D Aug 1857' (Plate 5.7).[33] However, after collecting and reading through all their records on this topic, we discovered that Bitumen C, is in fact another heat-processed material made by the company with a complex list of ingredients of its own; it is not a simple raw material at all. A fixed database would not allow users to correct assumptions nor to annotate and link records, leaving future researchers to potentially repeat the misinterpretation of Bitumen C ad infinitum.

## The authority field and the field for interpreted terms

Regarding the interpretation of terminology, while in many cases researchers can resort to commonly available dictionaries of historical terms, this is not always possible. As the W&N Database revealed, the company used general terms, such as 'Spar' very specifically; in their case 'Spar' consistently refers to magnesium carbonate. Other terms appear to be idiosyncratic: 'White Earth' as produced by W&N consisted of aluminate.[34] The need to track the reasoning for the interpretation of the company's terms and to cite both internal and external references led to the introduction of the 'Authority Field'.[35]

To provide a controlled consistent vocabulary with normalized spelling, and to identify arcane and obsolete terminology (e.g. 'Sacrum' and 'Sugar of Lead', both of which were synonymous with lead acetate), the W&N database includes fields for 'Interpreted terms' which appear in the recipe summary page alongside the company's original names. The explanation for interpreted terms is held in the Authority Field. As new insight and expertise is developed, both the 'authority field' and 'interpreted terms' need to be capable of revision.

33 'Weig[h] out 8 Parcels of Asphaltum of 4lbs each composed of 3# of Egyptian bot of S D in Aug 57 and 1# of the brown sort bot of S D in Aug[t] 1852 then let each parcel be broken down roughly to ab[t] the size of hazle nutts' (W&N Researcher's Edition V2P465). The identity of S D was established by other references in the W&N database to S Druke selling asphaltum (e.g. P2P115A), then by cross-referencing S Druke to the Roberson Archive Bought Ledgers where they purchased similar materials from a Sarah Druke on 5 Cloak Lane (Carlyle database, see Carlyle, 'Using a Portable Computer'; Carlyle, 'A Critical Analysis of British Handbooks'). Her identity was fully revealed in Jacob Simon's database on British Artists Materials: https://www.npg.org.uk/research/programmes/directory-of-suppliers.php. Checked on 4 2019.

34 Vitorino et al., 'Nineteenth-Century Cochineal'; Otero, 'Historically Accurate'

35 Introduced by Mark Clarke along with his design for allowing forward and backward movement through each of the 85 books in the archive. Another of his many innovations was a programme to automatically convert old British measures to metric equivalents for ease of interpretation and for comparison with similar recipes (Clarke and Carlyle, 'Page-Image Recipe Databases' (2005a&b))

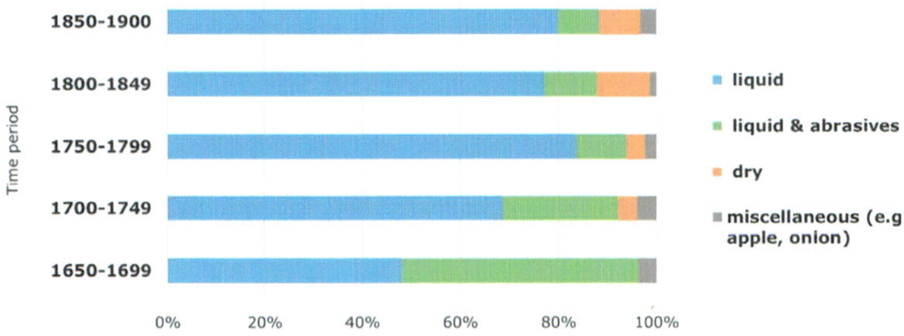


Plate 5.8: Graph generated from Joana Devesa's database on historical methods for cleaning oil paintings. Depicted are cleaning methods by date range (1650-1900) showing that the use of liquids was common throughout. Dry methods (e.g. abrasives) appear less frequently, however, references to dry methods do increase in the 19th century. Note that abrasive materials used with liquids appear more often in the early records (Devesa et al. 'Early Cleaning'). Table: Devesa et al., used with permission.

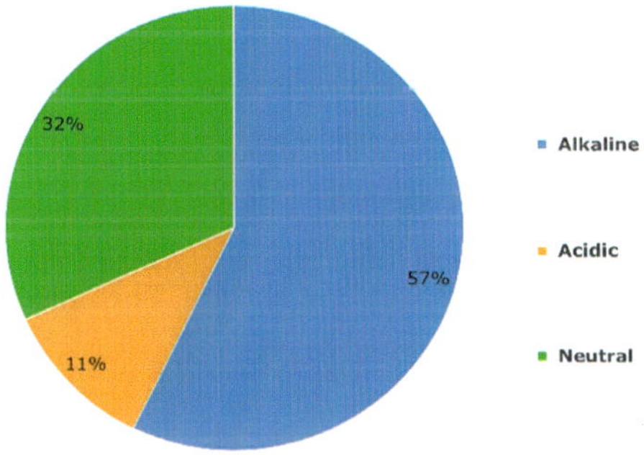


Plate 5.9: Pie chart generated from Devesa's database (see Plate 5.8) showing the high percentage of historical references to cleaning with alkaline ingredients out of the total records for aqueous-based cleaning (Devesa et al. 'Early Cleaning'). Figure: Devesa et al., used with permission.

## Information extraction through recursive refinements

Current work to extract information from a collection of over 900 records relating to historical methods for the cleaning of oil paintings (see endnote 30) offers a good example for describing the procedure of successive topic 'refining'. This is such an important aspect of carrying out a critical analysis of historical records that it deserves explanation. After all bibliographic details were placed in appropriate fields (author, title, date of publication, location of publication etc.), records were sorted into broad categories: e.g. recipes which identify materials and amounts; general instructions; and comments/observations. General categories were then subdivided by introducing yes/no fields. For example, to investigate liquid cleaning, further refinements were introduced, e.g. aqueous-based versus solvents

(Plate 5.8). Then, new subdivisions were introduced to break down aqueous ingredients into alkaline or acidic, and in the case of cleaning with solvents, to track their identity (e.g. alcohols or hydrocarbons).

At this point it was possible to show that of the aqueous-based cleaning records, over fifty percent involved cleaning with alkalis (Plate 5.9) and that a frequent source of the alkali was wood-ash lye. Damaged and abraded paint on historical oil paintings was thought to have resulted from the use of lye, but only based on anecdotal evidence. For the first time rigorously investigated evidence demonstrated that this type of cleaning has been described over an extended period of time and in a variety of locations.[36]

To identify a representative recipe to reconstruct lye with historically appropriate materials and methods, new fields were introduced to effect the extraction of information in more and more detail: was the wood type for making the lye identified? If so, was any particular wood mentioned more frequently than others? Each query then involved going back through previous records to tag them accordingly. Finally, a drop-down search field listing all wood or plant matter described in the sources for lye making could be built. The evidence was compiled through successive, increasingly refined queries of large sets of text-based data. The database ensured that meaningful reconstructions of historically appropriate cleaning agents could be accomplished within a reasonable time frame.

## The 'End Use' field

Another refinement to extract information from recipes is a keyword field entitled 'End Use'. It revealed that a single ingredient, such as 'mastic resin', most often associated with a final picture varnish, was not restricted in use to the top varnish layer of a painting. The collection of keywords showed that this material could be present in-between layers of paint (as an isolating varnish), on top of the preparation layer or ground (to seal the ground) or in the paint itself, as part of a painter's Medium meant to modify the paint's flow properties.[37]

36 Ashes from oak and beech woods are being characterized in terms of their chemistry and the pH of their lye solutions made with a number of variables, such as the source and age of the ashes, the age of the lye and the exposure of the lye to evaporation. A variety of oil painted surfaces (naturally aged and model reconstructions) will be 'cleaned' with the lye to determine whether there is a visual/chemical correlation between the surface damages seen on old paintings and that caused by wood-ash lye.

37 Carlyle, 'British Nineteenth-Century Oil Painting'.

Similar multi-functions were revealed for Copal Oil Varnish which, along with Mastic Varnish, was sold throughout the nineteenth century by artists' suppliers.

As the W&N database revealed, the company incorporated residues from a variety of products in multiple ways, e.g. oil varnish 'bottoms' or residues in their varnish-making pots were used in their preparation of artists' grounds,[38] and more recently it has become evident that the Purple Lake colour used in their Bitumen Oil Tube paints, was produced from residues of other lake manufacture.[39] An 'End Use' field would quickly capture W&N's incorporation of manufacturing residues in their other products, and 'links/annotations' between records would make the original passages from the manuscript (MS) containing this information immediately accessible.

## Manual transcription versus digital page-imaging

With the W&N Database the necessity of a full typed text transcription was eliminated by providing a high-resolution image of each MS page from the 85 MS books available (e.g. see Plate 5.7). Researchers can find individual page images through a series of summary and keyword fields, such as recipe title, topic (pigment or varnish manufacture, etc.) and by individual ingredients. Once a given page is accessed, the user can then page through the MS forwards or backwards as one would with a real book, such that the information on the page of interest can be seen in the context of the MS as a whole.[40] The incorporation of page images avoids transcription errors and allows abbreviations, line breaks, corrections, cross-outs and annotations to be easily viewed, enhancing interpretations. The Authority Field can be used to explain interpretations of abbreviations and conventions in the handwritten texts.

## Report function

An essential feature for using a text-based database as a research tool is the 'Report' function. This can entail customized print-outs of fields containing for example, varnish and Medium recipes, to reveal patterns and commonalities as well as chronological developments. Report options should include the ability to show content from selected fields in a variety of chosen orders and layouts (e.g. in tables by column or in a paragraph format). While summary tables can be produced by exporting information

38 Carlyle et al, 'Historically Accurate Ground Reconstructions'.

39 Vitorino, 'Nineteenth-Century Cochineal'.

40 Clarke and Carlyle, 'Page-Image Recipe Databases' (2005a&b)

to spreadsheets or word documents, this is unnecessarily time consuming when a database can be designed to produce tables and reports directly.[41]

## Historical sources databases for multi-topic research

The W&N Database was designed primarily for materials historians, conservation scientists and oil painting researchers, to identify how pigments, binders and paints were manufactured by a colourman in the nineteenth century, and what range and type of materials were used. Records describing pigment manufacture have been thoroughly investigated, analysed, and reconstructed in work carried out for Masters and PhD theses, for example, on W&N's lead chromate pigments[42] and their cochineal lake pigments.[43] An evaluation of binders and additives used in W&N paints was crucial to various projects for reconstructing nineteenth century oil paints[44] and as seen, their production of Bitumen paint is under investigation in the work of Marques (see endnote 32).

During the construction of the database, it was evident that the W&N manuscripts also contain details of the labour involved ('man's time') and other aspects of the company's operation, including the cost of producing their products as well as diagrams of specific manufacturing equipment. The original W&N Database was designed with a number of 'open fields' in anticipation of queries for this information and on other topics which were not the focus of the initial database designers. Open fields were intended to allow future researchers to populate their own fields and create their own tags and links to other records.

At the time this was a novel concept, that once all data had been entered, the ability to search and identify topics in the digital information was not restricted to a particular focus or unconscious bias built in by the original designers. There was also the intention that the database could receive updates and new expert information based on the findings of users.[45]

41 In her research with the HART Project, and later in her PhD on documentary sources covering artists' ground preparations, Maartje Stols-Witlox introduced the use of graphs to illustrate information by topic and chronology (Stols-Witlox, 'Historical Recipes'; Stols-Witlox, *A Perfect Ground*). This has been further refined in the work of Devesa (Devesa et al., 'Early Cleaning').

42 Otero et al., 'Chrome Yellow'; Otero, 'Historical Accurate Reconstructions'.

43 Vitorino et al., 'Nineteenth-Century Cochineal'.

44 Carlyle, 'HART Report'.

45 Unfortunately, these aspirations were not realized in the final Researcher's Edition since at that time the W&N company chose to impose limited access and restrictions on certain records (detailed in Otero, 'Historically Accurate'). The Winsor & Newton Nineteenth-century Archive Database project was initiated by Leslie Carlyle as part of the De Mayerne Programme funded

## Conclusion

The central role of performative research within the HART model has been highlighted by recounting discoveries that can only be found through the physical act of making. The need for intellectual rigour in choosing the route and developing the means to explore past practices has been expressed in the details provided which explain how large collections of historical information can be mined with interactive databases, and how historically appropriate materials can be identified and sourced.

The HART model rests on the aspiration to achieve some degree of historical accuracy, but as argued here, it is recognized that 'accuracy' can only ever be judged in relative terms: from wildly inappropriate material choices (e.g. modern silica coated pigments to represent historical materials) to the more appropriate option of using pigment reconstructions based on historical recipes. The need for thorough documentation of each reconstruction was argued with examples of how these products of re-enactment can be used in both visual and chemical analyses in comparison with historical oil paintings.

One aspect of the investment involved in the HART Model, which was not discussed directly, is the problem of sustaining the collection of the HCRs produced in terms of institutional commitments to providing long-term storage and accessibility. The value of the physical samples produced only increases as they age naturally and undergo the 'slow chemistry' that their historical counterparts in actual painted objects have already experienced.

A similar challenge awaits with regard to the sustainability and development of text-based databases, not only to make them widely available online, but also to build into their design methodologies for users to extract and interrogate the data beyond the expectations of the original designers. The HART three-part model for investigating historical oil painting methods and materials has proven successful thus far, but its future success rests on

by The Netherlands Organisation for Scientific Research (NWO). The initial database design was carried out in partnership with Mark Clarke, who then led its continuation and further development in the UK with a Resource Enhancement Grant from the Arts and Humanities Research Council (AHRC). The NWO funded database, known as the Pilot Project, was then supplanted by the AHRC funded Researcher's Edition completed in 2008. Portals for access to the Researcher's Edition are available at RKD Netherlands, Tate Britain (Conservation Department) and at the Department of Conservation and Restoration, Faculty of Sciences and Technology, University NOVA of Lisbon. The Hamilton Kerr Institute, University of Cambridge currently hosts a limited and partial edition of the Researcher's Database which is available online but is not currently fully functional for researchers.

the improvements and refinements that will be accrued as new researchers explore and extend its potential.

## Bibliography


Berg, Klaas Jan van den, Jerre van der Horst, and Jaap J. Boon, 'Recognition of Copals in Aged Resin/Oil Paints and Varnishes', in *ICOM-CC 12th Triennial Meeting Preprints*, (Paris: International Council of Museums, 1999), Vol. II, pp. 855-861.

Carlyle, Leslie, 'A Critical Analysis of British Handbooks, Manuals and Treatises on Oil Painting Published in the Nineteenth Century – An Interim Report of Work in Progress', in *Conservation today: papers presented at the UKIC 30th Anniversary Conference, 1988*, ed. by Vitoria Todd (London: United Kingdom Institute of Conservation, 1988), pp. 54-58.

Carlyle, Leslie, 'Using a Portable Computer in Art Historical Research', in *Computers and the History of Art*, ed. by Anthony Hamber, Jean Miles and William Vaughan (London: Mansell Publishing Limited, 1989), pp. 189-199.

Carlyle, Leslie, 'British Nineteenth Century Oil Painting Instruction Books: A Survey of Their Recommendations for Vehicles, Varnishes, and Methods of Paint Application', in *Cleaning, Retouching & Coatings* (London: International Institute for Conservation of Historic and Artistic Works, 1990), pp. 76-80.

Carlyle, Leslie, *A Critical Analysis of Artists' Handbooks, Manuals and Treatises on Oil Painting Published in Britain Between 1800-1900: With Reference to Selected Eighteenth-Century Sources*, Unpublished PhD dissertation (London: University of London, 1991).

Carlyle, Leslie, *'The Artist's Assistant': Oil Painting Instruction Manuals and Handbooks in Britain, 1800-1900. With Reference to Selected Eighteenth-Century Sources* (London: Archetype, 2001a).

Carlyle, Leslie, *MOLART Fellowship, Historical Reconstructions of Artist's Oil Paint: An Investigation of Oil Processing Methods and the Use of Medium-Modifiers* (Ottawa: Canadian Conservation Institute, 2001b), Revised Edition.

Carlyle, Leslie, with contributions by Maartje Stols-Witlox, Kathrin Pilz, Meta Chavannes and Brian Baade, *HART Report 2002-2005, The De Mayerne Programme* (unpublished, 2005a). Available from FOM-AMOLF, ICN, CCI and the author.

Carlyle, Leslie, 'Representing Authentic Surfaces for Oil Paintings: Experiments with 18th and 19th Century Varnish Recipes', in *Art of the Past: Sources and Reconstructions, Proceedings of the First Symposium of the Art Technological Research Study Group*, ed. by Mark Clarke, Joyce H. Townsend and Ad Stijnman (London: Archetype, 2005b), pp. 82-90.

Carlyle, Leslie, 'Historically Accurate Reconstructions of Oil Painters' Materials: An Overview of the Hart Project 2002-2005', in *Reporting Highlights of the De Mayerne Programme: Research Programme on Molecular Studies in Conservation and Technical Studies in Art History*, ed. by Jaap J. Boon and Ester S.B. Ferreira (The Hague: Nederlandse Organisatie voor Wetenschappelijk Onderzoek, 2006), pp. 63-76.

Carlyle, Leslie, Jaap J. Boon, Ralph Haswell and Maartje Stols-Witlox, 'Historically Accurate Ground Reconstructions for Oil Paintings', in *Preparation for Painting, The Artist's Choice and Its Consequences*, ed. by Joyce Townsend, Tiarna Doherty, Gunnar Heydenreich and Jacqueline Ridge (London: Archetype, 2008), pp. 110-122.

Carlyle, Leslie, 'Practical Considerations for Creating Historically Accurate Reconstructions', in *Fatto d'archimia: Historia e identificación de los pigmentos artificiales*, ed. by Marián del Egido and Stefanos Kroustallis (Madrid: IPCE; Ministerio de Educacion Cultura Y Deporte, 2010), pp. 105-117.

Carlyle, Leslie, and Diana Conde, 'Moisture Treatment and Flattening of a Severely Distorted Large Format Oil Painting on Canvas Without Facing or Looming', *The Picture Restorer*, 39 (2011a), pp. 9-12, 23-26.

Carlyle, Leslie, Pedro Caetano Alves, Vanessa Otero, Maria João Melo and Márcia Vilarigues, 'A Question of Scale and Terminology, Extrapolating from Past Practices in Commercial Manufacture to Current Laboratory Experience: The Winsor & Newton Nineteenth Century Artists' Materials Archive Database', in *ICOM-CC 16th Triennial Meeting Preprints* (Paris: International Council of Museums, 2011b), pp. 1-9.

Carlyle, Leslie, 'Towards Historical Accuracy in the Production of Historic Recipe Reconstructions', in *Tempera Painting 1800-1950, Experiment and Innovation from the Nazarene Movement to Abstract Art*, ed. by Patrick Dietemann, Wibke Neugebauer, Eva Ortner, Renate Poggendorf, Eva Reinkowski-Häfner and Heike Stege (London: Archetype, 2019), pp. 81-86.

Clarke, Mark, and Leslie Carlyle, 'Page-Image Recipe Databases: A New Approach to Making Art Technological Manuscripts and Rare Printed Sources Accessible', in *Art of the Past: Sources and Reconstructions, Proceedings of the First Symposium of the Art Technological Research Study Group*, ed. by Mark Clarke, Joyce Townsend and Ad Stijnman (London: Archetype, 2005a), pp. 49-52.

Clarke, Mark, and Leslie Carlyle, 'Page Image Recipe Databases, a New Approach for Accessing Art Technological Manuscripts and Rare Printed Sources: The Winsor & Newton Archive Prototype', in *ICOM-CC 14th Triennial Meeting Preprints* (London: James and James, 2005b), Vol. I, pp. 24-29.

Clarke, Mark, *Tricks of the Medieval Trades, The Trinity Encyclopedia: A Collection of Fourteenth-Century English Craft Recipes* (London: Archetype, 2018).

Devesa, Joana, Leslie Carlyle and Maartje Stols-Witlox, *Early Cleaning of Rembrandt's Paintings: What Materials Could Have Been Used and When?*, poster presentation at the Rembrandt Conservation Histories Symposium (Amsterdam, 2018).

Haswell, Ralph, and Leslie Carlyle, 'Van Gogh's Painting Grounds: Quantitative Determination of Bulking Agents (Extenders) Using SEM/EDX', *Microchimica Acta*, 155 (2006), pp. 162-169.

Haswell, Ralph, Leslie Carlyle and Kees T. J. Mensch, 'Quantitative Determination of Van Gogh's Painting Grounds using SEM/EDX', in *Microscopy & Microanalysis* (Cambridge: Cambridge University Press, 2010), pp. 1-5.

Haswell, Ralph, Leslie Carlyle, Kees T. J. Mensch, Ella Hendriks and Muriel Geldof, 'The Examination of Van Gogh's Grounds Using Quantitative SEM-EDX', *Van Gogh's Studio Practice, Exhibition Catalogue*, ed. by Marije Vellekoop, Muriel Geldof, Ella Hendriks, Leo Jansen and Alberto de Tagle (Brussels: Mercatorfonds, 2013), pp. 202-215.

Heinicke, Susanne, and Peter Heering, 'Discovering Randomness, Recovering Expertise: The Different Approaches to the Quality in Measurement of Coulomb and Gauss and of Today's Students', *Science & Education*, 22 (2013), pp. 483-503.

Hendriks, Laura, Irka Hajdas, Ester S.B. Ferreira, Nadim C. Scherrer, Stefan Zumbühl, Markus Küffner, Leslie Carlyle, Hans-Arno Synal and Detlef Günther, 'Selective Dating of Paint Components: Radiocarbon Dating of Lead White Pigment', *Radiocarbon*, 61 (2018), pp. 1-21.

Hendriks, Laura, *The Art of Microscale Radiocarbon* ($^{14}C$) *Dating of Paintings*, PhD dissertation (Zurich: ETH Zurich, 2019).

*European Paintings 15-18$^{th}$ Century: Copying, Replicating & Emulating*, ed. by Erma Hermens and CATS Proceedings (London: Archetype, 2014).

Langley, Allison, Kimberly Muir, and Kenneth Sutherland, 'Scenes from the Life of Picasso's *Still Life* (1922): History, Materials, and Conservation', *Springer Nature Applied Sciences Journal*, 2 (2020).

Marçal, Hélia, *From Intangibility to Materiality and Back Again: Preserving Portuguese Performance Art from the 1970s*, Unpublished PhD dissertation (Lisbon: NOVA University of Lisbon, 2018).

Oltrogge, Doris, 'The Cologne Database for Painting Materials and Reconstructions', in *Art of the Past: Sources and Reconstructions, Proceedings of the First Symposium of the Art Technological Research Study Group*, ed. by Mark Clarke, Joyce Townsend and Ad Stijnman (London: Archetype, 2005), pp. 9-15.

Oomah, B. Dave, and Edward O. Kenaschuk, 'Cultivars and Agronomic Aspects', in *Flaxseed in Human Nutrition*, ed. by Lilian U. Thompson and Stephen C. Cunnane (Champaign: AOCS Press, 1995).

Otero, Vanessa, *Historically Accurate Reconstructions of Amadeo's Chrome Yellows: An Integrated Study of Their Manufacture and Stability*, Unpublished PhD dissertation (Lisbon: NOVA University of Lisbon, 2018).

Otero, Vanessa, Leslie Carlyle, Márcia Vilarigues, and Maria João Melo, 'Chrome Yellow in Nineteenth Century Art: Historical Reconstructions of an Artists' Pigment', *RSC Advances*, 2 (2012), 1798-1805.

Salis Gomes, Carolina, Cátia Ferreira, Brenda Rossenaar, Ineke Joosten, Inez van der Werf, Leslie Carlyle and Klaas Jan van den Berg, 'Pigment Surface Treatments: 20th and 21st Century Industrial Techniques and Strategies for their Detection', *The Conservation of Modern Oil Paintings*, ed. by Klaas Jan van den Berg, Ilaria Bonaduce, Aviva Burnstock, Bronwyn Ornsby, Mikkel Scharff, Leslie Carlyle, Gunnar Heydenreich, Katrien Keune (Cham: Springer International Publishers, 2019), pp. 39-60.

Stols-Witlox, Maartje, *Historical Recipes for Preparatory Layers for Oil Paintings in Manuals, Manuscripts and Handbooks in North West Europe, 1550–1900: Analysis and Reconstructions*, Unpublished PhD dissertation (Amsterdam: University of Amsterdam, 2014).

Stols-Witlox, Maartje, *A Perfect Ground. Preparatory Layers for Oil Paintings 1550-1900* (London: Archetype, 2017).

Vitorino, Tatiana, Vanessa Otero, Leslie Carlyle, Maria João Melo and Marcello Picollo, 'Nineteenth Century Cochineal Lake Pigments from Winsor & Newton: Insight into their Methodology through Reconstructions', in *ICOM-CC 18$^{th}$ Triennial Conference Preprints*, ed. by Janet Bridgland (Paris: International Council of Museums, 2017).

## About the Author

**Leslie Carlyle** is a paintings restorer/conservator and Associate Professor of Paintings Conservation in the Conservation and Restoration Department of the Faculty of Sciences and Technology at the Universidade NOVA de Lisboa. Her doctorate from the Courtauld Institute of Art, University of London (1991) focused on 19$^{th}$-century British artists' oil painting handbooks and manuals. She specialises in the study of historical recipes for the preparation of artist's oil paint materials and techniques. She introduced the concept of 'historically accurate' reconstructions of painting materials to the fields of conservation and technical art history.

# 6. Imperfect Copies. Reconstructions in Conservation Research and Practice[1]

*Maartje Stols-Witlox*

**Abstract**

In paintings conservation, reconstruction plays a role as a treatment, when it serves to complete the structural image of a lost or damaged area or object, and in research, where reconstructions are employed, for instance, to investigate the impact of conservation treatments or for their innovation. The conservation field has developed several strategies to deal with questions of validity, truthfulness and relevance. By example of a number of recent conservation and conservation research projects, this chapter discusses terminology adopted for reconstruction practices within the field, strategies developed by conservators to deal with ethical and practical issues surrounding the use of reconstructions, and relates these strategies to the ethical framework that guides conservators in their daily work.

**Keywords:** conservation treatment, conservation research, code of ethics, terminology, retouching, digital reconstruction

## Introduction

Reconstruction plays an important role in the conservation of cultural heritage. While most of us are familiar with reconstructions of missing parts of statues or missing areas in paintings, the role of reconstructions in the investigation of degradation processes, treatment methods and

1 Acknowledgements: Gwen Tauber, Laurent Sozzani, Leslie Carlyle, Joana Devesa, Ella Hendriks, Agnes Brokerhof, Kees van der Meiracker, Tonny Beentjes, Rozemarijn van der Molen, Lambert Baij, Joen Hermans, Katrien Keune, Piet Iedema.

Dupré, S., A. Harris, J. Kursell, P. Lulof, M. Stols-Witlox (eds.), *Reconstruction, Replication and Re-enactment in the Humanities and Social Sciences.* Taylor & Francis Group
DOI 10.5117/9789463728003_CH06

insight into future changes to objects is less well known. Conservators investigate and care for works of art of the past and the present that are often unique, i.e. of which only a single or very few copies exist. It is this uniqueness that makes conservators resort to reconstructions in research, as all interfering with or altering 'the original' in conservation or restoration places this original in danger of losing part of its uniqueness or significance.

Art conservation[2] makes use of a range of reconstruction types, the choice of which depends on the purpose of the reconstruction. In this chapter I describe the different uses of reconstructions within the field, discuss the methods developed by conservators to deal with the use and limitations of reconstructions in conservation research, and reflect on the role and practices of reconstruction in this field. Reconstruction is the term used throughout, as this is a term that is generally employed within the field of conservation.[3] Yet, the cases discussed in this chapter show that re-enactment or replication are an integral part of the performative processes used in this field.

## The conservation Code of Ethics and the invisibility of the conservator

As a paintings conservator, I follow the professional Code of Ethics of the European Confederation of Conservator-Restorers Organizations (E.C.C.O.). In this document, a conservator's obligations towards Cultural Heritage have been formulated (see Appendix 1). These obligations reflect current thoughts about ethically responsible conservation and reconstruction practice. Some of the principles described relate directly to the use of reconstructions in this field. Reconstructions can help minimize danger to the physical integrity when they replace an original during tests and they play a role in restoring symbolic meaning and aesthetic significance when they are employed to fill in missing pieces or areas in damaged objects (article 5). When objects are kept in an unsuitable environment, replacement with a reproduction will remove the object itself from danger of being damaged

2 While conservators in cultural heritage care for art objects, they also conserve and restore applied art objects or structures, and artifacts that primarily carry historical value. For ease of reading, the term 'art' has been chosen here and should be seen to include these other categories as well in this chapter.

3 Although the term 'visualization' is preferred for colour 'reconstructions'.

(article 16).[4] The conservation Code of Ethics also implicitly warns against limitless reconstruction. Articles 6 and 8 draw attention to the principle of minimal conservation treatments. If reconstruction or 'improvement' of an object by reconstruction goes too far, it harms the spiritual significance, physical integrity, historic, and often aesthetic values contained within it.[5]

The Code of Ethics places the main focus on the object and not on the practitioner, in this case the conservator. This emphasis reflects the attitude that many conservators, including myself, were infused with during training. We were trained to be minimally 'visible' as a person in our work, because a conservator is supposed to put the interests of the object first with the goal of making the object speak for itself. The idea of the conservator pushing aside his own personality to follow the artist has a long history. For example, in 1829, when French author Jacques Nicolas Paillot De Montabert wrote about retouching, he stated that 'the tint, the tone and the touch should resemble the work by the master; while the taste, the ideas and the manner of the repairer should not be at all apparent'.[6]

As discussed by Étienne, investigating painting restoration in France 1750-1850, historic references such as these, to the restorer imitating the 'hand' or capturing the 'spirit' of the master are numerous. While one might assume that such an attitude would lead to modest interventions, this was not necessarily the case. Out of one and the same spirit completely different restorations could flow, drastic reconstructions that would 'help' the artist or work of art by 'improving' certain areas. Étienne includes a quote by Italian restorer Bedotti (1837), who wrote: 'It sometimes happens to even the most skilful painters to make mistakes that are too gross and too visible. In that case, one must not fear to seek and to remedy; whenever possible, and to improve the painting by eliminating or concealing the most shocking mistakes.'[7]

Recent conservation theory justly challenges the notion that a neutral intervention is possible and that such an intervention can be achieved by placing the artist or the work of art before the individuality of the conservator. Conservation and restoration influence the life trajectory of an object,

4 Although of course, removing an object to a different location in order to keep it safe may in fact remove meaning and identity from the object depending on its meaning in its original setting.

5 An interesting discussion on the difficulties of applying such Ethical Codes in conservation practice is given by Smith, 'A Role'.

6 'La teinte, le ton et la touche doivent sembler l'ouvrage du maître; et le goût, les idées, la manière du réparateur ne doivent apparaître aucunement'. De Montabert, *Traité complet*, p. 717.

7 See Étienne, *The Restoration*, p. 82; quoting Bedotti, *De la restauration*, p. 32.

changing the story contained within it. Therefore, being invisible or 'neutral' as a conservator is simply impossible. Even though not always visible at first sight, I, as a conservator, cannot prevent that my choices, my interventions, change the life trajectory of an object. If I remove a varnish, I may reveal more of the colours the artist used in his painting, but I also take away part of the history of the painting.

In modern conservation discourse, the term 'legibility' is often used in this context, the argument being that conservation is needed to make the object understandable for an audience. However, as conservator and philosopher Salvator Muños Viñas writes in his *Modern Theory of Conservation*, while reconstructions or retouching in conservation are often defended as a method of increasing the legibility of an object, the conservator is in fact not increasing legibility. Instead, the conservator is deciding which legibility prevails over different possible ways to read or understand an object.[8] For why would an object that before conservation is read as very damaged or broken, be *more* legible when it is less damaged or completed? The legibility does not increase, it only changes.

## Reconstruction terminology

Notwithstanding the impossibility of being a neutral party, the profound way in which our aim to be invisible in conservation interventions is integral to the profession may in fact be why conservation terminology to describe methods that replicate artistic processes equally focuses on the object. Conservators will immediately recognize terms like reconstruction, replication, model paint, test panel or dummy – the first two terms employed mainly for illusionistic reconstructions that replicate the pictorial effects created by the artist and the last three used as alternative terms in reconstructions in conservation research. However, the conservation field is less familiar with process-related terms like re-enactment or reworking, terms that are employed in other fields discussed in this publication, fields that investigate human history with performative methods. And yet, reconstruction research within the conservation field does often include some form of performance or enactment, as we are also studying historical actions and processes in order to reconstruct historical materials or objects. An example is research into the effects of historical recipes to clean oil paintings that I am currently

8 Muños Viñas, *Theory of Conservation*, p. 100.

involved in.[9] In this research, in which we investigate the visual, chemical and physical consequences of the use of historical methods for cleaning oil paintings, we use reconstructions to test and evaluate what actually happens when a historical cleaning agent like wood ashes and water touches a paint surface. And while we use modern methods to document the effects of such exposures (e.g. microscopy, UV-induced fluorescence photography, electron microscopy and analysis to investigate chemical changes), we investigate the circumstances of the use of historical methods described and base our approach to tools or the production methods of ingredients on those available at the time of writing of a recipe. Equally, in our field notes and publications, we document and describe the sensations that we experience while using such methods (the time it takes and the difficulty of making cleaning solutions, the feel of the surface during application, the sounds made by the ashes touching a paint surface, etc.). In short, the process is as important as the final effects.

## Where the difficulties lie

As any other researcher who employs re-enactment, replication or reconstructive methods to learn about the past, researchers in conservation are keenly aware of the challenges of bridging the gap between the present and the past in reconstruction projects. It is only very rarely that we know exactly under which circumstances certain materials or methods were employed in conservation treatments. This uncertainty forms the basis of the development of visible retouches or replacements of missing areas, methods that go from 'neutral' retouchings that aim to make damage less obvious without filling in for the artist, to integral retouches that aim to be invisible to the naked eye but still detectable with additional techniques (see discussion below).

The same challenge also influences the way we set up research reconstructions and how we describe the outcomes of reconstruction experiments in publications. Authors reporting on research that involves reconstruction may choose formulations like 'the painting may well have been subjected to this type of handling',[10] or indicate general connections, as in: 'comparison with reconstructions make it possible to read the surface of a painting

9 'From Wood Ashes', PhD research at the New University of Lisbon by Joana Deves, supervized by Dr. Leslie Carlyle and Dr. Stols-Witlox, provides a description of the project.
10 Mengshoel, *In Artists' Footsteps*, p. 136.

with enhanced understanding of what that paint must have been like in order to behave as it has',[11] or the description of reconstructions as offering 'flavours and insights'.[12] Verissimo Mendes et al. pointed out in their research into the cleaning of unvarnished contemporary oil paintings that their reconstructions 'are not necessarily representative when compared to the variability of "real, natural" dirt and naturally aged painted surfaces'. As they had available real objects for their research, in this case deaccessioned fragments of original and naturally aged paintings, they were able to draw comparisons between their tests performed on reconstructions and on a real aged object, concluding that 'tests performed with the same cleaning material show that the well cured and aged deaccessioned surface suffered less abrasion and polishing than the relatively young albeit light aged surfaces of the prepared panels'. Yet, they concluded that 'cleaning test results on these [reconstruction] surfaces give a good insight on the potential use of the cleaning materials' on specific young surfaces.[13] I myself have described reconstructions as 'approximations' and pointed out that 'today's researchers lack long-term experience of handling historic materials'.[14]

A specific difficulty in undertaking research into the effects of conservation methods using reconstructions, is the fact that we not only need to consider the distance between the present and the past but must multiply that distance by two. For example, when my colleague Joen Hermans investigates the way solvents used in conservation treatments interact with – or maybe damage – seventeenth-century oil paints, he needs to perform his tests with reconstructions that firstly, represent the composition of the original seventeenth-century oil painting, and secondly, also compare chemically to this same oil painting after three hundred years of ageing, i.e. he requires a reconstruction of an *aged* original. And chances are that a centuries-old painting may even have been cleaned or restored during its lifetime, further complicating the chemical and physical characteristics of the object that he wishes to replicate.

Regarding reconstructions used in conservation research, chemical and physical comparability are of particular importance because these reconstructions function as stand-ins for tests of conservation methods, usually followed by application on original objects themselves, necessitating

11 Carlyle, 'Exploring the Grammar', p. 37.

12 Bucklow, *In Artists' Footsteps*, p. 25.

13 Veríssimo Mendes et al., 'New Approaches', p. 387. The 'deaccessioned' in the first quote from this publication refers to the fact that the tests were performed on a painting that was first in a public collection, deaccessioned, i.e. removed from the collection, and donated to science.

14 Stols-Witlox, *A Perfect Ground*, pp. 224-225.

sufficient chemical and physical resemblance to make the tests on the replicate relevant for actual conservation practice.

Different approaches have been developed to deal with the complexity of producing reconstructions or test samples that are comparable to *aged* oil paintings. Research into the topic of cleaning water-sensitive paints by Klaas-Jan van den Berg and Aviva Burnstock, among others,[15] shows the complexity of, and the necessity to, include reconstruction processes in research that aims to design suitable conservation methods for such fragile paintings.

## Reconstruction as conservation treatment: Aesthetic enhancement

Replication processes are introduced into conservation trajectories for two reasons: Firstly, to replace an original, for instance when the decision is made to remove an object from harmful weathering, and when an original is lost, and secondly, to fill in damaged or lost areas and segments of works of art, in a process that is referred to by Muños Viñas as 'aesthetic enhancement'. A famous example of the first category is the lower left panel of the altarpiece by Jan van Eyck in the St. Bavo Cathedral in Ghent. The copy panel of the *Righteous Judges*, painted by Jan van der Veken, art dealer, artist and painting restorer, after the theft of the original in 1934, is still in place.[16]

In processes primarily aimed at aesthetic enhancement, visual comparability is considered more crucial than chemical and physical similarity with the original. In fact, chemical and physical dissimilarity is even considered a positive trait in the filling of lost or damaged areas, as is discussed below.[17] The evaluation of reconstructions as conservation treatments rests on a number of primary criteria: the thoroughness of the research that precedes

15 Both authors have published widely on this topic, with various other researchers. See for example *Issues in Contemporary Oil Paint*, ed. by Van den Berg et al. A full overview of publications is outside the scope of this paper, but a list of publications by Van den Berg can be found at http://www.uva.nl/en/profile/b/e/k.j.vandenberg/k.j.vandenberg.html, checked on 17 February 2019. See for publications on the topic by Burnstock: https://courtauld.ac.uk/people/aviva-burnstock, checked on 17 February 2019.

16 http://closertovaneyck.kikirpa.be/ghentaltarpiece/#home/sub=panel23, checked on 17 February 2019.

17 This comment about the emphasis on visual similarity generally applies to paintings produced within a Western European and North American tradition. In particular, in situations where objects have an important spiritual or religious meaning, the balance between material and visual similarity that is aimed at during treatment may be entirely different. See for instance the discussion of the restoration of Thai Stupa in Byrne 2014, p. 80.

the reconstruction, the ethical and aesthetic qualities of the result, and a balanced approach to notions of authenticity. In the field it is felt that a replicate(d area) is only acceptable if sufficient evidence can be found to support its visual characteristic, and when the reconstruction is aesthetically and structurally suitable; ideally it should also age in a manner that is comparable to that of the original in order to prevent future dissimilarities.

Historically, aesthetic enhancement was the domain of artists who were hired to restore paintings because of their painting skills. Depending on the context of an object, aesthetic enhancement was not always restricted to reconstruction, but could go much further and cover original surfaces. It could even include the addition of certain elements, for example to adapt a work of art to a new role, a new aesthetic or changed morals.[18] Because for cases in this last category the term reconstruction does not apply, these are not considered in this chapter.

However, with the rise of the conservation profession in the eighteenth century the role of professional artists in conservation slowly decreased, although both professions remained closely related for a long time.[19] With the establishment and professionalization of conservation came the development of conservation theory, and conservation professionals developed conservation ethics. This had its effect on the degree of reconstruction in conservation, as conservation theory started to question the ethical validity of reconstruction as a conservation treatment. In fact, the conservation field has struggled and is still struggling to define the role of reconstruction and its principles as a method. In particular the tension between concept and material, the connection between a reconstruction and its original, the subjective versus the objective, the role of the conservator in the interpretation of the original, and the way in which the public may be informed about replicated areas are central to this discussion.

In the case of severely damaged paintings, conservators have sought and are seeking methods to reconstruct missing areas without falsification of the object: they feel that ideally, the audience should be able to distinguish

18 One of the well-known examples of such interventions are the loincloths to cover the nudity of the figures in Michelangelo's *Last Judgement* (1536-1541) in the Sistine Chapel. Added by Daniele da Volterra *c.* 1565, a number of them were removed in the restoration campaign of 1980-1994.

19 See Conti, *History of the Restoration*, for a discussion of historical restoration treatments and restorers from the sixteenth century onward, with emphasis on Italy and the UK, and Étienne, *The Restoration*, about the rise of the restoration profession in France. Varnish removal and retouching were estimated nobler than structural treatments of painting supports (tear mending, lining, transfer, etc.), and were often the domain of artists, although Étienne's sources did show that restoration was considered by many a last resort for professional artists who had too little success to earn their living with the selling of paintings.

original from restoration. Nadolny provides a complete overview of the retouching – or 'visual compensation' – systems the field has developed for this purpose and discusses the historical context of their development. Methods range from fully mimetic or integrated retouches to the application of a 'neutral' tone, neutral in this case being an even tone that is considered suitable for the surrounding painted areas. Between these extremes we find a number of techniques that mimic the original to a different degree, but that can be distinguished visually from close by. Such methods include several stippling (or *pointillism*) and stripe (*tratteggio*) techniques, or the use of fully mimetic or integrated retouches that have a slightly lower surface level than the object's surface, etc.[20] The choice in material used for aesthetic enhancement also plays a role; modern retouching binders having different solubilities than those of the original. This is considered particularly important ethically, because such differences allow for later removal of the retouches without destruction of or damage to original material.[21]

The degree to which the effects of ageing are reconstructed during aesthetic enhancement varies. In fully integrated reconstructions, signs of age such as craquelure in paintings or surface dirt are mimicked using different techniques. Craquelure patterns can for instance be scratched into the paint, a cast can be made of an intact area and used as a mould to impress texture in a filling. The effect of dirt can be created using semi-transparent brownish or greyish paints.

An interesting case illustrating the effects of different approaches to reconstructing severely damaged paintings, is the *Holy Kinship* by Geertgen tot Sint Jans (c. 1495). This painting had suffered severe water damage in the past, and was restored by Rijksmuseum Amsterdam paintings conservators Luitsen Kuiper between 1983-1989 and by Gwen Tauber from 1991-2001. Overall, approximately fifteen percent of the painted layers was lost, and in particular the painting's bottom half was very damaged, with many areas of exposed wooden support and ground (Figure 6.1). As filling in the losses would have required extensive reconstruction, Kuiper and Tauber started exploring some of the techniques mentioned above, Tauber using as a test panel a fragment of a fifteenth-century panel painting given to the museum for research (Plate 6.1). On this test panel, different techniques with stripes and stipples were explored, each executed to different degrees of fineness and colour resemblance. After careful deliberation and discussion with studio colleagues, museum curators and an advisory commission including

20 Nadolny, 'History of Visual Compensation', pp. 573-585.

21 Digney-Peer et al., 'The Imitative Retouching', p. 613.

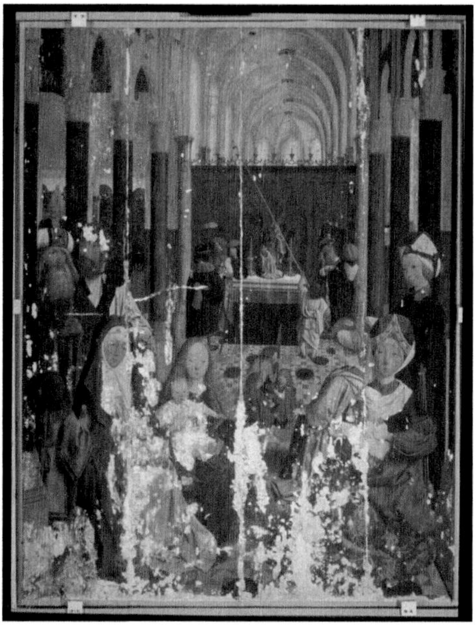

Figure 6.1: Geertgen tot Sint Jans (studio), *The Holy Kinship*, oil on panel, c. 1495. Collection: Rijksmuseum Amsterdam. After removal of non-original paint (overpaint). Photo: Rijksmuseum Amsterdam.

external conservators, the choice was made to reconstruct missing areas using a stripe and stipple technique that would only be visible to the trained eye from very close by. The more visible techniques were considered too disturbing, as the conservator and curators thought that they would draw the public's attention away from the preserved original areas. The idea was to restore the losses to such a degree that the unity of the whole composition would be regained, guiding primary attention to the original (intent), before the damage. Craquelure patterns were not imitated. Tauber explained that as such patterns were not very visibly dominant in the original, she did not find their absence in reconstructed areas visually disturbing.[22] The lengths conservators go to in such a difficult reconstruction process is illustrated by the reconstruction of the brocade pattern of saint Elizabeth in this painting. So much of the original gold brocade was lost that it was extremely difficult to reconstruct its folds on the basis of what remained. Tauber finally made her reconstruction using a visual aid fabricated by a colleague, a cloth marked with horizontal and vertical lines, stiffened with glue and pushed into shape following the remnants of the original cloth (Figure 6.2 and Plate 7.2).

22 Gwen Tauber, paintings conservator Rijksmuseum Amsterdam, oral communication, 23-10-2018.

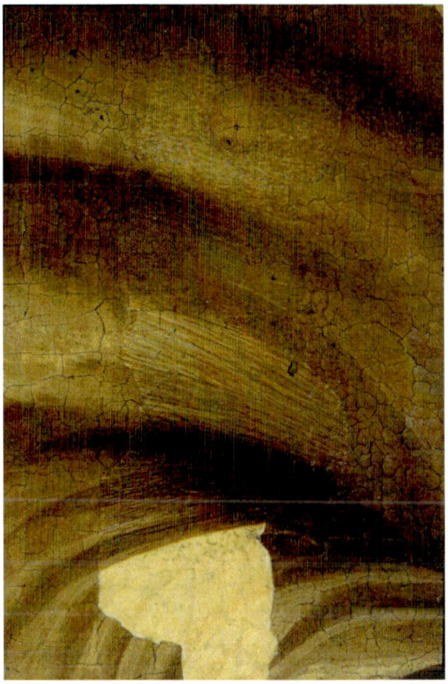

Plate 6.1: Testing panel with two different retouching options, stippling that imitates the 'worn' details and the colours of the adjoining areas (top), retouching that imitates details and colour, but in a more distinct striped brush stroke (lower area). From left to right, the texture of the craquelure was imitated to different degrees. Photo: Gwen Tauber, Rijksmuseum Amsterdam.

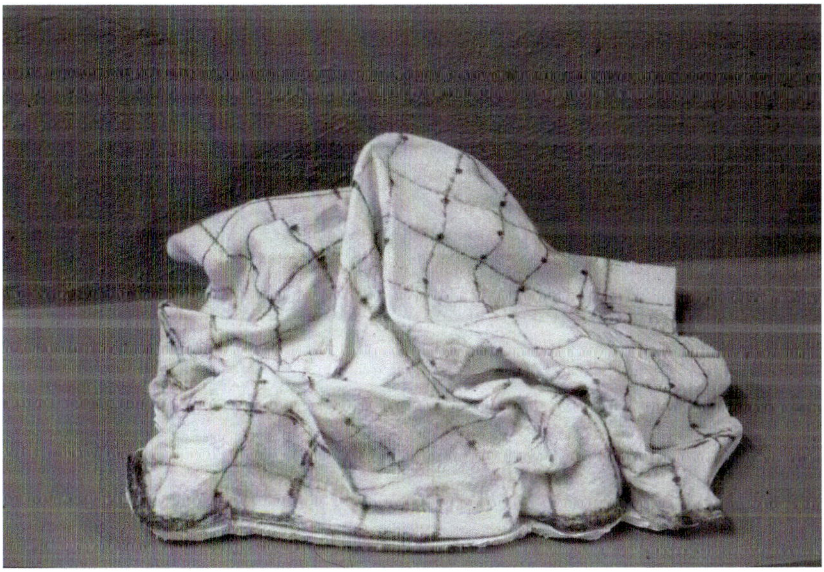

Figure 6.2: Checked cloth made to assist in the reconstruction of folds in drapery by Gwen Tauber's colleague Laurent Sozzani. The cloth was folded to imitate the folds of remaining areas of the skirt of Elizabeth, seated in the lower right of the painting. Photo: Laurent Sozzani, Rijksmuseum Amsterdam.

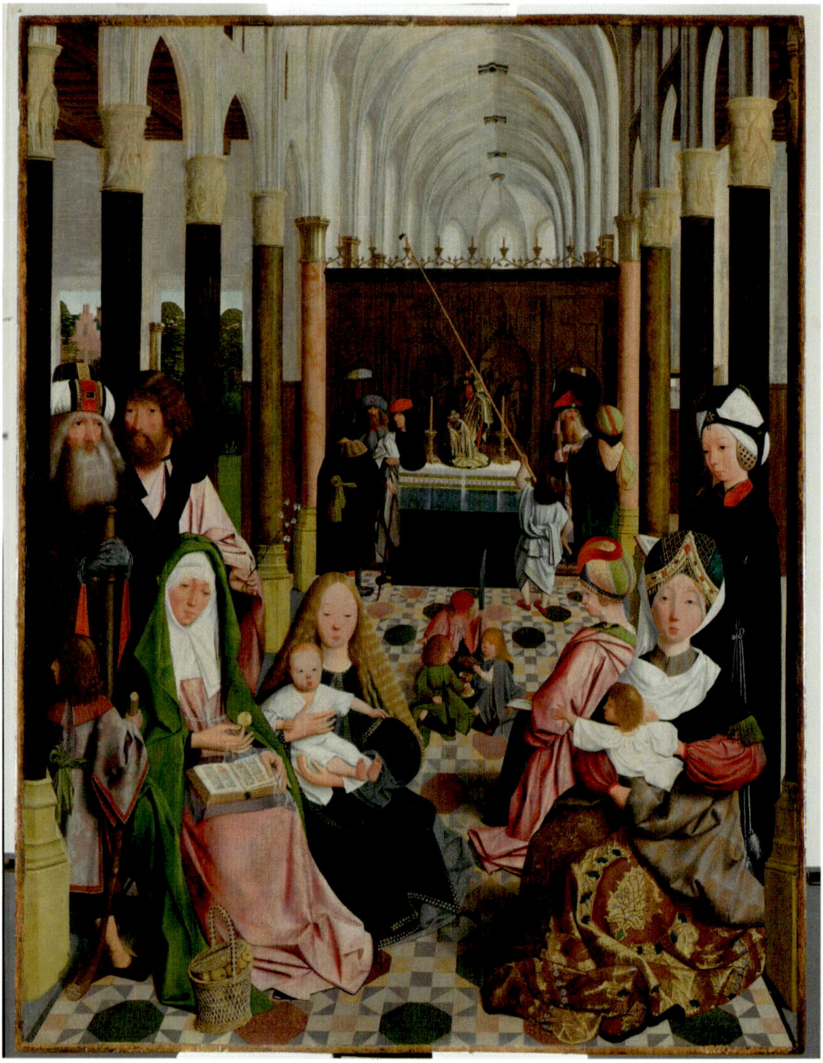

Plate 6.2: Geertgen tot Sint Jans (studio), *The Holy Kinship*, oil on panel, c. 1495. Collection: Rijksmuseum Amsterdam. After restoration and reconstruction of missing areas by Gwen Tauber. Photo: Rijksmuseum Amsterdam.

Tauber still feels only partly satisfied with some of the reconstructed areas: 'It is better than it was, but the chance that you choose exactly the method of the artist is near zero. You may get close, but never reach it completely. This is a heavy responsibility.'[23] These mixed feelings are why

23 Gwen Tauber, paintings conservator Rijksmuseum Amsterdam, oral communication, 23-10-2018.

Tauber emphasizes the fact that the museum audience can at least learn about the degree of loss in this painting and its restoration treatment in a museum tour and in a book dedicated to this painting and its restoration, as well as the importance of using materials which can be removed later without damage to the original paint.[24] This means that Tauber's interpretation of the painting can be changed in a new conservation or restoration treatment if information surfaces that necessitates a change, or if our general vision of such paintings or their restoration changes. The term 'retreatability' is commonly used in this context. As discussed by Appelbaum, retreatability can be read both as 're-treatability' or as 'retreat-ability', both terms indicating that reversing or withdrawing should remain possible. The term was adopted by practitioners in the field to replace the earlier term 'reversibility' in conservation ethics, the idea being that current conservation measures should not diminish future options for restoration or interpretation.[25] The issue with reversibility is that absolute reversibility is unattainable in practice, as nearly every conservation measure removes, adds or changes an object in such a manner that it is impossible to remove every trace of this measure. Retreatability is seen as a more practical aim, which also takes into account the fact that the idea of the conservator as a neutral mediator for the object (as an independent entity) is oxymoronic.

The choice of a retouching or reconstruction approach is inherently a personal choice, the restorer deciding which characteristics of the object should be revealed to the audience, which values of the object should be most noticeable, of course within the borders of what is considered ethically defendable and the limitations of what is practically possible, as skill plays an undeniable role. Should the audience know exactly which parts of the object have the same age as the original, or should the audience not know this, as visible retouches are considered too disturbing for the visitor experience? Muños Viñas argues that conservation is in essence a creative profession, exactly because of this subjectivity,[26] something that is confirmed by Tauber's account. Even when based on much scientific research, aesthetic enhancement depends as much on the conservator's feeling about what is *right for* or *needed by* an object and on the collection's curatorial policy, as on the conservator's research of the original materials, the state of preservation of the object and on the perceived intention of the artist.

24 Wallert and Tauber, *The Holy Kinship*. Unfortunately, and ironically, currently out of print.

25 Appelbaum, *Conservation Treatment*, pp. 353-359.

26 Muños Viñas, *A Contemporary Theory*, p. 147.

## Reconstructions in conservation research

Physical reconstructions in conservation research follow completely different trajectories. Visually, they are also a separate category. In conservation research, such reconstructions often look like long series of squares or rectangles of paint, without any reference to a pictorial image. Sometimes they are not applied by paint brush but with a paint applicator called a drawdown-bar that ensures paint layers of equal thickness (Figures 6.3 and 6.4). While reconstructions executed as aesthetic enhancement are physically connected or part of the original they are reconstructing/completing, reconstructions in conservation research are generally separate objects.

As explained earlier, reconstructions are needed in conservation research because conservators and scientists require surfaces for testing conservation materials and methods, materials and methods that may in the future also be used on original paintings. Also, in degradation studies investigating the way materials change through time, and in preventive conservation studies focused on improving the environment of an object, reconstructions are employed to help understand material degradation or change through time and to determine the factors that influence this change.[27]

A major problem that researchers identify in this approach is dealing with 'the chemical complexity of oil paint and the great variation in paint formulation and paint history', as Hermans writes. In order to find 'true explanations of the phenomena that are observed in real paintings', chemists like Hermans work with model systems and vary a single factor. According to Hermans, such model systems cannot be classified as reconstructions. Yet, the approaches share similarities, as in using such model systems researchers do try to reconstruct the steps in chemical processes that occur in art objects themselves, even though their experiments are performed out of context, in a situation that differs enormously from that surrounding the creation of a real oil painting. And always the chemist is left with the immense challenge to 'relate this knowledge of paint models to actual paints', which Hermans attempts 'by gradually reintroducing complexity'.[28]

The time gap between the present and the past, and back again from the past to the present needs to be bridged in some way in conservation studies, as replicates of originals are relevant only if they possess certain ageing and degradation features that are similar to those of the original. Two approaches have been developed for the creation of such reconstructions, one making

27 See the chapter by Carlyle in this publication.

28 Hermans, *Metal Soaps*, p. 14.

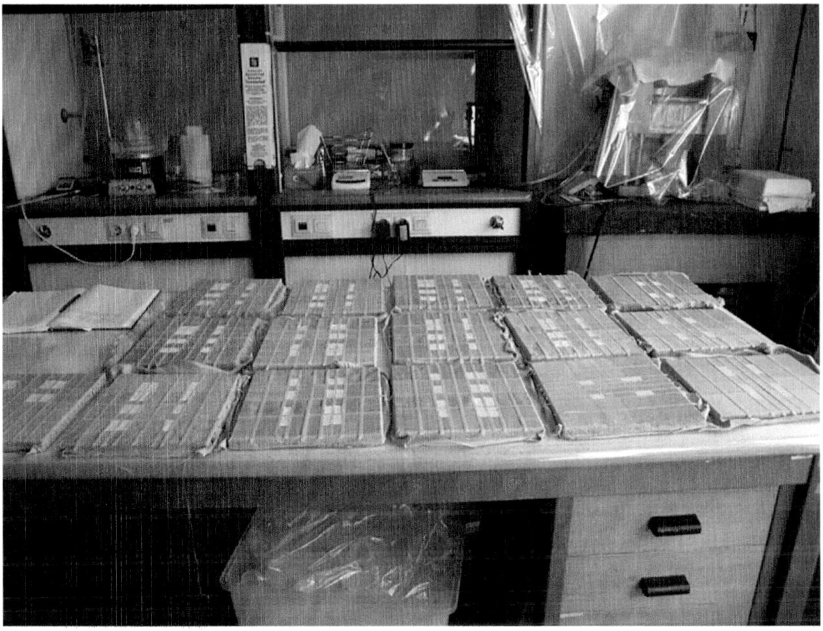

Figure 6.3: Set of reconstructions to examine the influence of a size layer on canvas. Figure: reproduced from Stols-Witlox, *A Perfect Ground*, p. 224.

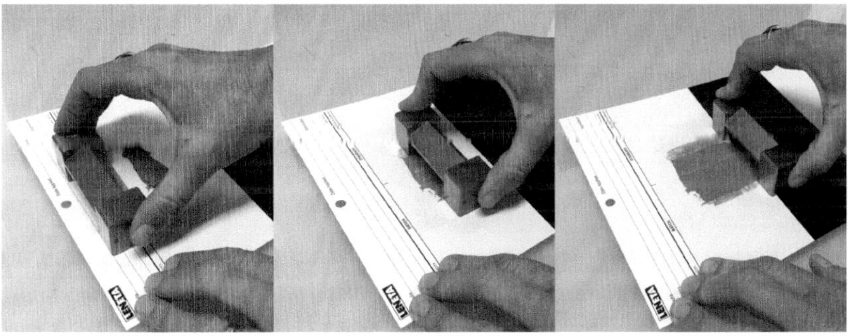

Figure 6.4: Use of a paint applicator, also called a drawdown bar, to generate paint layers of even thickness. The drawdown bar is pulled down to spread the paint (from left to right in this image). In this case, the paint is applied to a black-and-white chart. Such charts are used to check colour and opacity of the layer. Photo: author.

use of a process described as artificial or accelerated ageing, the other directly preparing materials using chemicals that have compositions similar to those that have been detected in the aged original by chemical analysis.

Artificial or accelerated ageing uses extreme environmental conditions to speed up degradation processes that would normally take much longer. Depending on the type of accelerated ageing, different effects are produced.

Increased light levels result in the formation of radicals, which initiate further chemical reactions. By raising the temperature, all chemical processes accelerate. Exposing samples or reconstructions to cycles of high and low humidity and/or of high and low temperature creates physical stress, as objects respond to the changes in temperature and humidity through shrinkage and expansion. This leads to the formation of tears and cracks or to flaking paint, visually comparable to the kinds of physical changes seen to naturally aged objects. Accelerated ageing has long played an important role in the paint industry, where it is used to compare materials and their stability under different circumstances. Standards and protocols that have been developed for such industrial applications[29] and equipment built for these standardized ageing tests finds application in conservation research.

While accelerated ageing can be used to explore possible reactions that may occur in materials when exposed to extreme conditions, the use of accelerated ageing to age paint to a level where it resembles an ancient oil paint is problematic. The reason is that extreme environmental conditions do not necessarily result in the same chemical and physical degradation processes that would occur naturally. Different energy levels may lead to different reactions. Therefore, one cannot re-create Rembrandt's paint by simply making a fresh paint according to a seventeenth-century recipe and exposing this paint to high levels of light and temperature. The resulting paint will not necessarily resemble a real seventeenth-century paint, even more so because this seventeenth-century paint has probably been restored a number of times during its lifetime, which has added more uncertain factors to the equation.

Much is still unknown about natural ageing. Therefore, we cannot fully imitate natural ageing with accelerated processes. Robert Feller concluded in his 1994 overview of contemporary approaches to accelerated ageing that 'it is really too soon to expect to find extensive, well-founded recommendations for specific testing procedures'. He was hopeful that 'this may be possible in the not-too-distant future for specific materials such as paper, dyed textiles, and artists' pigments'.[30] Progress has been made since Feller's publication. For example, new insights have been obtained into degradation processes of oil paint. More knowledge about the chemical reactions that may occur during the natural ageing of oil paint can be used to improve protocols for accelerated ageing, which can aim to create circumstances under which such reactions are most likely to occur.

29 See for instance the accelerated aging tests described by the American Society for Testing and Materials: www.astm.org, checked on 25 October 2018.

30 See Feller, *Accelerated Aging*.

An example of conservation research using artificial ageing is found in the work of Van den Berg with different co-workers on the cleaning of modern oil paintings. Within this context, Volk and Van den Berg published on cleaning methods suitable for surfaces that were covered with epsomite crystals. Epsomite crystals occurring on modern oil paints are thought to be the result of a chemical reaction between magnesium carbonate, a paint constituent, and sulphurous gases that were abundant in city air during periods of high pollution in the past.[31] Epsomite is very water-sensitive, and severely limits the conservator's ability to clean non-varnished paint surfaces. Volk and Van den Berg worked with reconstructions, or as they call them, 'artificial test paints' in which epsomite crystals were created by exposing paints containing magnesium carbonate to sulphurous gases inside a custom-built box. For their subsequent surface-dirt removal tests, they applied 'artificial soiling', dirt made up of powdered pigments and fillers (ivory black, ochre, silica, kaolin, cement) and organic materials (gelatin powder, starch, olive oil and mineral oil).[32] They explain that reconstructions allow for 'well defined systems', which, for this research is considered an advantage, but they also see limitations and note problems. For instance, in their experiments, the raised temperatures used to speed up degradation during accelerated ageing led to a softening of the relatively fresh paint films, which had the result that the artificial soiling was adhered too strongly to the paints.[33]

The second approach to the problem of bridging the age gap in conservation reconstructions is to eliminate accelerated ageing from a reconstruction protocol altogether and directly create aged paints, as done by Baij and colleagues in their model-system-based experiments into the penetration of solvents used in conservation into paint layers. This penetration of solvents is an important concern of conservators, as solvents-on-the-move may 'drag along' small molecules in paint layers that are thus pulled out of place. This may play a role in promoting unwanted short- and long-term chemical reactions within paint layers, and may result in swelling of paint films, which in turn can make paint layers more vulnerable to damage when rubbed or pressed when swollen. To answer the question of how quickly solvents penetrate into aged paint layers, and what effects such penetration may have, Baij et al. chose not to use artificially aged paint

31 Magnesium sulphate is used as an additive in certain modern oil paints. See Silvester et al., 'A Cause of Water-Sensitivity' for a detailed discussion on the formation of epsomite in oil paints.

32 The recipe for this artificial dirt was introduced by Bronwyn Ormsby, who explains the reasoning behind its composition in Ormsby, 'An Empirical Evaluation', pp. 77-87.

33 Volk and Van den Berg, 'Agar – A New Tool', pp. 390, 393, 402.

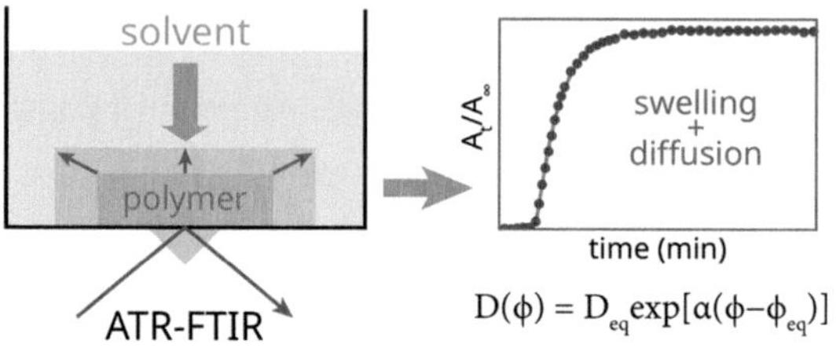


Figure 6.5: Schematic drawing of the set-up used by Baij and Hermans to investigate solvent penetration through paint layers. Figure: reproduced from Baij et al., 'Time-dependent ATR-FTIR', abstract / American Chemical Society, used with permission.

films. They rejected this approach for their research because they felt that such paint films would be too complex, obscuring insights into the process they wish to investigate, as every additional variable could play a role in changes, thus confusing their results. Secondly, they asked themselves, why follow the indirect approach of starting with new materials and ageing them to make our test materials (which leads to problems itself), if we can combine the molecules we are interested in investigating directly in the lab? Thus, they set about combining chemical compounds that are found within aged paint films and created their model system. They worked with linseed oil co-polymerized with lead or zinc sorbate complexes, as they found in previous studies that these compounds gave readings in instrumental investigations that are comparable to those of original aged paint films.[34] Their 'paint films' were exposed to different solvents in an ingenious set-up: they placed an ATR-FTIR detector (Attenuated Total Reflection – Fourier Transform Infrared Spectroscopy) on one side of their paint film and distributed the solvent on the other side. The solvent would only become detectable with the ATR-FTIR after having travelled through the paint film ('polymer' in figure 6.5). By taking measurements with the ATR-FTIR equipment at different intervals, they were able to determine how long it took the solvent to travel through a paint film. They measured its effects on the paint film, while the same set-up also allowed them to determine the swelling of the paint film (Figure 6.5).[35]

The above examples underscore the difference between reconstructions in conservation practice that need to visually replace or complement an original, and in conservation research, where a high degree of abstraction

34 A model system developed in the context of PhD research by Joen Hermans (defended in 2017) which aims to provide insights into chemical reactions between oil and pigments that take place in oil paintings and lead to degradation phenomena.

35 Baij, et al. 'Time-Dependent ATR-FTIR', p. 7137.

is sometimes accepted, and is even considered crucial by researchers who wish to get to the bottom of the chemical or physical changes that underlie degradation effects known to occur in paintings. These researchers lift out single processes from the complexity of the paint layer and re-enact and analyse them.

In this simplification lies the main complicating factor of the approach, for in order for such experiments to have relevance for actual paintings conservation, the results need to be translated to the more complex reality of the conservation studio. Indeed, researchers find this translation extremely challenging and acknowledge that at the moment it is often only partly possible. Due to this issue, the main value of such performative experiments is that they help paintings conservators understand the principles that lie at the basis of processes we experience in conservation practice. This type of experiment does not lead to tailor-made solutions for individual cases. Therefore, conservation practice is still characterized by extensive testing procedures, practical tests where original paint layers are exposed to different chemicals in order to test their suitability. Trial-and-error still plays a big role in conservation practice, even though trials are executed according to protocols that are designed to minimize risk to the original object, and that rest on an ever-growing scientific foundation.

## Recent developments

The last decades have seen the introduction of new reconstruction methods in conservation, both in research and in conservation practice. In particular, digitization plays an important role. Trumpy et al. describe early uses of digital image processing methods to virtually reconstruct what paintings look like without signs of age or degradation, by digitally removing discoloured varnishes or craquelure patterns. Their work aims to check the reliability of digital reconstructions as tools to visualize the effects of the removal of yellowed varnishes. While earlier studies did this by correcting the blue, red and green values, Trumpy et al. train neural networks to make predictions. As the paintings involved are actually undergoing varnish removal at the same time, they can compare their virtual varnish removals with actual varnish removal, concluding that while their approach gives good results, the match is not perfect. They believe that this is due to small local variations in the degree of discolouration and craquelure formation of the varnish, and differences in surface roughness of the paint that influence light absorption and scattering. This indicates the difficulty of digitally modelling or reconstructing variations that exist in the real world.[36] More recently, Kirchner et al. used a different approach, applying the Kubelka Munk theory, in a study focusing on varnish removal from Van Gogh's *Field with Irises*. They had more promising results when comparing the digital varnish removal and the actual varnish removal.[37]

Comparable attempts have also been made to reconstruct areas in paintings that have suffered discolouration due to the instability of particular pigments, such as the blue pigment smalt, organic yellow and red lakes.[38] The effects of such discolourations can be huge, as for instance in the paintings of Vincent van Gogh, who is known to have worked with a red pigment called geranium lake, a pigment that discolours from exposure to daylight and UV-radiation. In paintings like Van Gogh's *Bedroom*, the role of complementary colours is described by the artist in his letters, like the red floor against the green windows and the purple walls against the yellow bed. However, currently, these purple walls have faded to light blue and the red floor has turned pink. While the current state does not reflect the artist's intent, re-colouring the painting (and certainly scraping off the discoloured surface of the paint!) is considered highly unethical because of the invasiveness of such a treatment. Therefore, reconstructions have been chosen to inform the public about the painting's changed state. Described as 'approximations' or 'visualizations' by the researchers involved (Berns et al.), digital reconstructions were carried out, based on what the researchers could find out about paths of discoloration through recipe reconstructions,[39] measurements of the colours in areas that did not discolour – as they had been protected from light by the frame, and the colours of pigments present below the surface, as seen in paint samples. The reconstructions were illustrated in museum catalogues, published in scientific papers and also incorporated into museum displays (Plate 6.3a and 6.3b). Researchers have also used these digital reconstructions to explore the future discolouration of the *Bedroom*. For, having approximate knowledge of the influence of fading in the past, they project these changes into the future to digitally simulate further colour changes expected in the painting.[40] The Van Gogh museum used

36 Trumpy, et al., 'Experimental Study'.

37 Kirchner et al., 'Digitally Reconstructing, Part 1'.

38 Dik, *Scientific Analysis*; Wrapson, *In Artists' Footsteps*; Kirchner et al., 'Digitally Reconstructing, Part 2'; Kirchner et al. 'Digitally Reconstructing, Part 3' and Geldof et al., 'Reconstructing Van Gogh's Palette'.

39 Burnstock, 'Comparison'; Kirby, 'The Reconstruction'.

40 Researchers acknowledge that their approximations of future change are no more than that, noting about the fading experiments with painted reconstructions that they used for their

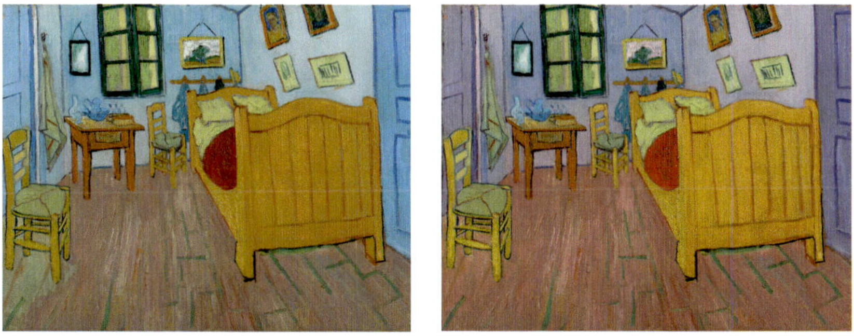

Plate 6.3 (a-b): Vincent van Gogh, *The Bedroom*, oil on canvas, 1888. Collection: Van Gogh Museum, Amsterdam (Vincent van Gogh Foundation). A digital visualisation of the work's original appearance (6.3a) is shown beside an image of the painting as it looks now (6.3b). Digital visualization: Roy S. Berns (Munsell Color Science Laboratory, Rochester Institute of Technology, U.S.A.). Figures: reproduced from Hendriks, Brokerhof, Van den Meiracker, 'Valuing Van Gogh's colours', n.p., used with permission.

these approximations as part of their move towards a new museum lighting system, based – amongst other reasons – on what were considered acceptable future discolouration levels (Fading continues and its rate depends on the amount of light the painting receives.) In an experiment, they showed these simulations with progressive colour fading to several groups of museum staff members and stakeholders and asked them what level of change they would consider acceptable, and within which time range. This information was taken into account in the decision-making process towards new LED museum lights as well as the formulation of new lighting instructions for paintings that go on loan to other museums, and played a role in raising staff awareness and acceptance of a new museum lighting system which lowered the light dosage received by the paintings in the Van Gogh Museum.[41]

In the case of Van Gogh's *Bedroom*, colour change was different for each area, depending on the way the artist mixed and applied his paints. For more homogeneous and well-defined colour changes, another approach has been introduced into conservation: digital retouching with light. This was the method used by Stigter (2016) in the conservation treatment of Dutch artist Ger van Elk's sculpture *Roquebrune* (1970, Frans Hals Museum Haarlem).[42] This triangular sculpture consists of a chromogenic print of a rock on the right

predictions: 'Obviously, the behaviour of freshly prepared paint is not the same as that of paint more than a century old, but this was accepted as indicating a worst-case scenario'. Hendriks et al., 'Valuing Van Gogh's Colours', p. 2.

41 Hendriks et al., 'Valuing Van Gogh's Colours'.

42 Stigter, *Between Concept and Material*, p. 117.

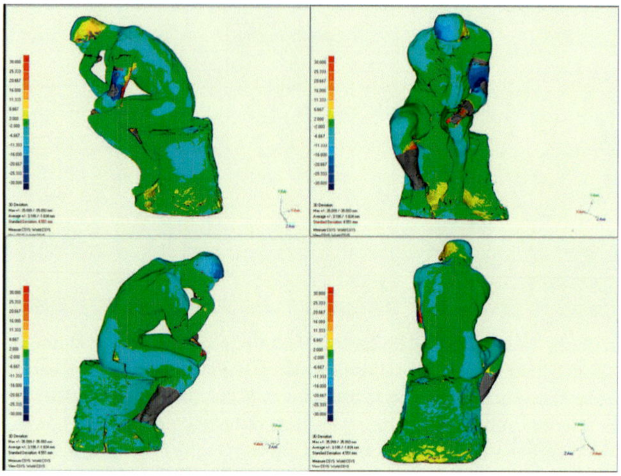

Plate 6.4: 3D scan of Auguste Rodin's *Thinker*, 1889, bronze (foundry mark Alexis Rudier/Fondeur Paris), digitally overlain with a scan of the casting mold, after the original statue in the collection of the Singer Museum Laren had been vandalized, before restoration. The colours show the level of overlap (green) and difference (from yellow and blue to red) between the casting mold and the mutilated statue, thus indicating areas of distortion. Image reproduced from Beentjes, Van de Molen, '3D scanning & printing', p. 149. Photo: Introtech.

side, and a painted version of the same pattern and colour on the left side. Because the photo print had faded, the relationship between the two elements was no longer as originally intended by Van Elk. Stigter describes the decision-making process around the conservation, which involved conservators, curators and the artist himself, discusses the ethical, conceptual and practical consequences of different options, ranging from a complete reproduction or reconstruction of the work to a non-intervention. The action chosen was to use a ceiling spot with coloured light to illuminate the faded plane, light that added such colour to the photographic print that both sides were visually similar again, thus 'retouching' the statue with light to reconstruct its unity.

In the case of Van Gogh's *Bedroom*, 2D reconstructions were used as a tool to restore or reconstruct original colour, and for *Roquebrune* light 'retouching' was introduced. Also, 3D, digitally printed reconstructions are being explored in conservation.[43] An example is research carried out by Beentjes et al. in relation to the restoration of August Rodin's 'Thinker' of the Singer Museum

43 This chapter focuses on conservation applications, not on 3D scanning and printing of art in general. However, the techniques developed by Joris Dik at Delft University may have interesting conservation applications. Dik et al., together with Océ Technologies, have performed research into 3D scanning and printing of paintings, resulting in 3D prints of paintings by Rembrandt and Van Gogh, amongst others. See also Elkhuizen et al., 'Topographical Scanning'.

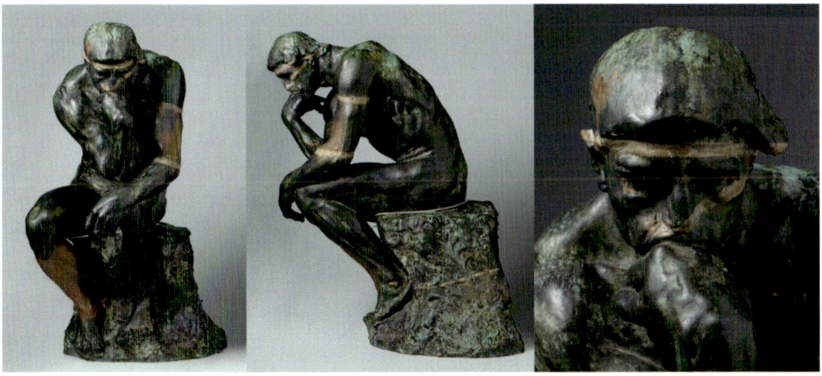

Plate 6.5: Auguste Rodin's *Thinker* of the Singer Museum in Laren. The statue is seen her with the brass powder/synthetic resin inserts and fillings, modelled after the 3D scans, before retouching was carried out. Image reproduced from Beentjes and Van de Molen, '3D scanning & printing', p. 152. Photo: T. Beentjes, R. van der Molen and M. Svenson.

in Laren. When recovered after a brutal theft from the museum garden, the bronze was severely damaged, having suffered attempts to cut it into pieces with a chainsaw and having received blows with a sledgehammer. Distortions, saw cuts and a missing lower leg needed to be dealt with in the conservation treatment. Research led to the discovery of the original casting mould in the Musée Rodin in Paris. A full 3D scan of this mould and of the mutilated statue were made and digitally overlain to identify changes (Plate 6.4). These scans were used by metal conservators as a guide during the re-bending of distorted areas. The 3D scan of the missing lower leg was used to 3D print a mould to cast the replacement part in a mixture of brass powder and synthetic resin (Plate 6.5). Colouring and retouching were carried out by hand, as this was considered the most practical option by the restoration team.

## Conclusion

As the cases discussed in this chapter exemplify, in conservation some kind of reconstruction is often part of a treatment. During the act of conserving, a conservator may attempt to improve or optimize the relationship between the current state of an object and its intended appearance as perceived by the conservator and stakeholders. Conservators do not always feel comfortable with this creative aspect of their profession, as demonstrated by the example of the *Holy Kinship*. Here the conservator in particular stressed the importance of telling the whole story to the public, including the reasoning behind the choices and illustrations of the damaged original to avoid falsification. Digital reconstruction techniques that have recently entered the field offer new possibilities to inform the public about what remains of an original and what has been added through reconstruction. This is an area that will no doubt develop further in the future, also because of the interest of museum audiences in this type of information.

Will digital methods replace 'traditional' reconstruction in conservation? It is expected that in some areas their role will increase. For example, do-it-yourself 3D scanning is developing rapidly and future availability of simple tools for scanning objects will offer interesting possibilities for the field. Applications such as computer-designed shaped fillings that can be printed, inserted into an object and subsequently finished on the spot do not seem far away.

In conservation research the role of digital aid is also increasing. Steps are being taken towards the digital modelling of degradation, such as the recent work of Piet Iedema in the Predagio project, modelling oil paint networks and their degradation,[44] and by Akke Suiker, who models the ageing and degradation of wooden objects and their adhesives.[45] While much progress is being made towards representative models in these projects, the complexity of the different chemical and physical processes at play makes computer modelling a great challenge. Designing computer models that replicate all instances that influence degradation of actual objects also remains challenging. An interesting aspect of the approach employed by these researchers is the fact that they make use of physical model systems, i.e. reconstructions, to validate their modelling approach. They compare their computer predictions with laboratory tests to check the outcomes of their mathematical models. In return, the mathematical models aid in guiding the direction of laboratory tests. So, while the balance between reconstruction in the lab and reconstruction behind the computer may change, they play a complementary, mutually supportive role in such research.

Reconstruction projects within conservation are more interdisciplinary now than in the past. In particular the cooperation with philosophy, history

44 Iedema et al., 'Mathematical Modeling'. Prof. Iedema is project leader of the project 'Predicting Ageing of Oil Networks', a project that uses multi-scale mathematical modeling to study the microstructure of oil paintings. See http://www.nicas-research.nl/full-projects/predagio/predagio.html, checked on 26 November 2018.

45 'A multi-scale and uncertainty approach for the analysis of the ageing of timber art objects adhesively bonded by animal glues'. Project leader Akke Suiker, TU Eindhoven, Netherlands. This project uses modelling to investigate the physical changes that occur during the degradation of such objects. See: https://www.nwo.nl/en/news-and-events/news/2018/01/eight-projects-receive-funding-for-collaborations-between-scientists-and-museums-with-the-application-of-data-science.html, checked on 26 November 2018.

of science and anthropology has increased, for instance within the RRR network. In the conservation of contemporary art such connections were established earlier,[46] as evidenced by the application of methods from anthropology, such as the interview, in conservation trajectories[47] or auto-ethnography inspired reflective trajectories.[48]

Such crossings of the disciplinary boundaries lead to a more profound articulation of discipline-related paradigms. Connections with other fields play an important role, as they result in increased attention to methodology that is required for reconstruction to fulfil its role as a mature scientific approach in conservation research and practice. These fields are welcome partners in discussing the way reconstructions themselves influence the way we experience art, in particular the questions they raise about authenticity and object value. For with all current advances towards in-depth understanding of the ageing of oil paint and the development of ever more precise reproductive techniques, we may need to prepare ourselves for the following question: Which do we consider to be the original? The aged three-hundred-year-old Rembrandt or the research-based 'perfect' copy of the painting as it must have left the artist's studio?

## Bibliography

Appelbaum, Barbara, *Conservation Treatment Methodology* (Oxford UK, Burlington, MA: Butterworth-Heinemann/Elsevier, 2007).

Ashley Smith, Jonathan, 'A Role for Bespoke Codes of Ethics', in *ICOM-CC 18th Triennial Conference Preprints*, ed. by Janet Bridgland (Paris: International Council of Museums, 2017).

Baij, Lambert, Joen J. Hermans, Katrien Keune and Piet D. Iedema, 'Time-Dependent ATR-FTIR Spectroscopic Studies on Solvent Diffusion and Film Swelling in Oil Paint Model Systems', *Macromolecules*, 51 (2018), pp. 7134-7144.

Bedotti, Giovanni, *De la Restauration des Tableaux, Traité Spécial sur la Meilleure Maniere de Rentoiler, Nettoyer et Restaurer les Tableaux Anciens et Modernes* (Paris: by the author, 1837).

46 See for instance Hummelen and Sillé, *Modern Art*, the book published at the occasion of a seminal symposium on the conservation of modern art in 1997.

47 Beerkens et al., *The Artist Interview*.

48 Stigter, *Between Concept and Material*.

Beentjes, Tonny, and Rozemarijn van der Molen, '3D Scanning and Printing as Conservation Tools: an Innovative Treatment of a Vandalized Bronze Statue, the Thinker by Rodin', *Lacona IX* (2011), pp. 146-153.

Beerkens, Lydia, Paulien 't Hoen, IJsbrand Hummelen, Vivian van Saaze, Tatja Scholte and Sanneke Stigter, *The Artist Interview. For Conservation and Presentation of Contemporary Art. Guidelines and Practice* (Amsterdam: Japsam, 2012).

Berg, Klaas Jan van den, Aviva Burnstock, Matthijs de Keijzer, Jay Krueger, Tom Learner, Alberto de Tagle and Gunnar Heydenreich, *Issues in Contemporary Oil Paint* (Cham: Springer International Publications, 2015).

Burnstock Aviva, Ibby Lanfear, Klaas Jan van den Berg, Leslie Carlyle, Mark Clarke, Ella Hendriks and Jo Kirby, 'Comparison of the Fading and Surface Deterioration of Red Lake Pigments in Six Paintings by Vincent van Gogh with Artificially Aged Paint Reconstructions', in *ICOM-CC 14$^{th}$ Triennial Conference Preprints*, ed. by I. Vergier (London: James and James, 2005), Vol. 1.

Bucklow, Spike, 'Housewife Chemistry', in *In Artists' Footsteps: the Reconstruction of Pigments and Paintings. Studies in Honour of Renate Woudhuysen-Keller*, ed. by Lucy Wrapson (London: Archetype, 2012), pp. 17-28.

Byrne, Dennis, *Critical Perspectives on Heritage Conservation in Asia*, Routledge Studies in Heritage (New York, London: Routledge, 2014).

Carlyle, Leslie, 'Exploring the Grammar of Oil Paint through the Use of Historically Accurate Reconstructions', in *The Conservation of Easel Paintings*, ed. by Joyce Hill Stoner and Rebecca Rushfield (London/New York: Taylor & Francis Group, 2012), pp. 33-38.

Conti, Alessandro, *History of the Restoration and Conservation of Works of Art*, trans. by Helen Glanville (Amsterdam: Butterworths Heinemann, 2002).

Montabert, Jacques Nicolas Paillot de, *Traité Complet de la Peinture* (Paris: Bossange père, 1829), Vol. 9.

Devesa, Joana, Leslie Carlyle and Maartje Stols-Witlox, 'From Wood Ash to Pearl Ash: Historic Alkaline Cleaning Agents and their Impact on Oil Paint', in *ICOM-CC 18$^{th}$ Triennial Conference Preprints*, ed. by Janet Bridgland (Paris: International Council of Museums, 2017).

Digney-Peer, Shawn, Karen Thomas, Roy Perry, Joyce Townsend, Stephen Gritt, 'The Imitative Retouching of Easel Paintings', in *The Conservation of Easel Paintings*, ed. by Joyce Hill Stoner and Rebecca Rushfield (London/New York: Taylor & Francis Group, 2012), pp. 607-634.

Dik, Joris, *Scientific Analysis of Historical Paint and the Implications for Art History and Art Conservation. The Case Studies of Naples Yellow and Discoloured Smalt*, Unpublished PhD dissertation (Amsterdam: University of Amsterdam, 2003).

*E.C.C.O. Professional Guidelines* (II). Code of Ethics. Promoted by the European Confederation of Conservator-Restorers' Organisations and adopted by its

General Assembly (Brussels, 7 March 2003). Checked on 21 October 2018: http://www.ecco-eu.org/documents/.

Elkhuizen, Willemijn, Tim Zaman, Wim Verhofstad, Pieter P. Jonker, Joris Dik and Jo M.P. Geraedts, 'Topographical Scanning and Reproduction of Near-Plane Surfaces of Paintings', in *Measuring, Modeling, and Reproducing Material Appearance*, Proceedings of SPIE-IS&T Electronic Imaging, ed. by Maria V. Ortiz Segovia, Philipp Urban and Jan P. Allebach (2014), Vol. 9018, 901809-1.

Étienne, Noémie, *The Restoration of Paintings in Paris, 1750-1815* (Los Angeles: Getty, 2017).

Feller, Robert L., *Accelerated Aging. Photochemical and Thermal Aspects*, Series Research in Conservation (Los Angeles: Getty Institute, 1994).

Geldof, Muriel, Art Ness Proaño Gaibor, Frank Ligterink, Ella Hendriks and Eric Kirchner, 'Reconstructing Van Gogh's Palette to Determine the Optical Characteristics of his Paints', *Heritage Science*, 6 (2018) n.p.

Hendriks, Ella, Agnes W. Brokerhof and Kees van den Meiracker, 'Valuing Van Gogh's Colours: from the Past to the Future', in *ICOM-CC 18th Triennial Conference Preprints*, ed. by Janet Bridgland (Paris: International Council of Museums, 2017), n.p.

Hermans, Joen, *Metal Soaps in Oil Paint. Structure, Mechanisms and Dynamics*, Unpublished PhD dissertation (Amsterdam: University of Amsterdam, 2017).

Hummelen, IJsbrand and Dionne Sillé, *Modern Art: Who Cares?* (Amsterdam: SBMK/ICN, 1997).

Iedema, Piet D., Joen Hermans, Katrien Keune, Annelies van Loon, Maartje Stols-Witlox, 'Mathematical Modeling of Mature Oil Paint Networks', *ICOM-CC 17th Triennial Conference Preprints*, ed. by Janet Bridgland, Melbourne, 15-19 September (Paris: International Council of Museums, 2014).

Kirby, Jo, 'The Reconstruction of Late 19th Century French Red Lake Pigments', in *Art of the Past: Sources and Reconstructions*, Proceedings of the First Symposium of the Art Technological Source Research Study Group, ed. by Mark Clarke, Joyce Townsend and Ad Stijnman (London: Archetype, 2005), pp. 69-77.

Kirchner, Eric, Ivo van der Lans, Frank Ligterink, Ella Hendriks and John Delaney, 'Digitally Reconstructing Van Gogh's Field with Irises near Arles. Part I: Varnish', *Color Research and Application*, 43 (2017), pp. 150-157.

Kirchner, Eric, Ivo van der Lans, Frank Ligterink, Muriel Geldof, Art Ness Proaño Gaibor, Ella Hendriks, Koen Janssens and John Delaney, 'Digitally Reconstructing Van Gogh's Field with Irises near Ales, Part II: Pigment Concentration Maps', *Color Research and Application*, 43 (2017), pp. 158-167.

Kirchner, Eric, Ivo van der Lans, Frank Ligterink, Muriel Geldof, Art Ness Proaño Gaibor, Teio Meedendorp, Kathrin Pilz and Ella Hendriks, 'Digitally Reconstructing Van Gogh's Field with Irises near Arles, Part III: Determining the Original Colors', *Color Research and Application*, 43 (2017), 311-327.

Mengshoel, Karen, Mirjam Liu and Tine Frøysaker, 'Shocking a Mock-up: Recreating the Damages and Historical Treatments Found in Edvard Munch's Monumental Aula Paintings to Test Materials and Procedure for Marouflaging a Marouflage', in *In Artists' Footsteps: the Reconstruction of Pigments and Paintings. Studies in Honour of Renate Woudhuysen-Keller*, ed. by Lucy Wrapson (London: Archetype, 2012), pp. 129-140.

Muños Viñas, Salvador, *Contemporary Theory of Conservation* (New York: Elsevier Butterworth-Heinemann, 2005).

Nadolny, Jilleen, 'History of Visual Compensation for Paintings', in *The Conservation of Easel Paintings*, ed. by Joyce Hill Stoner and Rebecca Rushfield (London/New York: Taylor & Francis Group, 2012), pp. 573-585.

Ormsby, Bronwyn, 'An Empirical Evaluation of a Range of Cleaning Agents for Removing Dirt from Artists Acrylic Emulsion Paint', *AIC Specialty Group Postprints*, 23 (2013), pp. 77-87.

Silvester, Genevieve, Aviva Burnstock, Luc Megens, Tom Learner, Giacomo Chiari, Klaas Jan van den Berg, 'A Cause of Water-Sensitivity in Modern Oil Paint Films: the Formation of Magnesium Sulphate', *Studies in Conservation*, 59 (2014), pp. 38-51.

Stigter, Sanneke, *Between Concept and Material. Working with Conceptual Art: A Conservator's Testimony*, Unpublished PhD dissertation (Amsterdam: University of Amsterdam, 2016).

Stols-Witlox, Maartje, *A Perfect Ground. Preparatory Layers for Oil Paintings 1550-1900* (London: Archetype, 2017).

Trumpy, Giorgio, Damon Conover, Lionel Simonot, Mathieu Thoury, Marcello Picollo and John K. Delaney, 'Experimental Study on Merits of Virtual Cleaning of Paintings with Aged Varnish', *Optics Express*, 23 (2015), pp. 33836-33848.

Verissimo Mendes, Beatriz, Klaas-Jan van den Berg, Luc Megens, Ineke Joosten and Maude Daudin, 'New Approaches to Surface Cleaning of Unvarnished Contemporary Oil Paintings – Moist Sponges and Cloths', in *Issues in Contemporary Oil Paint*, ed. by Klaas Jan van den Berg, Aviva Burnstock, Matthijs de Keijzer, Jay Krueger, Tom Learner, Alberto de Tagle and Gunnar Heydenreich (Cham: Springer International Publications, 2015), pp. 372-388

Volk, Annegret, and Klaas Jan van den Berg, 'Agar – A New Tool for the Surface Cleaning of Water Sensitive Oil Paint?', in *Issues in Contemporary Oil Paint*, ed. by Klaas Jan van den Berg, Aviva Burnstock, Matthijs de Keijzer, Jay Krueger, Tom Learner, Alberto de Tagle and Gunnar Heydenreich (Cham: Springer International Publications, 2015).

Wallert, Arie, and Gwen Tauber, *The Holy Kinship. A Medieval Masterpiece* (Amsterdam: Rijksmuseum/Zwolle: Waanders, 2001).

## Appendices

### *Appendix 1: Excerpts from the E.C.C.O. Code of Ethics*

Article 5: The conservator-restorer shall respect the aesthetic, historic and spiritual significance and the physical integrity of the cultural heritage entrusted to her/his care.
Article 6: The conservator-restorer, in collaboration with other professional colleagues involved with cultural heritage, shall take into account the requirements of its social use while preserving the cultural heritage.
[...]
Article 8: The conservator-restorer should take into account all aspects of preventive conservation before carrying out physical work on the cultural heritage and should limit the treatment to only that which is necessary.
[...]
Article 15: The conservator-restorer shall not remove material from cultural heritage unless this is indispensable for its preservation or it substantially interferes with the historic and aesthetic value of the cultural heritage. Materials, which are removed, should be conserved, if possible, and the procedure fully documented.
Article 16: When the social use of cultural heritage is incompatible with its preservation, the conservator-restorer shall discuss with the owner or legal custodian, whether making a reproduction of the object would be an appropriate intermediate solution. The conservator-restorer shall recommend proper reproduction procedures in order not to damage the original.[49]

## About the Author

**Maartje Stols-Witlox** is Associate Professor Conservation and Restoration at the University of Amsterdam. She is an art historian and paintings conservator; she specializes in the investigation of historical artists' recipes, with particular emphasis on reconstructions with 'historically informed' materials. She is project leader of the NWO Free Competition Humanities project 'Down to the Ground, a historical, visual and scientific analysis of coloured grounds in Netherlandish paintings, 1550-1650, core-team member of the 'Irradiation Passport for Art' (seed money, NICAS/NWO) and is affiliated to the ARTECHNE-project (Utrecht University).

49 *E.C.C.O. Professional Guidelines*, part II, Code of Ethics.

# 7. Reworking Recipes and Experiments in the Classroom

*Thijs Hagendijk, Peter Heering, Lawrence M. Principe and Sven Dupré*[1]

**Abstract**

This chapter addresses the potential of reworking experiments or recipes in educational settings. We reflect on educational practice in several university settings, which span the Liberal Arts and Science Program in Utrecht and a course for historians of science and technology at Johns Hopkins University to physics teacher education at the Europa-Universität Flensburg. The classroom use of RRR methods serves to teach the exploratory nature of science, and focuses the attention of students on materials and the sensory dimensions of experiments. Together, the three cases argue that the use of RRR methods in the classroom allows teachers to engage students in new ways, and offers students the opportunity to participate more meaningfully in research into the history of science.

**Keywords:** experimental history of science, laboratory, science education, interdisciplinarity, two cultures, sensual experience

## Introduction

The reworking, reproduction, or re-enactment (RRR) of experiments has a long and rich tradition in the history of science. This methodology has revealed considerable information and insights for historians of science and

1 Thijs Hagendijk's and Sven Dupré's research for this chapter was realized with funding from the ARTECHNE project, a European Research Council (ERC) Consolidator Grant awarded to Sven Dupré under the European Union's Horizon 2020 research and innovation program (grant agreement No. 648718).

Dupré, S., A. Harris, J. Kursell, P. Lulof, M. Stols-Witlox (eds.), *Reconstruction, Replication and Re-enactment in the Humanities and Social Sciences*. Taylor & Francis Group
DOI 10.5117/9789463728003_CH07

technology that would otherwise have been unobtainable.[2] Historians of science and technology based in (history of) science and technology museums, use methods of reworking (including the 'restoration of behaviour') to access skills in the production and the use of instruments, tools and technological artefacts.[3] Experimental history of science, as it has come to be known, has also been deployed in science education. An influential moment in this regard was in the 1990s when a reform of physics teacher training at the University of Oldenburg added a new laboratory course that used replicas of historical scientific instruments. This pedagogical innovation encouraged analogous work with carefully-reconstructed instruments to address historical questions by historians of science, such as Otto Sibum's work with Joule's paddle-wheel device.[4]

Today, historians of science, most prominently Peter Heering and Hasok Chang, continue to re-enact historical experiments as an important element of science education. While Heering has advocated the role of reworking for teaching students about the nature of science, Chang has emphasized the potential of 'complementary experiments' for recovering lost scientific knowledge and for instilling a sense of wonder in students that might attract them to science.[5] Science educators like Elizabeth Cavicchi use comparable approaches to help students develop a better understanding of science and its exploratory nature.[6] As briefly discussed in this chapter, the latter aspect proves highly relevant from the perspective of science education, as recent education policies emphasize the development of concepts about science in addition to scientific content.

More generally, hands-on educational approaches have emerged within the arts and humanities. For example, at the beginning of the twenty-first century the Stanford Humanities Lab developed 'Artereality' as a new model for arts education within the academy. This new model helps shape the pedagogical environment and learning processes in ways that apply to reworking both historical experiments and recipes in the classroom: 'teamwork-based education as a complement to the traditional individualized studio; a scrutiny of process as an essential complement to product; the embrace of project-based and performance-based learning'.[7] Similarly, the use of a hands-on approach in science education, now widely applied at all

2 See Fors et al., 'From the Library to the Laboratory' and the introduction to this volume.

3 Boone et al., 'Histories of Use and Tacit Skills'.

4 Sibum, 'Mechanical Value of Heat'.

5 Heering,'Historical Approaches'; Chang, 'Historical Experiments'.

6 Cavicchi, 'Historical Experiments'; Cavicchi, 'Learning Science as Explorers'.

7 Schnapp and Shanks, 'Artereality', p. 143.

levels, from primary school to university, is based on similar insights about the benefits of learning by doing. However, an important distinction between such hands-on approaches and the use of RRR in the classroom is that the latter adds the historical dimension. The courses discussed in this chapter use historical experiments, recipes, and artefacts, and show how instructors also benefit from elaborating historical questions in collaboration with students.

Throughout this chapter, we use different RRR terms for two reasons. In the first place, different terms help us to subtly highlight different aspects of RRR methodologies. For instance, reworking puts emphasis on the *doing* by endeavouring to access and understand the manual, sensual, and bodily skills of an experiment or process for historical purposes. Reproduction, on the other hand, draws attention to the final products and the means of their production, working to explain the underlying processes or reactions involved as well as the material circumstances that help shape these products. Likewise, reconstruction addresses the process of making a device according to source information; a replica is the outcome of this process. In the second place, RRR terms are intended to point towards a richer historiographical tradition, and to underscore the importance of gaining specifically historical understanding through these practices.

While experimental history of science in its educational dimension has interdisciplinary potential, bridging what C.P. Snow famously diagnosed as the gap between the 'Two Cultures,' that is only one of its many benefits. We argue that the use of RRR methods in the classroom also creates fruitful ground for raising new and often unanticipated questions, in both the sciences and the humanities. We discuss here three cases of classroom RRR for contemporary education in science and the history of science and technology: A science teacher training programme in Flensburg (Peter Heering), a liberal arts and science programme in Utrecht (Thijs Hagendijk) and an interdisciplinary history of science and technology programme at Johns Hopkins University in Baltimore (Lawrence Principe and Yulia Frumer). In Flensburg, the study of the nature of science and the reflection upon students' experimental practices are central aims of the science teacher training programme. In the other programmes as well, classroom RRR serves to teach the exploratory nature of science. In the courses in Utrecht and at Johns Hopkins, they focus the attention of students (who have lost the material literacy of previous generations) on materials and the sensory dimensions of experiments. Moreover, the interdisciplinary programmes in Utrecht and Baltimore bring methods of the humanities to science students and scientific methods to humanities students through close engagement with material culture, historical texts, and their parallel investigation.

Together, the three cases make an argument that the use of RRR methods in the classroom allows teachers to engage students, and offers students the opportunity to participate in research, here specifically in the history of science.[8]

## I. Science teacher training at the Europa-Universität Flensburg

Starting in the mid-1980s, the University of Oldenburg group, led by Falk Rieß, played a key role in the development of the experimental history of science for educational purposes by implementing reconstructed historical experimental set-ups in a compulsory lab course for teacher training students in their third or fourth year of study.[9] The experiments included several that used canonical instruments, such as Coulomb's torsion balance and Ohm's balance, but also lesser-known devices such as a Gauss-Weber magnetometer and even a device historically rejected – Thomas Young's eriometer.[10] Besides this compulsory course, students had the option of writing a thesis on one experiment or instrument, and the thesis was also an option for students enrolled in a master's programme in physics. As a result, a number of instruments were reconstructed and several experiments analysed using the replication method.[11]

The Oldenburg model has been inspirational for the development of the science teacher training programme at the Europa-Universität Flensburg. Here, physics teacher students begin with a module on the history of physics. This compulsory module consists of a seminar on the history of ideas (in which some instruments are demonstrated) and a lab course; the latter

8 This is, of course, not unique to history of science; compare Ingold, 'The 4A's'.

9 See Heering, 'Approaches'; Rieß, 'Short History'. Other approaches include: Devons and Hartmann, 'History-of-Physics Laboratory'; Hartmann Hoddeson, 'Pilot experience'; Kipnis, 'Rediscovering Optics'; Chang, 'How Historical Experiments'; Chang, 'What History Tells Us'.

10 For a discussion of the instruments and experiments used, see Rieß, 'Teaching Science'; Heering, 'Role of Experiments'. The magnetometer allows the determination of the Earth's magnetic field in absolute measures, the eriometer measures the diameter of a sample of small objects through analysing the produced diffraction patterns.

11 For the results of some case studies, see e.g. Beneken, 'Robison'; Frercks, 'Fizeau'; Lühr, 'Ampere'; Nawrath, 'Newton', or Voskuhl, 'Herschel'. In the evolvement of the programme, the methodological approach developed further and was at the same time theoretically reflected. Relevant publications in the latter respect are Heering 'Grundgesetz', Sichau 'Replikationsmethode', and Frercks 'Forschungspraxis' who developed a somewhat different notion that is not the basis of this discussion. The most recent approach in this respect was carried out in collaboration with colleagues from Jena, see Breidbach et al. 'Experimentelle Wissenschaftsgeschichte'.

consists of five two-hour sessions. Prior to each session, students complete a reading assignment and afterwards write an essay reflecting upon their experiences. In the first session, making devices is the primary topic; each student makes a gnomon and uses it to determine solar noon in Flensburg (which poses an additional challenge considering usual autumn weather conditions). The second addresses Galileo's inclined plane experiment: The students are challenged to determine the movement of a ball rolling down an inclined plane. Here, the students are confronted with the twin challenges of analysing movement without a proper stopwatch and of measuring length as done prior to the existence of the meter. The former in particular challenges the students, and in the end they are guided towards using rhythm to determine equal time intervals.[12] The third session deals with eighteenth-century electrical experiments. Here, the interplay between experimental procedures, social settings, and conceptual development are central. Both the lecturer and illustrations from eighteenth-century publications depicting the experimental settings guide the experiments. The fourth session deals with Ampère's work in electromagnetism[13] – experiments using coils, spirals, magnets, and batteries are used in order to understand the structure of electromagnetic interactions. In contrast to these experiments (which are *exploratory* in terms of Steinle's analysis), the students watch a video showing Ampère's current balance[14] in order to understand the difference between exploratory and theory-driven experimentation. This part of the course concludes with a visit to a modern lab where the students learn about current physics research.

A central aim of this course is to address issues related to the nature of science (NOS).[15] Science educators agree on certain objectives extracted from science standard documents including the following (to name but a few):

- Scientific knowledge relies heavily, but not entirely, on observation, experimental evidence, rational arguments, and scepticism;
- There is no one way to do science (therefore, there is no universal step-by-step scientific method);
- People from all cultures contribute to science;

12 This approach is a modification of the interpretation provided by Drake, 'Renaissance Music'.

13 This session is structured according to Steinle's analysis of Ampère's experiments, see Steinle, 'Exploratory Experiments'.

14 This video can be found at https://youtu.be/_qw5FHjmZY8. Checked on 12 November 2018.

15 NOS is a construct that has become relevant in science education over the last three decades, see Allchin, 'Teaching'; Lederman, 'Nature of Science', and McComas, 'Nature of Science'.

- New knowledge must be reported clearly and openly;
- Observations are theory-laden;
- Scientific ideas are affected by their social and historical milieu.[16]

The experiments discussed here exemplify such aspects; students are expected to reflect on experimental standards as dependent upon time, place, and (scientific) communities. In so doing, they develop the understanding that experiments may have various different purposes and that there is not a single scientific method (or even experimental method). At the same time, they come to understand that experimental evidence is crucial (but not unique) in producing scientific knowledge, that communication plays an important role, and that contributions to science come not only from scholars.

These aspects are relevant (and taken up again) in the course on the nature of science which students take in their fifth semester. This course is aimed at all science teacher students. From the discussions taking place among the students, it is striking that the physics students have a significantly more profound understanding of the relevant concepts. Developing a thorough understanding of the nature of science is fundamental to science education as advocated in the German educational system. Developing competency is a key aspect of education, and such competency is not limited to content; process-based competency is equally relevant. Future science teachers' understandings can be developed through the history of science, and in particular through the reconstruction of instruments and the re-enactment of selected experiments.

This laboratory course is not only about addressing NOS. Students are also confronted with unfamiliar experimental practices and standards, forcing them to think about their own practices and standards and to develop a fresh perception of experimentation. The students have the opportunity to write a BA thesis (workload 300h) or MA thesis (workload 600h); one option for these theses is to carry out a study that applies the replication method. To do so, the students either build their own devices and carry out experiments or use an already existing device and study its experimental practice in more depth. They do not aim to 'check' the initial experimental report. Instead, they try to develop an understanding of the experimental specifics or try to characterize the device with which they are working by determining the relevance and effect of several parameters that affect the behaviour of the device.

16 These objectives are part of a list published by McComas, Clough and Almazroa, see McComas, 'Nature of Science', p. 6f. See also Lederman et al., 'Views'.

From an educational perspective, there are three relevant aspects to such work: First, the students are expected to carry out their research project in order to develop a hands-on understanding of research. This topic was already covered in seminars they took during previous semesters, however, it was then dealt with in a mainly theoretical manner. The thesis provides an opportunity to gain insight into inquiry-based learning.

Secondly, the students are enabled to gain insight into the interplay between experimenter, instrument(s), and conceptual understanding. To give an example: One student concluded in her analysis of Benjamin Thompson, Count Rumford's experiments on radiant heat, that 'Rumford's descriptions are formulated so illustratively and comprehensibly that initially I did not think the replication of his experiments would be problematic. A realization that I had to experience over and over again during my study is that becoming familiar experimentally with an apparatus can turn into a time-consuming process. I also initially underestimated the relevance of laboratory conditions.'[17] 'By using the replication method, I frequently had to put myself into a completely foreign way of thinking [...].'[18] Both aspects are typical for students working with such an approach. On the one hand, they begin to realize where problems can arise only when they are confronted with the task of reconstructing a set-up or of reworking an experiment. On the other hand, it is challenging for them to discuss apparently familiar phenomena in a conceptual manner that differs to their own. This quality is particularly valuable for science teachers who frequently talk to pupils who have different conceptions than the scientific ones; thus, being able to argue within a different conceptual frame is a substantial competence.

However, conceptual understanding and experimental practices are frequently not the only challenges. In addition, the reconstruction of the instruments may cause substantial difficulties. In the case study on Rumford, a substantial challenge turned out to be the reconstruction of a thermoscope. According to Rumford's specifications, the instrument consists of two blackened glass bulbs connected with a U-shaped capillary. A bubble of 'spirit of wine, tinged of a red colour' is to be inserted into the middle of the horizontal part of the capillary before the capillary is sealed hermetically.[19] Even though the glass of the thermoscope already existed, there were a number of questions to be answered: How did eighteenth-century people tinge spirit of wine red? How does one insert the bubble into the capillary?

17 Mercier, 'Rumford', p. 89.
18 Mercier, 'Rumford', p. 90.
19 Rumford, 'Enquiry', p. 101.

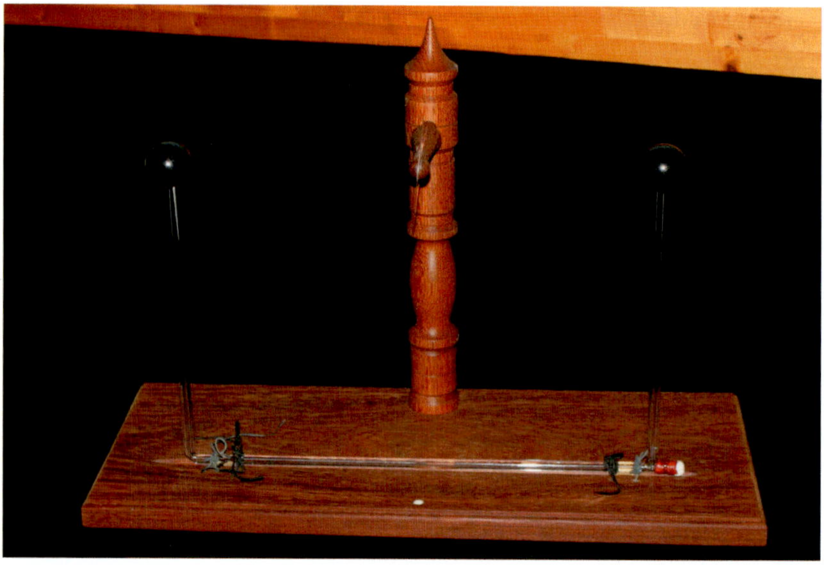

Plate 7.1: Reconstruction of Rumford's set-up to analyse radiant heat. Photo: M. Mercier, Europa-Universität Flensburg.

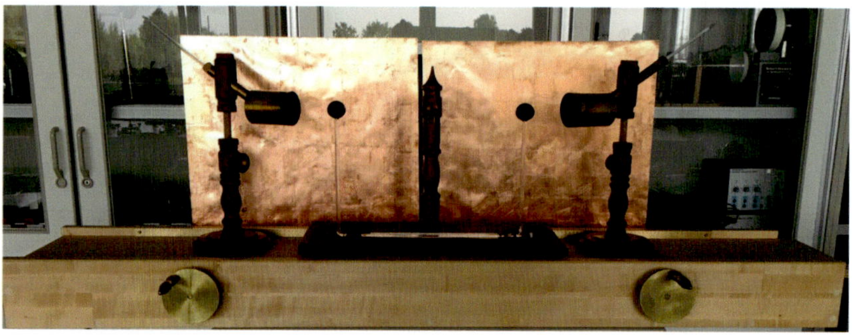

Plate 7.2: Reconstruction of Rumford's thermoscope. Photo: M. Mercier, Europa-Universität Flensburg.

How does one make sure it is in the middle when the air temperature in the two bulbs is equal? And, how does one seal the capillary hermetically? For questions such as these, the text does not provide answers. Further questions arose only during the practical work, and answers came only practically, some of which could not be adequately verbalized.

In this case study, the student aimed to develop an understanding of the historical practices through analysis with the replication method. This is not necessarily the case in all studies; others relate more to physics in the modern sense. One thesis analysed the electrical ignition of liquids, a popular electrical experiment of the mid-eighteenth century. The basis for this study was three letters by William Watson, published in 1746, describing his experiments in some detail. Watson reported that he ignited several substances: 'I have not only fired *Frobenius's* Phlogistion, rectified-spirit and common proof-spirit, but also Sal volatile Oleosum, Spirit of Lavender, dulcified Spirit of Nitre, Peony Water, *Daffy's* Elixir, *Helvetius's* Stiptic, [...].'[20] Watson used a poker connected to the conductor of an electrical generator to ignite the fluids that were held in a spoon, and identified several parameters that he considered crucial for success.

The student undertook a variety of experiments to identify parameters that may affect the outcome: The kind of liquid, its temperature, the amount to be ignited, the discharge voltage, the length of the spark gap, the air pressure and humidity, the form and material of the spoon, and the position of the poker with respect to the spoon. From his experimental analysis, the student identified several criteria that facilitate the ignition of the substance. He reflected upon his research process and wrote that, 'all in all, the development of an experimental set-up that combines reliable measurements with reliable spark production turned out to be surprisingly difficult, as shown by the number of failed experimental setups'.[21] Yet, despite these failures, his summary of his work was positive and included reflections upon the research process: 'The experimental set-up initially produces unexpected data, yet these data can be explained using recent understandings of physics. In composing this study, it was not only the development of the set-up, but also the performance of the process that provided an idea of how much practical knowledge researchers such as Watson must have developed and which does not appear explicitly in their texts.'[22]

Reconstructing and working with historical instruments provides several learning opportunities. In reflecting on their laboratory experiences with the historical set-up, students realize that experimentation is not done by blindly following a protocol but instead is to be understood as a human activity localized in time and place. This understanding cannot be effectively communicated to students by means of lectures and seminars that address the topic predominantly cerebrally. Being confronted with

20 Watson, 'Experiments', p. 19f. Watson stressed that the spark ignites the vapours of the liquids, not the liquids themselves. One challenge in the case study lay in identifying these liquids from their now unfamiliar names.

21 Holländer, 'Entzündung', p. 41.

22 Holländer, 'Entzündung', p. 41. Holländer uses the term 'Handlungswissen', which I translate as practical knowledge – it is not clear from the context which epistemological concept he is addressing in this comment.

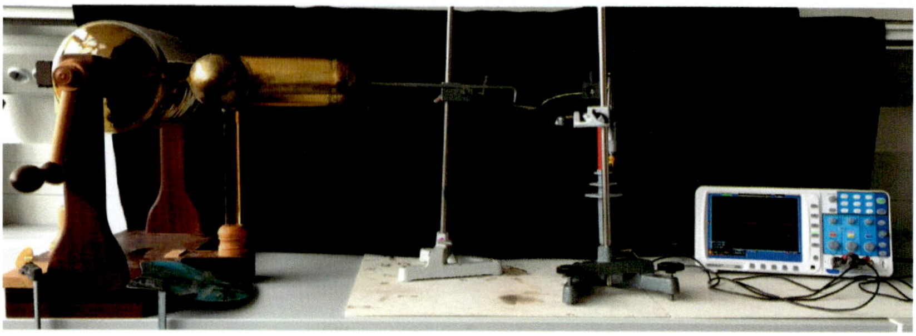

Plate 7.3: Experimental set-up to analyse the ignition of liquids by sparks. Photo: R. Holländer, Europa-Universität Flensburg.

unfamiliar experimental practices and standards helps students to reflect upon their own understanding and thus gain an awareness that may lead to a different perception of the role of experiments in science. In contrast to modern science experiments (which are not to be confused with teaching lab experiences), students can also reflect on different standards in the historical situation and what these standards mean for their practice. Thus, on a meta-level, they can reflect on experimentation.[23]

## II. Exploring chemistry and art in the Liberal Arts and Science Programme in Utrecht

Understanding the nature of experimentation and lab work is also one of the aims of the course *Chemistry and Art* which was offered for the first time in January 2017 at University College Utrecht, a Liberal Arts and Science Programme at Utrecht University.[24] Students work with a selection of sixteenth-century recipes for artisanal materials and are asked to transcribe, rework and further investigate them in the laboratory.[25] Reworking early modern recipes not only fosters an interdisciplinary approach, it also turns out to be an excellent starting point for reflecting on the role of interpretation, failure, improvization, and trial-and-error in laboratory work. After running a successful pilot, the decision was made to continue this course annually.

23 Just to clarify: This is a discussion of a programme for teacher students, not within a history, philosophy, or sociology programme.

24 Developed in collaboration with Gert Jan Vroege (Utrecht University) and Dominique Thies-Weesie (Utrecht University).

25 Comparable courses (but with a different student audience drawn primarily from graduate studies in the humanities) are offered by the Making and Knowing Project at Columbia University, New York. See: Bilak, 'The Making and Knowing Project'.

The course was initially designed as a practical module in which students develop basic laboratory skills, such as working safely, keeping a notebook, and mastering basic lab techniques. The contents came however from the history of art, science and technology, such that the course could be incorporated into an already existing Cultural Heritage programme offered by University College Utrecht. In principle, the course is open to students of all academic backgrounds, yet the vast majority of students have a background in the sciences, while occasionally an art history student enrols as well. The course runs full-time for two weeks during which the students rework a selection of different recipes, including for iron-gall ink, a red brazilwood pigment, imitation pearls, and silver-plated copper. Students visit the Rijksmuseum Research Library in Amsterdam to inspect the early modern sources they work with, and receive lectures and demonstrations by historians, chemists and conservators, who teach them about the boundary between chemistry and art. In the end, students submit concise experimental reports on three different recipes and write an elaborate research report on a recipe of their choice.

The recipes come from an English translation (1595) of Alessio Piemontese's *De' Secreti* (1555).[26] This book of secrets proved immensely popular in the sixteenth century and contains recipes ranging from alchemy and cooking to medicine to the decorative arts.[27] The book was chosen because it soundly illustrates the intimate connection between alchemy and the arts in the early modern period. It groups together alchemical, medicinal, metallurgical and artisanal recipes, which help students realize that the boundaries between these fields were, and in fact still are, very fluid.[28] This choice was further informed by the fact that the book went through multiple editions and translations, thus enabling students to compare recipes in different languages. Comparing translations was no mere luxury; it uncovered incongruities that significantly altered interpretations.

Interdisciplinary and boundary-crossing situations occur at several levels during the course. First, students are confronted with the fluid line between alchemy and the arts, as demonstrated, for instance, in a recipe for counterfeiting pearls. The recipe instructs the reader to take the shells of white mussels, grind them, and then heat them in a vessel covered with *lutum sapientiae*. The use of *lutum* and the practice of luting were rather common to alchemists. They covered their glassware with specific types of

26 Piemontese, *The Secrets.*

27 Eamon, *Science and the Secrets of Nature.*

28 Dupré, *Laboratories of Art.*

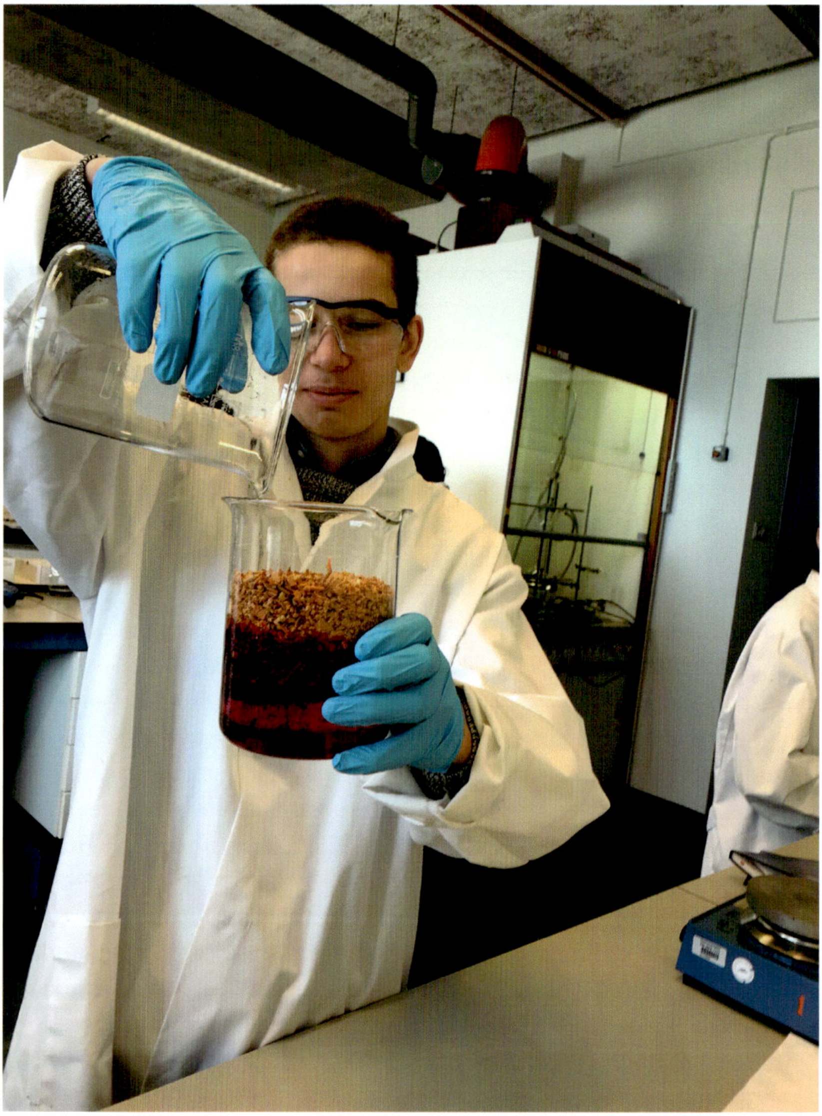

Plate 7.4: Students work with raw materials and simultaneously have to rely on their senses for reworking the recipes. The brazilwood in the beaker is processed into a red pigment. Photo: Thijs Hagendijk.

clay to prevent them from cracking when exposed to extreme heat. Students are shown that techniques like luting, but also ingredients such as vitriols or alum, commonly travelled between different workshops, and they are introduced to the idea that alchemy was not a 'dark and occult' occupation but can also be understood as an archetypical material science.

Secondly, the luting technique immediately poses a new interdisciplinary problem for the students. Because it would be too dangerous to work with luted vials in an open fire, students are asked instead to carefully consider what happens during this procedure and to start working with the resulting compound. An understanding of modern-day chemistry becomes important. They learn that shells are composed almost entirely of $CaCO_3$. The question remains whether the heat would merely burn off organic contaminations, or whether it would lead to thermal decomposition of the $CaCO_3$ into CaO. It is left to the students to make a chemically-informed decision.

Thirdly, after the students have decided how to substitute their processed shells, they combine the powder with egg yolk to shape the paste into pearls, as instructed by the recipe. To date however, every attempt to create the much-desired pearls has failed. Disappointed by the results, the students are encouraged to take another look at the recipe and to compare different translations. Doing so, they learn that the English translation 'yelks of egges' is diametrically opposed to an earlier French edition of Piemontese's *Secrets* that prescribes egg white, or *la glaire d'oeuf* instead of yolk.[29] Eventually, most attempts with egg white fail too, but by then, students have already critically compared editions and traced the genealogy of recipes to decide on an interpretation. The ambiguity of recipes or mistranslations thus serves as an invitation to deploy humanistic methods in the laboratory.

The richness of an application of humanistic methods in the laboratory is even more evident when students transcribe their recipes prior to reworking them. To understand and rework a material procedure, they first need to work through typographical peculiarities such as the 'long S', deal with early modern orthography, and learn how to use lexica to understand terms like 'ciche pease' (chick pea). They discover how volumes and weights were measured before the metric system, and how recipes give indications like 'the bignes of a Walnutte'. Finally, they learn how materials were named before the era of IUPAC nomenclature, with examples as 'unsleaked lime' and 'roche alome'.[30] In short, a recipe does not provide clear-cut instructions but requires interpretation. Meanwhile, students learn that familiar and apparent unequivocal chemical concepts and nomenclature are relative and can be historicized.

Transcribing, interpreting, and reworking sixteenth-century recipes thus requires a combination of historical, linguistic, and chemical approaches. After the students have worked with each recipe, they are asked to pick one

29 Piemontese, *Les Secrets*, pp. 817-818; Piemontese, *The Secrets*, p. 252.
30 Piemontese, *The Secrets*, pp. 58-59.

recipe on which to perform further research. Performing further research on a recipe can be understood as what Hasok Chang has described as 'extension'. He argues that once a historical experiment or procedure has been performed, it is 'difficult to resist the natural curiosity'.[31] In other words, historical recipes not only evoke historical questions, but can also trigger scientific interest and chemical wonder, even today. Students are encouraged to use state-of-the-art equipment to find answers to their problems.[32] Sometimes problems lie in observed discrepancies between historical procedures and modern chemical intuition.[33] An example is a recipe to silver-plate copper employing common salt, wine lees, alum, water and silver.[34] When the recipe was tried for the first time, it was expected to fail. To silver-plate copper, a redox-reaction has to occur in which the silver must be dissolved first, which is unlikely to happen given the other ingredients. When after a few attempts the copper started to look like silver nonetheless, the question remained how this sixteenth-century recipe was able to get silver in a dissolved state, counterintuitively to modern chemical theory. Several possible explanations can be tested by the students. For example, the mechanism might rely on a natural tarnishing process in which silver reacts with sulphides from the air, prior to the reaction prescribed in the recipe. More than once, the unmistakable smell of rotten eggs was noted, indicating that sulphides were actively involved in the process. Another possibility is that common salt is more corrosive to silver than initially assumed. To date, the definitive mechanism has not been found, but the recipe's ability to puzzle the minds of both chemistry students and experienced chemists alike is fascinating. It demonstrates that it pays to keep an open mind, even when modern chemistry curbs expectations.

Another advantage of this course's RRR approach is that students learn to work with raw materials and simultaneously learn to rely on their senses for reworking the recipes. They become familiar with historical laboratory practices and modes of chemical experimentation, often characterized by a high degree of sensual experience and practical knowledge.[35] Recent scholarship has, for instance, emphasized the historically important role that the senses played in the classification and recognition of chemical substances.[36] However, such practical and sensually-rich modes of investigation are

31 Chang, 'Historical Experiments', p. 320.

32 E.g. (electron) microscopy, UV-vis spectroscopy, pH-instruments, rheometers.

33 Principe, 'Chemical Translation'.

34 Piemontese, *The Secrets*, p. 246.

35 Principe, *Secrets of Alchemy*, p. 208.

36 Ragland, 'Chymistry and Taste'; Tillman Taape, 'Distilling Reliable Remedies'.

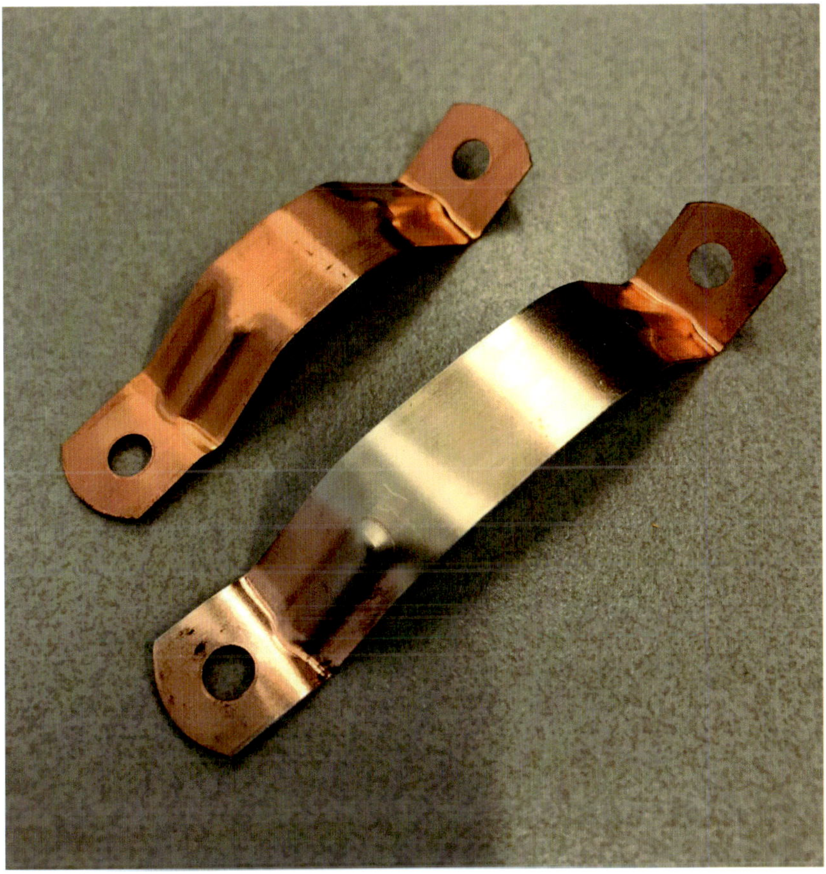

Plate 7.5: The copper piece on the right has been silver-plated based on a sixteenth-century recipe and against the expectation of modern chemistry. Photo: Thijs Hagendijk.

no longer obvious for contemporary students of chemistry.[37] In practical training, the importance of the senses is significantly downplayed and substituted by a wide range of analytical equipment. Moreover, materials in the laboratory are often no more than a series of greyish bottles containing extremely purified white powders and working from predetermined protocol is the rule rather than the exception. Indeed, each time students enrol in this course, they exhibit an overreliance on text and theory, are easily daunted by failure and have to be pushed to observe more closely what happens to their materials.

An illustrative example of overthinking and overreliance on theory and texts is the following. When the students start working on the recipe for

37 Chang, 'What History Tells Us'.

iron-gall ink, they are first asked to transcribe the recipe, do some research, and write a chemical translation that can be taken into the laboratory. All this happens in the lecture room. They are allocated about two hours for this task, which they spend frantically writing and looking up information on the internet. They usually gather more than enough information to start their practical work. The students are taken to the laboratory, asked to suit up accordingly and start their experiments. What usually happens instead is that they look around for the nearest table, sit down and continue their theoretical discussion, oftentimes not even looking at the raw materials prescribed by their recipes. A major aim during this course is therefore to try to make students stop thinking *about* materials and instead think *with* the materials by engaging with them.[38] The outcome of this approach is illustrated in the following example. Once, after about thirty minutes and two failed attempts at reproducing the pearls from the pearl recipe, one group simply gave up, stating that their protocol obviously did not work. After they were encouraged to forget the protocol for a while and to just continue experimenting, they had fun for another two hours and came up with something that, much to their own surprise, started to look like a pearl. This shows the importance of a creative and improvizational attitude. The students learned that there is no harm in forgetting the protocol for a while, and that much can be learned by simply following the materials. The final results from past courses demonstrate that students start to use their senses in a way they had not been doing before. Their reports contained sentences like this one, which concerns iron-gall ink: 'As a result, a viscous, sticky dark green mixture was created, that smelled like a mixture of vinegar and apple cider.'[39]

The *Chemistry and Art* course illustrates that reworking historical recipes is a valuable tool that helps students develop a set of epistemic attitudes that complement the skills taught in traditional lab courses. The ambiguity of early modern recipes means that there is no obvious best practice or correct way to solve the puzzle. The course trains students to deal with failure, use trial-and-error, and teaches them to trust not only in theories, but also in materials and their own senses. Moreover, these centuries-old recipes quite naturally invoke interdisciplinary approaches. Working these recipes successfully encourages students to integrate scientific reasoning with humanistic methods, and to bridge the 'Two Cultures' as defined by C.P. Snow.[40]

38 Ingold, 'Materials against Materiality'; 'The Textility of Making'.

39 Patricia Jäger and Sebastiaan Berschoor Plug, 'Recreation of an Ink Recipe by Alexis of Piemont'. Report written for *Chemistry and Art*, University College Utrecht, Summer 2016-2017.

40 Snow, *Two Cultures*.

## III. Thinking through things for historians of science at Johns Hopkins University

Similar to the *Chemistry and Art* course at Utrecht University College, exposing students to the sensory experiences of materials is one of the central aims of courses taught at Johns Hopkins University. However, these courses further demonstrate the utility of reworking experiments for students in the humanities as well as for those in the sciences. The reproduction of experiments and the engagement with material objects has been used, for example, to supplement the traditionally lecture-based history of science surveys for undergraduates. In addition, two recent upper-level courses – *Thinking Through Things and Thinking Things Through* (designed and taught by Professor Yulia Frumer) and *Practical History of Science* (Professor Lawrence Principe) – involved a wide range of practical engagement with objects, historical artefacts, and experimental reproductions. The subject areas of these interactions ranged from physics and astronomy to chemistry and technology, and from Late Antiquity to the twentieth century.

Some exercises were intended simply to expose students to the sensual experience of materials and objects. Today's students increasingly lack the direct experiences that were normal and expected for students of previous generations. The current world of the computer and the recent, and ill-advised, substitution of 'virtual' simulations for traditional laboratory, anatomy, and other training experiences, has led to this state of affairs. The unfortunate result is that today's students are increasingly separated from meaningful direct contact with the real material world, do not develop adequate manual skills (except perhaps for pressing buttons on a portable device), and are thus disabled from coming to understand the observations and thoughts of those authors and workers who *did* engage directly with materials on a daily basis.

One of the professors (Principe) found that even advanced chemistry students had no concept or experience of the natural origins of metals, salts, and other chemical substances. Students were initially befuddled, for example, as to why early workers would have thought to put one stone rather than another into a fire (thus discovering how to extract metals). But after having actually handled chunks of metallic ores, they understood sensually, based on the striking colour, density, gleam, and other features of these minerals, why attention was fixed on these stones rather than on other, less unusual ones. In another exercise, students were given an array of different salts – all of them white, crystalline substances and thus indistinguishable by sight alone – and asked to identify them using the type of fire tests (i.e. throwing them on burning coals) described by the ninth-century Persian author al-Rāzī and other medieval authors. This experiment underscored the difficulties and ambiguities that premodern workers faced in classifying and identifying individual chemical substances (before the time reliable materials arrived in neatly-labelled bottles) and taught them the properties of several of these substances. Such work drove home the crucial role of the senses in scientific endeavours, and the value for historians in re-accessing some portion of that sensual experience in order to better understand the historical actors and their thinking.

Other exercises dealt with the role of instruments and the engagement with technological materials. Students explored, for example, the inherent difficulties of visual perception and discovery by using inexpensive telescopes of approximately the quality and magnification of Galileo's first instruments, and this hands-on experience occurred alongside their more traditional reading of Galileo's 1610 *Sidereus nuncius* that announced and described his telescopic discoveries.[41] They were thus able to read Galileo's descriptions (which of course embody his interpretations) in parallel with seeing for themselves approximately what he and his contemporaneous readers saw. Weather and celestial positions permitting, students were able to watch the nightly movements of Jupiter's moons (the Galilean satellites), see the phases of Venus, and observe the surface of the moon at different phases, and compare their own observations with those of Galileo. One aspect that students found particularly instructive was their encounter with the same sort of unexpected difficulties faced by early observers, such as how to find objects in the telescope, where exactly to position one's eye for observations, how to keep the instrument steady and focused long enough to make reliable observations, and most of all, how to interpret what they were seeing. Whereas modern accounts of telescopic observations overlook the initial difficulty of using the instrument, through this exercise students came to understand some of the difficulties and uncertainties encountered by early observers at the start of the seventeenth century when faced with using this new instrument of vision. Some students noted how, when looking at a distant street light through a telescope with poor optics, a series of smaller, weaker spots of light sometimes appeared on both sides of the light, initially similar in appearance to the moons seen around Jupiter. With this observation, they were better able to understand the claims by some of Galileo's critics that the tiny 'stars' he saw around Jupiter, and which he interpreted as orbiting satellites, were actually optical artefacts of the instrument. This realization in turn explained why Galileo

41 Galilei, *Sidereus Nuncius*.

spent so much of the *Sidereus nuncius* (seemingly *too much* from a modern perspective) displaying the regular and changing movements of those 'stars' from night to night.

Both graduate and undergraduate students also worked with astrolabes (in simple cardboard format), using them as observing devices for surveying, taking astronomical measurements, and time-reckoning as well as calculating devices for predicting the rising and setting times of the Sun and stars. Some students in the *Thinking Through Things* course engaged in the process of 'making' by building their own scientific instruments using contemporaneous descriptions and depictions, such as the single bead-lens microscope devised by Antoni van Leeuwenhoek in the 1670s. This endeavour required them to deal not only with constructing a technological device from written descriptions, but also in acquiring, shaping, and engaging with the necessary materials, in this case, brass, steel, and glass.

The *making* of objects and materials was also inverted in order to study *already-made* technological artefacts. The goal was to tease out the manufacturing or technological know-how embodied in them, as well as to consider their original use and impact on users. Thus, students in the *Thinking Through Things* course had the opportunity to study and inspect an original and functioning 1918 Model T Ford touring car. They identified physical traces of its method of production, after having read about Ford's moving assembly line.[42] They were asked to put themselves in the position of the original owners and operators, to think about using the automobile and consider what drivers and passengers had to know, do, and encounter in order to operate and maintain it. The presence of the physical object itself inspired many questions that would never have been thought of otherwise. Without gauges, how did drivers know how much gasoline was left? How many gas stations were there in 1918 and how could you be sure of finding one? Why is there no door for the driver's seat? After actually feeling the physical strength and technique required to pull the crank in order to start the engine, students wondered what less robust people could do if they wanted to drive. One student observed that several normal tasks (like raising and lowering the top or starting the car when cold) generally required two people working together, and therefore cooperation between driver and passenger must have been essential at times. In this way, they explored questions of how human beings use, interact with, are changed by, and come to accommodate a complex, life-changing, and originally revolutionary technology.

42 Mahoney, 'Reading a Machine'. The course title is a tribute to the late Prof. Michael Mahoney, who taught a graduate seminar by this title at Princeton.

Some revealing results for both students and professor came from the reproduction of Robert Boyle's famous 1661 experiment on the 'reintegration' of saltpeter (potassium nitrate).[43] This process was performed as a demonstration before the *Practical History of Science* seminar, after the students had read and discussed Boyle's essay. Because Boyle's main purpose in writing his *Essay on Salt-Petre* was to call attention to the variety and changes of observable qualities in chemical processes (thus undermining the Aristotelian concept of substantial form), this experiment was especially valuable as a reconstruction since it exposed students directly to those observables. In the experiment, small pieces of ignited charcoal are dropped sequentially into a crucible of molten potassium nitrate. The mixture deflagrates violently with each addition until all the potassium nitrate has reacted, leaving behind a residue of 'fixed nitre' (potassium carbonate). The slow, dropwise addition of spirit of nitre (nitric acid) to this residue produces vigorous effervescence and mild heat – which can be heard and felt – at the end of which time 'regenerated' nitre crystallizes out of solution. In this case, the professor had not himself done this experiment nor practiced it prior to doing it with the seminar students, which led to a learning experience for all involved. In particular, the professor (who had assigned Boyle's essay to classes for many years) had long puzzled over Boyle's claim that the 'fix'd Niter' left in the crucible 'was of a deep colour betwixt blue and green', since the product to be expected from modern chemical knowledge, potassium carbonate, should be white.[44] Nevertheless, he was very surprised to find that Boyle's experiment, conducted as described, did in fact yield a deep turquoise-coloured product. (He was inspired thereafter to conduct further experiments that identified the source of this colour as trace amounts of iron present naturally in the wood from which the charcoal had been made.) One student who was not experienced in chemical operations reflected that he was surprised, confused, and slightly frustrated by the implicit choices the professor made spontaneously during the demonstration – how much heat to apply to melt the nitre, what size pieces of charcoal to add, when to conclude that the experiment was finished, how quickly to add the spirit of nitre, as well as how exactly to manipulate the tongs, crucible, burner, and other tools necessary for the operation. His observations revealed more vividly (even to the professor) the depth and variety of unarticulated experiential

43 Boyle, '*A Physico-chymical Essay*'.

44 Boyle, '*A Physico-chymical Essay*', p. 96.

knowledge an experimenter brings implicitly and unconsciously to any experiment, and also the various forms that such knowledge can take.

These courses at Johns Hopkins make a special point of bringing together both textual sources and experimental reconstruction or the direct engagement with objects. The purpose here is to ensure that the exposure to the sensual experiences is clearly directed towards gaining a deeper understanding of historical texts, in short, to set up a dialogue between text and object or experience.[45] The greatest benefit of reworking experiments for students of history accrues when historical questions spur engagement with materials, processes, and objects and when such engagement answers historical questions and proposes new questions for consideration.

## Conclusion

Why is it important to bring RRR approaches to the classroom? The cases discussed above show that engaging with historical instruments, materials and recipes allows teachers and students to address interdisciplinary questions and to activate different epistemic attitudes in the classroom. Still, bridging the gap between the 'Two Cultures' and the creation of an interdisciplinary learning environment in which science students are exposed to humanities questions, and vice-versa, while a valuable side-product, is only one of the goals. Why then should science students be introduced to the vivid sensual experiences of materials, instruments and experiments of the past? What is the added value of exposing students to the reworking of historical experiments and recipes? In this chapter we have argued that the pedagogical benefits of the use of RRR methods in the classroom are twofold.

First of all, reworking experiments in the classroom is a powerful methodological tool to encourage science students to reflect on the connection between the past and the present and to gain insight into their own laboratory work. One of the most important aims of the courses in both Flensburg and Utrecht is to teach science students about the nature of experimental science. Explicit reflection on the experiences and the difficulties encountered in reworking a historical experiment on the basis of a text, such as a recipe from the past or historical experimental account, confronts science students with the intricacies of interpretation typical of approaches in the humanities, allowing them to question the protocol-like nature of experimental science as it is now commonly taught.

45 See Fors et al., *'From the Library to the Laboratory'*.

Yet a further innovative power of RRR approaches in the classroom is to create collaborative learning environments in which teachers learn together and alongside their students. In such a setting, teaching is not only research-driven, as one would expect in a university classroom, but research in the history of science and technology is also nurtured by teaching. Teachers learn about past historical practices of science from the inquiries of their students as students learn from their teachers. The use of RRR in the classroom is not unique in this regard, but the reworking of historical experiments and recipes creates a classroom setting which seems particularly apt in this regard. We have seen examples in which students reworking historical recipes are confronted with questions in science which for their teachers are also new and often unanticipated. Curiosity can be tremendously contagious if teachers are as interested in – or surprised by – the outcomes of an experiment as students. These reworkings of experiments and recipes are instantiations of the 'extensions' and 'complementary experiments', of which (as we have discussed in the introduction) Hasok Chang has argued for the usefulness to science students. However, as especially the pedagogical experience at Johns Hopkins University shows, this is equally true for students in humanities classes. Students' inquisitive questioning of the teachers' experimental skill and knowledge can help to make them articulate and address the historical questions which, for the historians of science and technology teaching the class, were the point of departure for reproducing the experiment. Thus, reworking experiments and recipes in the classroom allows teachers to engage students in their research and to offer students the opportunity to participate in research in the history of science. Students and teachers become co-producers of knowledge, especially regarding the historical practices of science.

## Bibliography

Allchin, Douglas, *Teaching the Nature of Science: Perspectives & Resources* (Saint Paul: Ships Education Press, 2013).

Beneken, Heiko, 'Die Bestimmung des elektrostatischen Grundgesetzes durch John Robison', in *Im Labor der Physikgeschichte – Zur Untersuchung historischer Experimentalpraxis*, ed. by Peter Heering, Falk Rieß and Christian Sichau (Oldenburg: BIS, 2000), pp. 71-106.

Bilak, Donna, Jenny Boulboullé, Joel Klein, and Pamela H. Smith, 'The Making and Knowing Project: Reflections, Methods, and New Directions', *West 86*$^{th}$, 23 (2016), pp. 35-55.

Boone, Tim, Roger Kneebone, Peter Heering, Klaus Staubermann and Yves Winkin, 'A Symposium on Histories of Use and Tacit Skills', *Science Museum Group Journal*, 8 (2017).

Boyle, Robert, 'A Physico-chymical Essay...Touching the Differing Parts and Redintegration of Salt-Petre [1661]', in *Certain Physiological Essays, Works of Robert Boyle*, ed. by Michael Hunter and Edward B. Davis, 14 vols. (London: Pickering and Chatto, 1999-2000), Vol. 14, pp. 93-113.

Breidbach, Olaf, Peter Heering, Matthias Müller, and Heiko Weber. 'Experimentelle Wissenschaftsgeschichte', in *Experimentelle Wissenschaftsgeschichte*, ed. by Olaf Breidbach, Peter Heering, Matthias Müller and Heiko Weber, (München: Wilhelm Fink Verlag, 2010), pp. 13-72.

Cavicchi, Elizabeth, 'Historical Experiments in Students' Hands: Unfragmenting Science Through Action and History', *Science & Education*, 17 (2008), pp. 717–749.

Cavicchi, Elizabeth, 'Learning Science as Explorers: Historical Resonances, Inventive Instruments, Evolving Community', *Interchange*, 45 (2014), pp. 185-204.

Chang, Hasok, 'How Historical Experiments Can Improve Scientific Knowledge and Science Education: The Cases of Boiling Water and Electrochemistry', *Science & Education*, 20 (2011), pp. 317-341.

Chang, Hasok, 'What History Tells Us about the Distinct Nature of Chemistry', *Ambix*, 64 (2017), pp. 360-374.

Devons, Samuel, and Lilian Hartmann, 'A History-of-Physics Laboratory', *Physics Today*, 23 (1970) pp. 44-49.

Drake, Stillman, 'Renaissance Music and Experimental Science', *Journal of the History of Ideas*, 31 (1970), pp. 483-500.

Dupré, Sven, *Laboratories of Art. Alchemy and Art Technology from Antiquity to the 18th Century* (Dordrecht: Springer, 2014).

Eamon, William, *Science and the Secrets of Nature* (Princeton: Princeton University Press, 1995).

Fors, Hjalmar, Lawrence M. Principe, and H. Otto Sibum, 'From the Library to the Laboratory and Back Again: Experiment as a Tool for Historians of Science', *Ambix*, 63 (2016), pp. 85-97.

Frercks, Jan, *Die Forschungspraxis Hippolyte Fizeaus: Eine Charakterisierung ausgehend von der Replikation seines Ätherwindexperiments von 1852* (Berlin: Wissenschaft und Technik, 2001).

Frercks, Jan, 'Creativity and Technology in Experimentation: Fizeau's Terrestrial Determination of the Speed of Light', *Centaurus*, 42 (2000), pp. 249-287.

Galilei, Galileo, *Sidereus Nuncius*, trans. by Albert van Helden (Chicago: University of Chicago Press, 2016).

Hartmann Hoddeson, Lilian, 'Pilot Experience of Teaching a History of Physics Laboratory', *American Journal of Physics*, 39 (1971), pp. 924-929.

Heering, Peter, 'Historical Approaches in German Science Education', *Eurasia Journal of Mathematics, Science & Technology Education*, 10 (2014), pp. 229-235.

Heering, Peter, 'The Role of Historical Experiments in Science Teacher Training: Experiences and Perspectives', *Actes d'Història de la Ciènca i de la Tècnica Nova època*, 2 (2009), pp. 389-399.

Heering, Peter, *Das Grundgesetz der Elektrostatik: Experimentelle Replikation, wissenschaftshistorische Analyse und didaktische Konsequenzen* (Oldenburg: Dissertation, 1995).

Holländer, Ruben, *Die Entzündung brennbarer Flüssigkeiten mittels elektrischer Funken – eine historische und physikalische Analyse*, BA-Thesis (Flensburg: Europa-Universität Flensburg, 2017).

Ingold, Tim, 'Materials against Materiality', *Archaeological Dialogues*, 14 (2007), pp. 1-16.

Ingold, Tim and Ray Lucas, 'The 4A's (Anthropology, Archaeology, Art and Architecture: Reflections on a Teaching and Learning Experience', in *Ways of Knowing: Anthropological Approaches to Crafting Experience and Knowledge*, ed. by Mark Harris (New York/Oxford: Bergbahn Books, 2007), pp. 287-305.

Ingold, Tim, 'The Textility of Making', *Cambridge Journal of Economics*, 34 (2010), pp. 91-102.

Kipnis, Naum, *Rediscovering Optics* (Minneapolis: Bena Press, 1993).

Lederman, Norman G., 'Nature of Science: Past, Present, Future', in *Handbook of Research on Science Education*, ed. by Norman G. Lederman and Sandra K. Abell (Mahwah: Lawrence Erlbaum, 2007), pp. 831–879.

Lederman, Norman G., Fouad Abd-El-Khalick, Randy L. Bell and Renée S. Schwartz, 'Views of Nature of Science Questionnaire: Toward Valid and Meaningful Assessment of Learner' Conceptions of Nature of Science', *Journal of Research in Science Teaching*, 39 (2002), pp. 497-521.

Lühr, Jörg, 'Die Geschichte eines Demonstrationsexperiments: Zur Geschichte der Ampère'schen Stromwaage', in *Im Labor der Physikgeschichte: Zur Untersuchung historischer Experimentalpraxis*, ed. by Peter Heering, Falk Rieß and Christian Sichau (Oldenburg: BIS, 2000), pp. 135-156.

Mahoney, Michael S., 'Reading a Machine'. Checked on 1 February 2019:
https://www.princeton.edu/~hos/h398/readmach/modeltfr.html

McComas, William F., *The Nature of Science in Science Education: Rationales and Strategies* (Dordrecht/Boston: Kluwer Academic Publishers, 1998).

Mercier, Michelle, *Ausgewählte Experimente Rumfords zur Wärmestrahlung: Physikalische und wissenschaftshistorische Analyse*, MA-Thesis (Flensburg: Europa-Universität Flensburg, 2018).

Nawrath, Dennis, 'Die Analyse von Newtons Prismenexperimenten zur Untersuchung von Licht und Farben (1672) – Ein Erfahrungsbericht', in *Experimentelle*

*Wissenschaftsgeschichte*, ed. by Olaf Breidbach, Peter Heering, Matthias Müller and Heiko Weber (München: Wilhelm Fink Verlag, 2010), pp. 73-105.

Piemontese, Alessio, *Les Secrets du S. Alexis Piemontois ...* (Lyon: Loys Cloquemin et Estienne Michel, 1572).

Piemontese, Alessio, *The secrets of the Reverend Maister Alexis of Piemont ... Newly Corrected and Amended, and also Somewhat Inlarged in Certaine Places, which Wanted in the First Edition*, trans. by William Ward (London: Peter Short, 1595).

Principe, Lawrence M., '"Chemical Translation" and the Role of Impurities in Alchemy: Examples from Basil Valentine's *Triumph-Wagen*', *Ambix*, 34 (1987), pp. 21-30.

Principe, Lawrence M., *The Secrets of Alchemy* (Chicago: University of Chicago Press, 2013).

Ragland, Evan, 'Chymistry and Taste in the Seventeenth Century: Franciscus Dele Boë Sylvius as a Chymical Physician Between Galenism and Cartesianism', *Ambix*, 59 (2012), pp. 1-21.

Rieß, Falk, 'Short History of the Use of Historical Experiments in German Physics Lessons', in *Constructing Scientific Understanding through Contextual Teaching*, ed. by Peter Heering and Daniel Osewold (Berlin: Frank & Timme, 2007), pp. 219-225.

Rieß, Falk, 'Teaching Science and the History of Science by Redoing Historical Experiments', in *Proceedings of the Third International History, Philosophy, and Science Teaching Conference*, ed. by Fred Finlay, Douglas Allchin, David Rhees and Steve Fifield (Minneapolis: University of Minnesota, 1995), Vol. 2, pp. 958-966.

Schnapps, Jeffrey T. and Michael Shanks, 'Artereality: Rethinking Craft in a Knowledge Economy', in *Art School: Propositions for the 21st Century*, ed. by Steven Henry Madoff (Cambridge, MA: MIT Press, 2009), pp. 142-157.

Sibum, H. Otto, 'Reworking the Mechanical Value of Heat: Instruments of Precision and Gestures of Accuracy in Early Victorian England', *Studies in the History and Philosophy of Science*, 26 (1995), pp. 73-106.

Sichau, Christian, 'Die Replikationsmethode: Zur Rekonstruktion historischer Experimente', in: *Im Labor der Physikgeschichte: Zur Untersuchung historischer Experimentalpraxis*, ed. by Peter Heering, Falk Rieß and Christian Sichau (Oldenburg: BIS, 2000) pp. 9-70.

Snow, C.P., *The Two Cultures* (Cambridge: Cambridge University Press, 1998).

Steinle, Friedrich, *Exploratory Experiments: Ampère, Faraday, and the Origins of Electrodynamics*, trans. by A. Levine (Pittsburgh: University of Pittsburgh Press, 2016).

Taape, Tillman, 'Distilling Reliable Remedies: Hieronymous Brunschwig's *Liber de arte distillandi* (1500) Between Alchemical Learning and Craft Practice', *Ambix*, 61 (2014), pp. 236-256.

Thomson, Benjamin, Count Rumford, 'An Enquiry Concerning the Nature of Heat, and the Mode of Its Communication', *Philosophical Transactions of the Royal Society of London*, 94 (1804), pp. 77-182.

Voskuhl, Adelheid, 'Recreating Herschel's Actinometry: An Essay in the Historiography of Experimental Practice', *BJHS*, 30 (1997), pp. 337-355.

Watson, William, *Experiments and Observations Tending to Illustrate the Nature and Properties of Electricity: In one Letter to Martin Folkes and Two to the Royal Society* (London: C. Davis, 1746).

## About the Authors

**Thijs Hagendijk** is a PhD candidate at Utrecht University and part of the ERC-funded ARTECHNE-project 'Technique in the Arts: Concepts, Practices, Expertise, 1500-1950.' His dissertation is about reading and writing practices in the early modern arts, with a specific focus on text usage in historical glassmaking, painting, and metalworking. He works on the intersection of technical art history and the history of chemistry, and is interested in performative methods such as reworking, re-enacting and reproducing historical techniques, materials and processes.

**Peter Heering** is Professor in the Institute of Mathematical, Scientific and Technical Literacy (Physics and its Didactics) at Europa Universitaet Flensburg. He specializes in the experimental history of science as developed at the Carl-von-Ossietzly-Universitaet Oldenburg, and its use and value for science teaching.

**Lawrence M. Principe** is Drew Professor of the Humanities in the Department of History of Science and Technology and the Department of Chemistry at the Johns Hopkins University. His research concerns the early history of chemistry, especially alchemy. The replications of alchemical processes that he has carried out in his laboratory have substantially revised our understanding of the history of alchemy and chemistry (or 'chymistry').

**Sven Dupré** is Professor of History of Art, Science and Technology at Utrecht University and the University of Amsterdam. He directs the project 'Technique in the Arts: Concepts, Practices, Expertise, 1500-1950' (ARTECHNE), supported by a European Research Council (ERC) Consolidator Grant. The project undertakes the experimental reconstruction of historical recipes to open the black box of the transmission of technique in the visual and decorative arts.

# 8. A Walk as Act / Enact / Re-enactment: Performing Psychogeography and Anthropology

*Jo Vergunst*

**Abstract**

This book's keywords of re-enactment, replication and reconstruction pose a distinction between an original on the one hand and some kind of copy on the other. The practice of walking and everyday life in general suggest alternatives. Using the series of terms 'act', 'enact' and 're-enactment', the chapter investigates creativity in ways that cannot be reduced to the dichotomy of original and copy. It begins with an account of some pedagogical experiments into walking and psychogeography, and then explores the act of walking in psychogeography. It moves on to the enactment of shared practice between psychogeography and anthropology, and finally the re-enactment of psychogeography as anthropology, and vice versa.

**Keywords:** anthropology, creativity, psychogeography, walking, re-enactment

## Introduction

The aim of this chapter is to question the widely-held distinction between creativity, originality and authenticity on the one hand, and transmission, repetition and copying on the other. This is a problem that seems most pertinent to the art world, but I want to explore it by way of research that crosses between contemporary art practice and anthropological fieldwork. At its heart is a simple question that has concerned me in research on walking and landscape: what makes a footstep? Is it a single bodily action that is repeated, and on sufficient repetition can be said to constitute a walk?

Dupré, S., A. Harris, J. Kursell, P. Lulof, M. Stols-Witlox (eds.), *Reconstruction, Replication and Re-enactment in the Humanities and Social Sciences*. Taylor & Francis Group
DOI 10.5117/9789463728003_CH08

Or should we understand the stepping motion of the feet to be submerged within the entire rhythmical progress of walking? To put it another way, can 'a walk' – that is, a particular person walking – be both a succession of repeated moments, and also an original act that is conceived of in its entirety? I will argue that just as every walk engenders both originality and repetition, it also re-enacts a series of relationships between walkers themselves, and their environments.

My contribution to this book is as follows. Where re-enactment, replication and reconstruction pose a distinction between an original on the one hand and some kind of copy on the other, walking, and in fact everyday life in general, suggests alternatives. By working through the series of terms 'act', 'enact' and 're-enactment', I want to inquire into the apparently simple divide between originality and copying invoked in our keywords of re-enactment, replication and reconstruction. To do this, I will investigate the kind of creativity that walking can invoke, both in an ordinary or everyday sense and through the artistic practice of psychogeography. I begin with an account of a couple of pedagogical experiments into walking, psychogeography and creativity, and then present some anthropological materials on the *act* of walking in psychogeography. I go on to discuss the *enactment* of shared practice between psychogeography and anthropology, and finally the *re-enactment* of psychogeography as anthropology, and vice versa. In writing this chapter I also aim to invoke something of a psychogeographical 'drift' in which the narratives unfold step by step, rather than sticking too closely to a predetermined path.

## Pedagogical experiments in creativity and the ordinary

For the 'Re-enactment, Replication, Reconstruction' workshop held in Leiden in June 2017 that forms the basis for this book, I brought together some techniques of anthropological fieldwork and psychogeography into a small scale event. The intention was to explore what they might tell us about our keywords, particularly through the lens of memory and remembering. Rather than defining psychogeography for participants in advance of the walk, I merely set the parameters of the walk itself and let it unfold in its own terms. In these written accounts I want to do the same.

The structure of the RRR event was developed from one held the year previously in Aberdeen. There, participants read a small part of Lord Byron's poem *Don Juan*, in which the poet fondly recalls his childhood spent in Aberdeen, and mapped their own personal memories onto a street map. In Leiden I chose one of the dozens of 'wall poems' (*muurgedichten*) that decorate the old city centre, painted on large gable ends and street corners. William Shakespeare's Sonnet 30 is on the corner of Houtstraat and Rapenburg. Like the excerpt of *Don Juan*, it tells of memory, loss and the sustaining power of friendship: 'When to the sessions of sweet silent thought; I summon up remembrance of things past [...]' The invitation to the RRR group was to begin on the street underneath the sonnet, and then wind towards a grassy bank of the Oude Rijn river on the other side of the city centre, pausing where we wanted to. We would allow the sonnet's invocation of memory to conjure up our own thoughts and observations of the city life around us, linking, I hoped, our own biographies with those of the city as they presented themselves on that morning.

In terms of anthropological techniques, I wanted to explore ways of creating fieldnotes, understood as notations in text or drawing of the observations and experiences of the researcher. Participants were given an A5 blank notebook in the anthropological tradition, but also some psychogeographical prompts. They had three different versions of the sonnet, including one derived from the original printing, another with the original language in a modern font, and a modern, plain English translation. They had a street map of the centre of Leiden with a transparent sheet of plastic over the top that they could draw on to make their own mapping, all held together on a clipboard. While fieldnotes in anthropology have been taken as a key symbol of professional identity, and the emphasis in methodological texts is often on content rather than form,[1] here the group were invited to make notes imaginatively and creatively rather than just observationally in a naturalistic mode. We used drawing and mapping as much as writing. The intention was, in other words, to engage with a psychogeographical tradition as much as an anthropological one.

What resulted were sketch maps tracing our route, marker pen lines and street names, and small drawings of things found or noticed along the way. Notes made in the A5 books moved from what was seen, heard and felt: 'Church bells are ringing – not the Pieterskerk', to the personal: 'disappearing books ... mother-in-law ... making ... pressed flowers ... loss,' to the reflexive: 'So much to know and completely impossible to encompass it all – that, too, is a loss to me'. The latter comment perhaps opens into the failing of any observational exercise, in that when one begins to be attentive, the realisation of the very partial view that one gets – indeed,

1 Jackson, 'I am a Fieldnote', pp. 3-33.

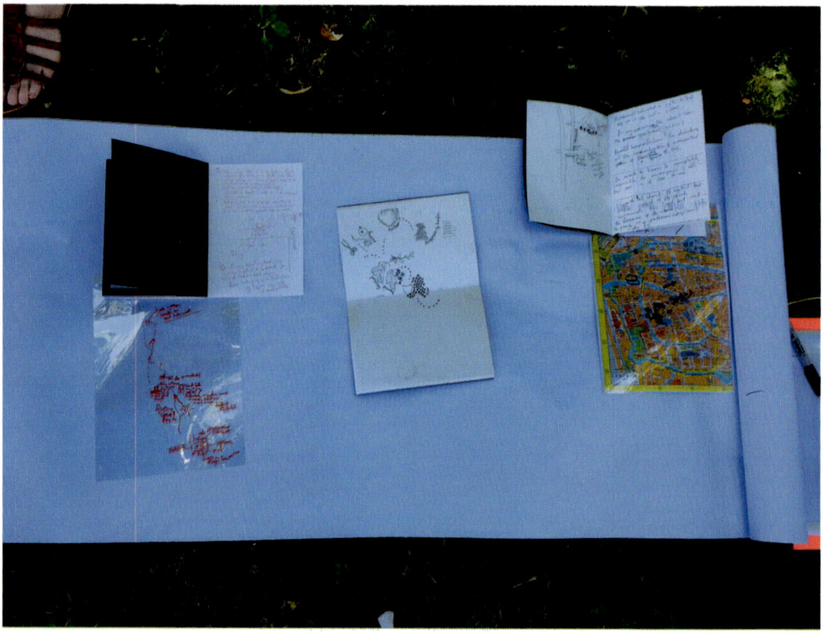

Plate 8.1: Fieldnotes from Leiden. Photo: Thijs Hagendijk.

the inability to see very much or even work out what is going on – can be painful. Anthropological fieldwork and psychogeographical drifting, are partial perspectives.

In the afternoon, we travelled from Leiden to the coastal town of Katwijk and immersed ourselves in another couple of hours of independent observations. This time we started with paintings associated with the nineteenth century Haagse School artists who specialized in working outdoors from 'real life'. We looked at these during a visit to Katwijk Museum, with collections of traditional fishing and artisanal life. Amongst our activities: we observed leisure time on the beach, now the dominant activity where fishing boats had been pulled up in the past ('abandoned place, towels, ball, bags, clothes, unclear where the people are'); a couple of participants got into conversation with a street seller; one sat outside a cafe with a drink, making notes.

A year following the RRR workshop, I was invited back to the Netherlands to work with a group of postgraduate students of art history. I wanted to pursue the idea of the ordinary itself, as the germ for this chapter. We were now in Delft and I could not resist starting with a Vermeer painting. *Het Straatje* (The Little Street) appears to be an entirely ordinary glance at a family group around a house and yet is at the same time extraordinary, simply for the feeling of realism, that one is actually there, seeing what is going on

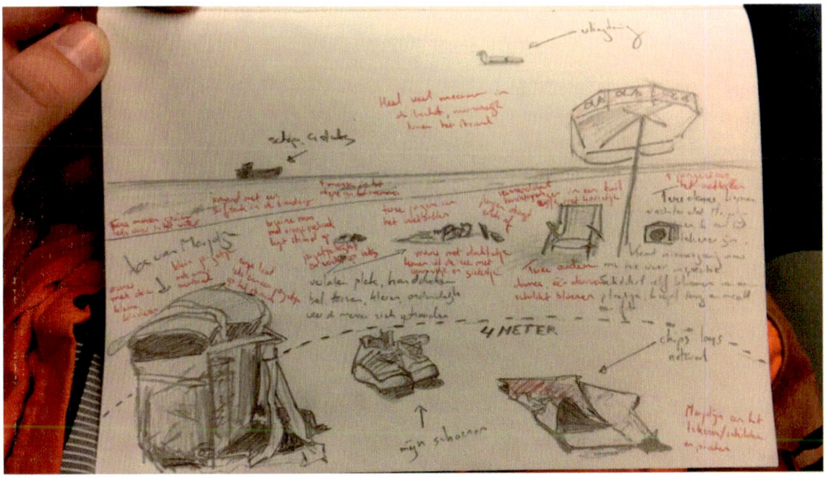


Plate 8.2: Fieldnotes from Katwijk. Photo by Thijs Hagendijk.

beside the artist. So, through attentiveness, I suggested to the students, we might slip between the ordinary and extraordinary, and be able to question a little more how things appear and how they came to be. The students left our TU Delft seminar room to find the location in Vlamingstraat identified (though not without dispute) as the site of the house painted by Vermeer.[2] I asked them to then look around the area for the things that an artist like Vermeer might notice today, and, as in Leiden, to generate fieldnotes on those topics. The students responded with a variety of notes – mostly text this time, though sometimes with diagrams and drawings, and a couple of found items: 'the space between houses – cracks, maintenance, materiality', a description of a house decorated for the birth of a baby, a thin twig with leaves picked up from the ground. In the next day's discussion, the students talked about the sense of themselves as being present in their fieldnotes, whether that was thought of as a problem in anthropology, and what they might do with the idea of subjectivity in all observation.

My aim in presenting these scenes is to connect, perhaps counterintuitively, the study of art and creativity with an anthropology of the ordinary or the everyday. In my 'standard' anthropological fieldwork on walking, as opposed to the pedagogical and psychogeographic episodes described here, I continue to explore people's senses of the ordinary: their perceptions of seasonal change through regular walks down the same city street, for example, or their developing confidence and familiarity in walking in the

2 Grijzenhout, *Vermeer's The Little Street.*

mountains, or their micro-gestures of sociality towards others passing by. Michael Taussig contends that the everyday entails a sensuousness that is tacit rather than explicit, an indirect perception such as might be experienced as the awareness of movement in the corner of one's eye.[3] In terms of walking it could be conceived as a 'pedestrian' knowledge, in that the sensory perception of movement on foot involves a tactile attentiveness and constant readjustment to conditions.[4] The everyday or ordinary is a sensory inhabitation, which is not to say it is not also 'constructed' by way of larger social forces that are often hidden from an individual's sight – in the classic Marxist reading of the everyday, as Highmore points out – but that it is not only such a construction.[5] The point is that every day walks are not simply repetitions of previous actions constrained by social forces, but are constantly made anew, each adding to the last. They may influence place-based politics and dynamics of power as well as being shaped by them.

Exploring the sensuousness of the ordinary through habitual actions that are repeated from day to day or moment to moment could be a task well suited to a pedestrian mode of anthropological fieldwork. But if we are content neither with a model of the endless mechanical replication of social life, nor of the spontaneous, innovative creation of new forms, we might ask instead how, then, is the ordinary improvised into being? And what slippages might we find between the ordinary and the extraordinary? In the following, I argue that the copying that takes place might belong to the realm of skill and engagement, rather than simple repetition.

## First steps to a psychogeographical perspective

To take these questions further, I want to pursue psychogeography as a particular strand of contemporary arts practice – but also one that is more than 'just' art. At this stage, the concern is with understanding the act of walking in both psychogeography and in ordinary life within an anthropological frame. My anthropological research on walking and landscape has touched on psychogeography over the years, and I now want to connect with its history and key writings as well as its diverse practices of walking and mapping. A backwards glance is a way of considering the possibility

3 Taussig, *The Nervous System*.

4 Ingold, 'Culture on the Ground'; Vergunst, 'Taking a Trip'; *Ways of Walking*, ed. by Ingold and Vergunst, pp. 105-121; Vergunst, 'Key Figure', pp. 13-27.

5 Highmore, *Ordinary Lives*.

of a relationship between anthropology and psychogeography. Without wishing to construct an art history account of psychogeography, exploring some key texts and further examples brings the relationship between the new and the copied, the enacted and re-enacted, to the fore.

If a definition of psychogeography would be useful at this stage, we could use the terms presented in the first edition of the journal of the *Internationale Situationiste* 'Situationist International' in 1958: 'The study of the specific effects of the geographical environment (whether consciously organized or not) on the emotions and behaviour of individuals.'[6] 'Situationist International' (SI) was a grouping of avant-garde artists, writers and political theorists who formally came together in 1957 in Paris and dissolved in 1972, but whose legacy continues in some of the artworks I describe in this chapter and, perhaps, in some anthropological practices too. Guy Debord is usually seen as the leading figure, and a variety of others contributed writings and practice to the movement.

Ken Knabb's translations of SI texts, including an essay written by pre-SI Letterist International member Ivan Chtcheglov, are a key source.[7] 'Formulary for a new urbanism' was written in 1953 when Chtcheglov was nineteen. He was committed to an asylum shortly after leaving SI but was recognized as influential by Debord and later had letters published in the SI journal. In the essay he rails against the 'banalization' of the world, in which 'everyone is hypnotized by production and conveniences – sewage systems, elevators, bathrooms, washing machines' which have become an 'omnipresent, obsessive image'.[8] His response was to seek a new form of architecture, which we might think of more broadly as urban fabric or even landscape: 'The architecture of tomorrow will be a means of modifying present conceptions of time and space. It will be both a means of *knowledge* and a *means of action*', while: 'The main activity of the inhabitants will be CONTINUOUS DRIFTING. The changing of landscapes from one hour to the next will result in total disorientation.'[9]

The urban 'drift', usually rendered in the original French *dérive*, is developed in the SI to encompass psychogeography not so much in the positivist and rather deterministic position set out in the 1958 SI journal definition cited above, but a practice of more open-ended attentive and reflective urban exploration. Debord writes in *Theory of the Dérive*, in the second edition of

6 Debord, 'Definitions', p. 52.
7 Knabb, *Situationist International Anthology*.
8 Chtcheglov, 'Formulary', p. 4.
9 Chtcheglov, 'Formulary', pp. 3, 7, emphasis and capitalisation as in the original.

the journal: 'In a dérive one or more persons during a certain period drop their relations, their work and leisure activities, and all their other motives for movement and action, and let themselves be drawn by the attractions of the terrain and the encounters they might find there.'[10] It seems important that the *dérive* is neither work (such as a survey), nor leisure (such as a stroll), nor any other obvious category. It is instead anti-categorical, in a movement against the preconceived divisions of time and function as they are reflected in the architecture of the city. Debord lists some activities of the group as examples of its 'general sensibility': 'slipping by night into houses undergoing demolition, hitchhiking nonstop and without destination through Paris during a transportation strike in the name of adding to the confusion, wandering in subterranean catacombs forbidden to the public, etc.'[11] Psychogeography comes to entail a set of practices that are enacted in ways significantly beyond Debord's initial, perhaps self-consciously positivist, definition.

In 2005 I was engaged in research on people's experiences of walking in and around the city of Aberdeen in north east Scotland, as I noted above. I spent my time finding people to share walking with, interviewing people about their personal histories of walking, and observing the ebb and flow of pedestrians and other walkers. In June of that year, artist Jim Colquhoun took up a week-long residency at Peacock Visual Arts, the city's most prominent contemporary arts venue. He conducted a series of walks, performances and talks with the promise that he would 'encourage others to cast themselves adrift with him into the collective unconscious of contemporary and historical Aberdeen' (quoted from the residency programme). The programme was to include 'organized "psychogeographical" walks', a term I had only recently heard of at that time.

I'll cut to the ethnographic (or perhaps psychogeographical) present tense. Jim explains the premise for his 'Peripatetic Randomiser walk' to a group of about a dozen of us, a few academics, a few associated with Peacock Arts, and a few others who are gathered during one afternoon of the residency. He has a small square folded paper chooser or fortune teller of the kind children sometimes have, held between thumbs and forefingers, that opens and can be unfolded to reveal messages. The messages are instructions that tell us to 'turn right', 'ask someone directions', 'cross the road' and so on, and each person in the group chooses something in turn so that the walk progresses with a random element through the city centre. It's fun

10 Debord, 'Theory of the Derive', p. 62.
11 Debord, 'Theory of the Derive', p. 65.

and sociable, and the ordinary sociability of walking became a theme in my work at this time.[12] But the walk ends more dramatically. We have crossed a dual carriageway, vaulting over railings in the process, and are then instructed by the Peripatetic Randomiser to find a corner. So, we are now perched and crammed precariously on a set of angular cobbles on the edge of an extremely busy junction. The cobbles seem to signal that we should simply not be there, this is not a pavement, not a place for moving through or pausing. The city, of course, is like this – distinguishing between places to be and places not to be – and the boundary has been traversed, in a small way, by the Peripatetic Randomiser. The turn-taking has ended, we disperse, and the walk is over. Jim gives us a 'thank you' for taking part, but there is no particular explanation for the walk available otherwise. We are left to reflect on it ourselves if we wish to, and my own thoughts turn to the distinctive reading of the city that the walk has thrown up.

Online in 2018, however, I read an artistic evaluation undertaken for the funders of Colquhoun's residency, the Scottish Arts Council (now Creative Scotland) that takes me down a different route. The reviewer found flaws with Colquhoun's work and rated it as merely 'competent': 'Artist walk/talk programme v. interesting but overall rating let down by the lack of visibility of events and marketing for the installation works.'[13] I am amused by the idea that the 'lack of visibility of events' was a problem. Of course, Jim's walks were visible only temporarily, and only really for those who took up his invitations to participate. There was no audience for the walks I did with him, other than perhaps those who saw us on the streets, but they would have had no idea what we were doing. 'Visibility' and 'marketing' could even be two of the watchwords for much contemporary art that walking art, including psychogeographical practices, rail against (though walking artists do also document and publish their work, as we will see later).

Perhaps the evaluation does nonetheless question the social reach of Colquhoun's events. The walks may have been art for those already 'inside' the art world, who heard about it and could imagine what it might involve. This critique might invoke social class as well as the specific networks surrounding the artist and host organization, and yet the intersection of

12 Lee and Ingold, *Fieldwork on Foot.*

13 Scottish Arts Council (2005) Artistic Evaluation – Visual Arts. Unpublished report, available at http://www.scottisharts.org.uk/resources/publications/Miscellaneous/Strategic%20Review%20papers/CFOs/M%020%20R/Peacock/2005%20Peacock%20Gangrel%20Bodies%20Jim%20Colquhoun.pdf. Checked on October 2nd, 2018.

social class with contemporary art is not straightforward. We could turn to Pierre Bourdieu, who links the consumption of art to cultural capital: 'To appropriate a work of art is to assert oneself as the exclusive possessor of that object, which is thereby converted into the reified negation of all those who are unworthy of possessing it.'[14] However, the psychogeographical walk cannot be owned or appropriated as such, and participants in the events I have joined and the artists who have instigated them have certainly varied in terms of being well off – although this has not been a research focus for me. In any case we might seek alternatives to a deterministic analysis of economic or cultural capital, while recognizing that contemporary art takes place within, and contributes to, a range of economic and cultural forms. Exploring the kinds of creativity entailed in these practices is one such alternative.

## The creativity of walking as art and art as walking

We could situate psychogeography as an artistic practice that is relevant to social science in its development of a critical and indeed subversive approach to the urban environment. The relevance to anthropology in particular might lie in the distinctive ways of knowing that it involves, which are based not in an objective interpretation of one's surroundings but through the shared, mobile practice of being attentive that is key to anthropological fieldwork. Yet the claim that a walk can be a work of art in any sense needs further consideration. Specifically, we need a notion of creativity that can incorporate more than just the spark of an individual artist and engage as well with the kinds of shared and collaborative practices described here. We could also seek to recognize the step of the walk as inherently creative, beyond the apparent replication that it involves.

Anthropological thinking on creativity has explored alternatives to the Western and modernist account of the individual mind as the seat and origin of creativity, and the body as the means by which it subsequently plays out. This has been a feature of mainstream thought on creativity since at least the time of Kant, who, as Eitan Wilf notes, set out the idea of the 'genius' as an aspect of individual spirit that is present from birth rather than learned (or indeed copied).[15] While Kant does not directly discuss creativity, the association of genius with the production of fine art provides a model for

14 Bourdieu, *Distinction*, p. 280.

15 Wilf, 'Semiotic Dimensions', pp. 397-412.

subsequent notions of how art is understood to come about, and the idea of the significance of art as contained merely within itself – art for art's sake. However, in *The Critique of the Power of Judgement*, Kant outlines the tension between the freedom from rules necessary for original genius to emerge, while at the same time deriving an expression for a work that must be recognizable to an audience.[16] So, 'there is no fine art in which something mechanical, capable of being at once comprehended and followed in obedience to rules, and consequently something *academic* does not constitute the essential condition of the art'.[17] Creativity then becomes, as Sassen puts it, 'a process of mediation between the realms of freedom and determination, both on the level of process and on the level of product'.[18]

This subtlety in Kant's work is, however, often overlooked. What results in modern times, according to Matthew Rampley, is the 'Kantian subject' of spontaneous and original creativity.[19] While artistic creation itself is understood by Kant as 'properly free of all external constraints, be they economic or any other', Wilf argues that the modern notion of creativity sits within a neoliberal framing of economic progress and individual responsibility in times of increased uncertainty.[20] This includes, for example, the popular framing of the 'creative class' as generators of wealth.[21]

Marxist approaches to creativity have explored the very opposite to the free-willed Kantian notion. Here it is instead labelled as a bourgeois theory based on the notion of an autonomous subject, which as Marx and Engels put it, 'depends wholly on demand, which in turn depends on the division of labour and conditions of human culture resulting from it'.[22] In other words, creativity is not based in the actions of an independent individual but on a particular system of production and consumption. In a later article Rampley summarizes this perspective drawing on Bourdieu's *Distinction* (described above): 'As a bourgeois ideological projection, aestheticized 'art' does not even exist in Western culture; the appeal to aesthetic judgement is primarily a strategy to disavow the social functions of art, beginning with the social stratification of taste.'[23]

16 Kant, *The Critique*.
17 Kant, *The Critique*, p. 171.
18 Sassen, 'Artistic Genius, p. 178.
19 Rampley, 'Creativity', p. 268.
20 Kant, *The Critique*, p. 267; Wilf, 'Semiotic Dimensions', p. 407.
21 Florida, *The Rise*.
22 Cited in Rampley, 'Creativity', p. 268.
23 Rampley, 'Art History', p. 548.

Anthropologists have sought to tread a path between this positioning of creativity between the poles of individual genius (a model of agency) and collective context (a model, we might say, of structure). Their means has commonly been through fieldwork that tracks the realities of practice, such as the making and circulation of works of art. To take a recent example, Alice Cant describes the forms of authorship used in artisanal workshops in a village in rural Mexico. There, many people making wooden sculptures for the international and tourist markets are employees of those who lead the studios and sign the works – in a manner reminiscent of paintings 'from the workshop' of an Old Master painter. The Mexican artisans negotiate authorship of the works through relations of kinship, employment and economy, sometimes striking out to launch their own workshop but more frequently staying tied to their existing network of labour and production.[24]

This connects with Alfred Gell's argument in *Art and Agency,* in which the 'enchanted' material property of art is held to be an active participant in social life, even standing in for people in relationships.[25] Social interactions and relations may be sustained by means of the art object. It is useful to consider what art does, or what it enables in the world, beyond the meaning that people might interpret from it. But as Ingold, Morphy and others have pointed out, the idea of agency can serve to differentiate in an unsatisfactory way between the qualities of some entities (for example those created as or touched by art) as opposed to others (those of everyday life).[26] The focus in Gell's work on adornment and decoration emphasizes the 'object' itself. Art is an object, which then does something to social relationships by means of the agency given to it by its maker – as opposed to art as a process emerging from the material grounding of all life. A model of creative action based in this generative field of everyday life, rather than just the 'moment' of individual genius, may be most helpful in tracking the way in which art operates in the world.

Ingold and Hallam find another register through which to discuss the creativity of social life, that of improvisation.[27] Where creativity has in modernity become hitched to the notion of innovation – the original and spontaneous production of a finished object – improvisation emphasizes instead the generative processes of making, the social and material relations through which they arise, and a temporality geared towards the continual emergence of present and future rather than the break with the past implied

24 Cant, '"Making" Labour'.

25 Gell, *Art and Agency.*

26 Ingold, 'Materials against Materiality'; Morphy, 'Art as a Mode of Action'.

27 Ingold and Hallam, 'Creativity and Cultural Improvisation'.

by innovation. Eitan Wilf's response to this is to note that there are forms of creativity that nonetheless seek to 'change a given set of constraints or field of possibilities', and in some ways it is the dynamic between continuity and change that gives life, and tension, to creativity in different circumstances – including perhaps the artisans' workshops in rural Mexico.[28]

If art is understood from the Marxist perspective to involve the production of an object that is separable from its maker in order to enter into circuits of exchange, and marks also an aesthetic distinctive to a social class, then a walk does not seem to fit very well. This would be the case even if we follow Roger Sansi's anthropological conception of the relationship between contemporary art and the gift.[29] For Sansi, what is being transferred when an art work moves from one setting to another – for example from its setting of production to its setting of display or consumption – is not just the object itself but the 'spirit' of the maker reproducing itself, much as a gift-giver sends something of themselves off with the gift towards the receiver.[30] But what can we make of art that creates no object, or leaves no trace?

If, alternatively, in the Kantian mode, art also springs forth as an idea, albeit enacted in a recognizable material form, that is informed by an original genius and given shape by an aesthetic perspective, a walk also does not work very well. This is because the walker almost always re-enacts a previous route or journey – by following a path – and yet also makes an improvisatory response to the particular conditions and circumstances of the individual walk itself. So, a walk is neither a commodifiable object nor an act of creative genius. In some ways it is in their opposition to both these categories that walking artists have worked, and in doing so, a useful critique of the categories emerges.

In their discussion of the generative aspect of improvisation, Ingold and Hallam go on to note that the reproduction of social life may frequently take place by way of copying or imitation, and these are not to be devalued according to what we might think of as the moral discourse of innovation and creativity. They write: 'Copying or imitation, we argue, is not the simple, mechanical process of replication that it is often taken to be, of running off duplicates from a template, but entails a complex and ongoing alignment of observation of the model with action in the world. In this alignment lies the work of improvisation.'[31] We might then find, in the improvisational and

28 Wilf, 'Semiotic Dimensions', p. 399.

29 Sansi, *Art, Anthropology*.

30 Cf. Mauss, *The Gift*.

31 Ingold and Hallam, 'Creativity and Cultural Improvisation', p. 5.

embodied activities of dance or calligraphy described elsewhere in Ingold and Hallam's edited collection – or indeed in walking – not examples of creative acts that transform the world, but explorations of creativity inherent in the progress of the beings that inhabit it and are continually changing it.[32] To imitate, or we might say, to follow in someone's footsteps, is no lesser a form of creative ingenuity.

## Walking Hamish Fulton's way

A useful formulation of creativity in walking comes from one of the senior figures of the walking art movement, artist Hamish Fulton. Here I will again consider contemporary art practice through anthropological fieldwork. Fulton visited Aberdeenshire to take part in another art residency with Deveron Arts (now Deveron Projects), in the town of Huntly in 2010. Fulton invited members of the public to join in a choreographed walk in Huntly. On the day, a group of around thirty people were asked to walk around one square block of this eighteenth-century planned town (about half a kilometre in total) continuously for two hours, evenly spaced, in silence. I took part in this and found it meditative (which was Fulton's intention), and slightly dizzying (which I guess was not). Each circuit of the block felt like not so much a repetition of the last as an accumulation, the conditions and the feel of one's body slightly different, attention being drawn to slightly different things – other people, window sills, the pavement, the breeze, one's own reverie. In a concentrated form, we learned the everyday sensory familiarity, and yet also diversity, of an ordinary walk.

The town walk was a prelude to one of Fulton's long 'wilderness' walks, undertaken to connect Huntly with the nearby Cairngorm mountains. He spent twenty-one days alone, camping and walking through the glens and mountains, and later publishing the results of the whole project in his book *Mountain Time, Human Time* with photographs and his characteristic clipped pieces of text that recount but do not explain the walk.[33] But before he set off from Huntly, Fulton gave a public talk for Deveron Arts. It was advertized with a strapline: 'Can walking be a kind of art?' and I remember turning this round in my head, as an anthropologist then interested more in walking than in art, to 'Can art be a kind of walking?' Rather than taking art as the universal category, I wondered whether the rhythmical, sociable

32 Ingold, 'Modes of Creativity'.

33 Fulton, *Mountain Time.*

and place-making activity of walking might have its counterpart in art. What I learned about Fulton's work, and that of others like him, was that both questions would probably be nonsensical from the inside of these practices. From this perspective, the art is the walk and the walk is the art. Each 'kind' is self-selecting of the other.

In the talk itself, one of Fulton's own comments stuck with me. He said (here I may be paraphrasing slightly through my fieldnotes): 'I appreciate that people enjoy painting watercolour scenes, but I don't want to make pictures like that. And I appreciate that people enjoy going on walks with the Ramblers Association, but I don't wish to walk in that way.' I have not had the opportunity to question Fulton directly about what he meant by this, and if I did, I suspect he might say that his words speak for themselves, just as he means his walks to stand for only themselves and do not in themselves require to be documented, analysed, or 'understood', as he also explained in his Huntly talk. Fulton's books could be said to document or, as I argue below, 'map' his walks but there is little that unpacks or contextualizes them within a wider frame of reference. Something of an exegesis of Fulton's statement can be attempted even so.

Taking the first part of what Fulton said, that he doesn't wish to paint watercolour scenes, speaks to the desire to avoid a representational response to the landscape as scenery that is common to a number of prominent environmental artists in Scotland (e.g. Tim Collins and Reiko Goto, Dalziel and Scullion, Alec Finlay, Elizabeth Ogilvie). It is to avoid the pointless creation of a 'better' version of the landscape itself, or one that translates it into an artefact that can be commodified. But it is possible still to locate Fulton's own response within the realm of the aesthetic and artistic, in the improvisational, embodied and emplaced sense that I have begun to outline in this chapter. The aesthetic response to the landscape that is given as a walk is also, however, not so far from the notion of alignment and attentiveness in Ingold and Hallam's discussion of copying.[34] The walking body in this sense maintains a kind of symmetry with the landscape in that the surfaces, substances, and textures being walked through entail specific reactions within the walker. An icy surface needs balance, rocks need flexibility to shift weight, while mud, sand, and everything else produce ways of walking that involve coming to know the material qualities of the landscape in particular ways. Hitting the hard pavements of Huntly at quite a fast pace for two hours, and constantly turning right, was at least as tiring as a hill walk, straining leg muscles and requiring considerable and uniform concentration. Familiar

34 Ingold and Hallam, 'Creativity and Cultural Improvisation'.

environments produce an everyday tactility of knowing that is no less skilful a 'copying' of the environment than a practiced artist's watercolour, even if it is, usually, less explicitly considered.

This leads into the second part of Fulton's avowal. That he doesn't wish to walk in the way that the Ramblers Association does, affirms that he does wish to walk in some way. But what would that be? The negative invocation of the Ramblers Association provides some pointers. In my fieldwork I have walked with the Ramblers Association and various other walking clubs, of which there are very many in the city and region of Aberdeen. These walks have a distinctive sociability and indeed often overflow with conversation and interaction between participants.[35] They have distinct and dedicated leaders, who decide upon timings, routes, resting places and have responsibility for the progress of the walk as a whole. The route is decided upon in advance, and indeed may even be 'pre-walked' by the leaders to ensure nothing will go wrong on the day. In all these respects, Fulton's walking, and we could add, psychogeographical walking, is not alike. The emphasis is instead on what the place stimulates in the walking participants and if there is conversation at all, it is usually a conversation of attentiveness to the surroundings. Fulton and a number of other walking artists usually walk without other people (Richard Long would be another example), in which any talk is simply with oneself and the surroundings. The walking is not so much carried out 'alone', which would emphasize the lack of human counterparts, but rather happens 'with' the landscape and all the entities and beings that comprise it.

At the same time, we should note that walking art can also take place in more overtly sociable ways. A major UK exhibition of work by walking artists, *Walk On: From Richard Long to Janet Cardiff. 40 Years of Art Walking*, detailed a wide variety of approaches by artistic partnerships as well as individuals, and acts of walking that purposely involved other people, as well as non-human others.[36] These stand as a response to the 'heroic', often male, artist who walks alone (and whose work did still form a significant part of the exhibition). In literature, perhaps the most lauded work of psychogeographical writing in English is Iain Sinclair's *London Orbital*, in which Sinclair describes a series of walks in the vicinity of London's M25 circular motorway as a counterpoint to the 'non-place' of the motorway itself.[37] The voices of people who are invited to share in the walks and those who are met along the way are an important presence in the book.

35 Lee and Ingold, *Fieldwork on Foot.*

36 Morrison-Bell, et al., *Walk On.*

37 Sinclair, *London Orbital.*

If being 'alone in nature' is not the significant feature of these walks, the improvisational, attentive and responsive character of much walking art, including Hamish Fulton's, is nonetheless very distinctive. In drawing a contrast with the Ramblers Association, Fulton may also have invoked a kind of walking that does not follow a map, although, like Colquhoun's walk in Aberdeen, it may also be circumscribed in other ways. But among the things it creates is a particular sense of place and engagement that acts as a mapping and can then be mapped – in other words, recreated in some kind of documentary form – if the walker chooses to do so. Fulton's exhibition and other published works could be seen in these terms. They often comprise of sequences of plain and unadorned descriptive words, sometimes numbers that count days, moons, or mountains, or one-word observations that follow the course of the walk. The walk becomes a mapping of the experience of the landscape, 'copied' into the body through careful attentiveness, in the same way that the intention to undertake an artwork becomes a walk. 'Walking art' means that the art is walked, and the walk is made as art. In these circumstances, the walk is the art, the art is the walk, and each maps onto – and is enacted in – the other.

## Psychogeography, maps and the politics of walking art

The enacting of walking and art in each other might thus mark a move beyond a reductionist approach to the 'act' of either, and the same might be true of anthropology and psychogeography. Rather than an anthropological account 'of' psychogeography and other art walking practices, in which walking is analysed by way of anthropology, according to – and in a sense reduced to – the latter's categories, thinking about the enactment of walking and art together could suggest a way that a primary act does not require a secondary analysis. They might find each in the other rather than being reduced to a subject-object relationship.

By analogy, I want now to pursue mapping in more detail, as an aspect of psychogeography that we might also see as enacted in anthropology. Mappings have been a significant practice in psychogeography over the years, including from both those identifying their work as psychogeography and those who we might connect more loosely with the tradition. We might seek a partial comparison with anthropological fieldwork techniques and sensibilities, and then explore where a fruitful common ground, or at least a fuzzy boundary, may be found.

The mappings that arose from the dérives of the SI took distinctive forms. Debord and collaborator Asger Jorn's well-known *Guide Psychogéographique de Paris* (1956) and *The Naked City* (1957) were, as Ford puts it, 'useless' maps. They split up Paris into regions connected with arrows denoting psychogeographic flows and spontaneous turns of direction, and so 'challenged traditional ideas of mapping related to scale, location and fixity'. Ford argues that they encouraged the sense of being a detective, 'stalking the city looking for clues to its significance and seductive power'.[38]

The notion of psychogeography as an enquiry, such as might be pursued by a detective and in contrast to simply an activity carried out for its own sake, seems important here. It is where we might distinguish the psychogeographic drift from other sorts of urban walking, such as the *flânerie* of nineteenth century Parisian arcades discussed by Walter Benjamin, or the surrealist deambulations of the avant-garde generation previous to the SI – though the transgressiveness of the *dérive* is, as Bonnett notes, a characteristic theme of the avant-garde.[39] Psychogeography involves an attentiveness to one's surroundings, and also an openness to the possibility that things may not be as they first appear. What appears as 'ordinary' may hide as well as reveal: indeed, *'Sous les pavés, la plage!'* as the famous May 1968 uprising slogan went ('Under the paving stones, the beach!). Walking occurs with a purpose beyond that of reaching a destination, and that is to make a difference in itself. A psychogeographical map would then not simply describe and document, but create an impetus to do something, to act differently upon and within the urban form. Practically useless for finding one's way through the streets, the *dérive* and its mappings – including text and other artistic forms as well as recognizable 'maps' themselves – would open on to other terrains that give shape to the emotional and perceptual aspects of experience.

All mapping practices are, of course, emergent from the particular circumstances and purposes of their makers, and this is in part what we might learn from the explicitly alternative and transgressive psychogeographical cases. The maps that anthropologists make could be an interesting comparative case. It is likely that most anthropological maps never make it to the published page, as they are created as part of fieldwork practice in the form of fieldnotes. But they may still be important in this regard. To take a couple of personal examples: during my first main fieldwork about the landscape of the Orkney Islands, situated just to the north of mainland Scotland, I spent some weeks visiting farms on one particular island to develop a case study of local farming and landscape histories. At each one, I took what became

38 Ford, *The Situationist International*, p. 59.

39 Benjamin, *The Arcades Project*; Bonnett, 'Walking through Memory'.

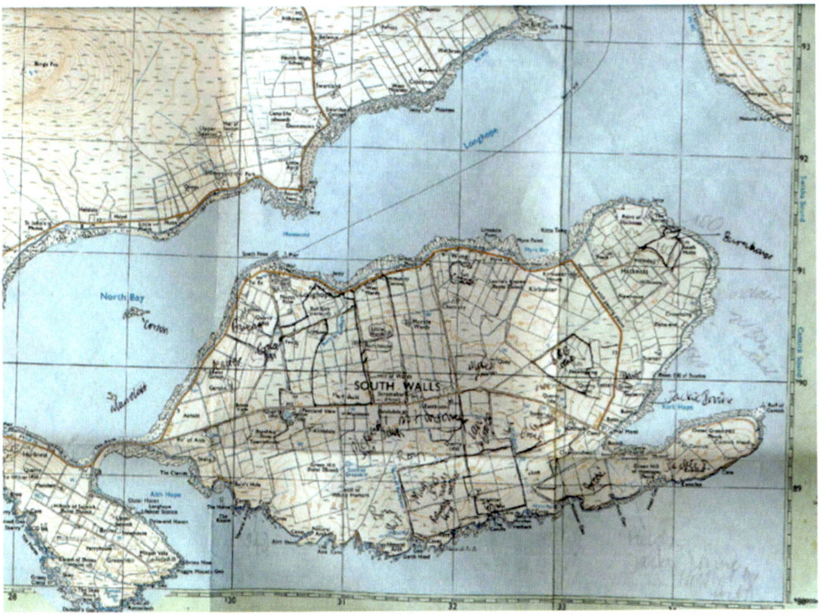


Plate 8.3: Annotated Ordnance Survey 1:25,000 map of Orkney. Photo: author.

a battered copy of a 1:25,000 Ordnance Survey map (the UK standard), and gradually wrote and drew layers of notes referring to place names, farm boundaries and land rental arrangements. After each farm visit, I wrote up careful descriptive fieldnotes in a separate computer document, and yet the physical map continued to accumulate its own notes as a palimpsest of the island's landscape and my own fieldwork. Today I can look at the map and feel – still be affected by – the intensity of my fieldwork as a PhD student that I can't maintain today, with more competing demands on my time.

In the same fieldwork period, I also produced a small, and clearer, set of plans of farm buildings in order to understand the shift from agriculture to a diversified agriculture and tourist business, as Orkney farms began to take in tourists as bed & breakfast businesses and reconfigure their buildings and the paths between them. Here the mapping of architectural forms allowed me to compare and consider the relationship between the buildings and a changing economy in a way that text alone would not. I saw the pattern of paths and driveways being moved away from byres and barns and towards the farmhouse or other accommodation directly. And more recently in separate fieldwork, I used another 1:25,000 scale map to record place names in a particular forest in the centre of Scotland. The forest is now designated and preserved as a 'natural' environment but actually has a whole series of interweaving personal and social histories running through

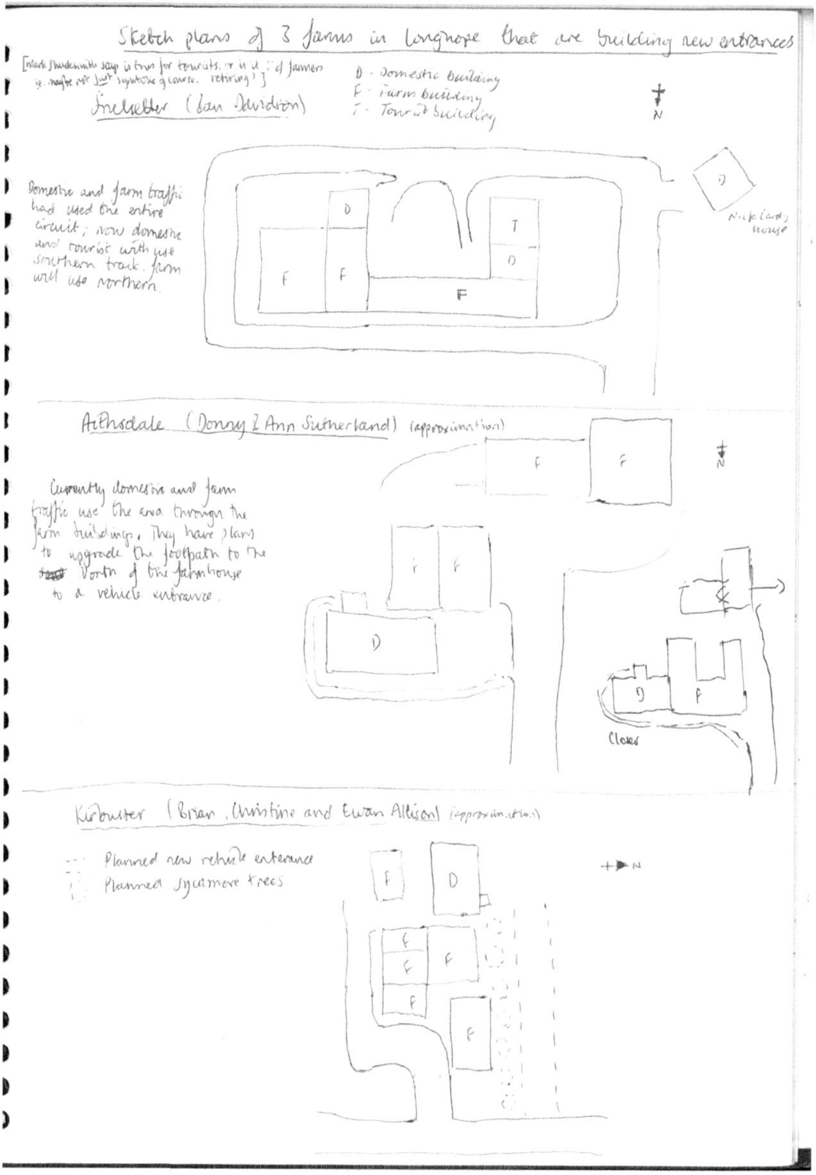


Plate 8.4: Fieldnotes from farms in Orkney. Photo: author.

it, as locally-named paths and other places. Again, I took the map to a series of interviews and added the names and histories in layers.

I suspect that many anthropologists, as well as geographers and others, would be able to tell similar stories of mapping practices as part of their fieldwork enquiries. Yet the methodological literature in anthropology has tended to focus for the most part on text rather than any 'drawn' material, other than the classic diagrams for understanding kinship.[40] Moreover the technological shift towards 'e-fieldnotes' could (although not necessarily will) increase the reliance on text rather than drawing – though digital images of other kinds are of course more and more preponderant.[41] Published examples of anthropological maps other than the topographic are not so common, but sometimes schematic maps of landscapes and places are used that are designed to convey a particular insight into a landscape. Simon Harrison explains with the help of a map how, along the Middle Sepik river in Papua New Guinea, the landscape is 'forgetful' as the constantly shifting meanders of the river are encouraged by ditch digging and tree planting, and forest re-growth quickly overtakes paths to a neighbouring village that fall out of frequent use. The inhabitants, though, are memorious of their social and kinship relations in circumstances of frequently shifting allegiances.[42]

More formal maps may also come to light during fieldwork and become the subject of it. Cristina Grasseni's informants in the Italian Alps decided to create a local mapping of their historic and cultural landscape, which became an exercise in cultural remembering rather than tourist-oriented planning.[43] While anthropologists have also come to use GPS technology to help accurately map landscapes and territories as part of indigenous rights movements, there is a sense in these examples of mapping that attempts something other than objective description. They may be no less carefully made than a topographic, scalar map, but serve to draw the attention (both of the maker and the viewer) to other phenomena and other relationships.

To return to the *dérive*, the shifting between formal and informal maps could be seen as analogous to the anti-categorical nature of the psychogeographical walk, which would not be stable or straightforwardly representable. As we saw above, the *dérive* does not easily fit into the dominant categories of work or leisure, but, for the SI at least, they did have a firmer footing within the scope of politics. But what sort of politics is being created here, and with what results? SI participants frequently concerned themselves with the future of the urban form, and what experiences in relation to it could be conceived in the future. Remaking the city in a freer, almost anarchic spirit of collective resistance to planned, structured forms of consumption was an important principle in SI texts – starting from Chtcheglov's open

40 Sanjek, *Fieldnotes.*

41 Sanjek and Tratner, *Fieldnotes.*

42 Harrison, 'Forgetful and Memorious Landscapes'.

43 Grasseni, 'Skilled Landscapes'.

statement 'We are bored in the city, there is no longer any Temple of the Sun' (in other words, no authentic and encompassing focus for the population's attention).[44] Subsequent psychogeographical practices (for example, as detailed in Richardson, as well as those described here) are perhaps subtler in their political framings, and the explicit fermenting of revolutionary ideals in the Situationist International now feels jarring.[45]

To find the politics of contemporary psychogeography and anthropology can be to seek both more subtlety and a more 'ordinary' mode of existence. As outlined in the above account of ordinariness and creativity, anthropologists have sought out political processes not just in the formal processes of legislation and public policy, but in micro-gestural relations, habitual actions and pedestrian ways of knowing. Politics can play out in an ordinary walk as a person makes decisions over routes, destinations, and things that happen along the way. De Certeau's identification of 'pedestrian rhetoric' in the city was in part to do with walkers' ordinary forms of resistance to the structures of planned urban forms, but it was also a reckoning of the attentiveness and sensuousness of the pedestrian as ordinary movement through a landscape.[46] As an example, we might look to the art walking practice of Misha Myers, who worked with asylum seekers to create new sketch maps of an English city superimposed with landmarks from the places they had left behind.[47] Hamish Fulton's town walk in Huntly, he explained later, referenced the repeated perambulations of Tibetan Buddhist monks around their temples, and was intended as a comment on the restrictions on 'ordinary' politics and everyday life in Tibet under Chinese rule. This sense of the political is less directly revolutionary than that of the SI's psychogeography, perhaps, but shares a grounding in the 'step-by-step' performance of ordinary life.[48]

## Re-enacting psychogeography and anthropology

While an important aspect of psychogeography is its explicit forms of action in the world, anthropology can appear to be rather subtle by comparison. Good intentions for the 'impact' of research can be subsumed by the act of immersion in the fieldwork itself. This is not to say that the outcomes of anthropology

44 Chtcheglov, 'Formulary', p. 1.
45 Richardson, *Walking Inside Out.*
46 De Certeau, *The Practice.*
47 Myers, 'Walking again Lively'.
48 Augoyard, *Step By Step.*

are not powerful, of course. But there may be things to learn from the practice in psychogeography that I have described here, formulated as interventions in – not documents of – the world that involve other people and places. This is where we can consider the value of the re-enactment of psychogeography as anthropology, and indeed anthropology as psychogeography, in which each is enfolded as a careful and attentive copy of the other. Not original, and not repetitive either, but emergent in a field of relations that involves researcher and participants (or co-enquirers) with the landscape.

The experimental walks I recounted at the start of this chapter, and others I have contributed to in recent years, were designed in different ways for different participants. Some include 'the public', some students or academics, or arts communities. But they have a common thread of purpose that goes beyond their immediate circumstances. They are not walks 'to' anywhere or to 'see' anything in particular. As a *dérive* would concern itself both with the experience of the walk and the wider critique of urban form, so these events are about walking but also the forms of everyday knowledge, perception and experience – the sense of the ordinary, in other words – that are normally taken for granted. In the events instigated, or co-instigated, by myself, I have been concerned to generate a sociable and collaborative form of walking and mapping. Personal and individual responses are possible but are created in social forms, through conversation and shared decisions about how to proceed. This collaborative mode is for me inspired by the collective psychogeographical *dérives* of mid-twentieth century Paris, but also figures in current accounts of anthropological practice that aim to critique and go beyond the authoritative figure of the lone ethnographer.[10]

I think of them, in fact, as re-enactments of anthropology as psychogeography, and vice versa. Where they have had a pedagogical purpose, it is to explore and communicate anthropology as a means of becoming attentive, rather than knowledge that needs to be learned. Instead of being introduced by way of a lecture, anthropology is seen straight away as a practice – not simply a methodology of ethnography to be bolted on to a separate research question, but an inquiry underpinned by ways of thinking discussed in this chapter. The idea of psychogeography lends a specifically pedestrian creativity to proceedings and enables a sociable collaboration. This creativity, however, is neither the Kantian spark of genius, nor is easily reduced by a Marxist critique to a commodity that reveals the class or identity of the consumer. It is instead improvized and attentive, yet still political in the unpacking and the questioning of the ordinary.

49 Lassiter, *The Chicago Guide.*

Another artist whom I work with in rural Scotland, Alec Finlay, describes his own practice as 'place-aware walking'. Making a virtue of a muscular condition that makes it hard for him to walk long distances, he explores local ecologies and place names – not aiming to 'bag' or conquer the summit of a hill. 'Rather than reeling in distant points of interest, I have had to accept seeing and naming them from where I am as ways to align places, and myself to them' he writes, in a poetic mapping of place names and landscapes.[50] In a world of almost constant distraction and continuing desire for acceleration and speed, the ability to move slowly and attentively needs to be valued too. What, then, of the footsteps with which this chapter began? My response now is to seek the enactment of each in the other: the walk enacts the footstep and the footstep enacts the walk. By this logic, the copy, or the re-enactment of the step and walk is not a mechanical repetition but an attentive, place-aware gesture. The 'act' of walking that might be amenable to an objective description disperses itself into a means to enact and re-enact ordinary life.

## Bibliography

Augoyard, Jean-François, *Step By Step. Everyday Walks in a French Urban Housing Project* (Minneapolis: University of Minnesota Press, 2007).

Benjamin, Walter, *The Arcades Project*, trans. by Howard Eiland and Kevin McLaughlin (Cambridge, MA/London: Harvard University Press, 1999).

Bonnett, Alastair, 'Walking through Memory. Critical Nostalgia and the City', in *Walking Inside Out. Contemporary British Psychogeography*, edited by Tina Richardson (New York: Rowman and Littlefield, 2015), pp. 75-87.

Bourdieu, Pierre, *Distinction. A Social Critique of the Judgement of Taste* (Cambridge, MA: Harvard University Press, 1984).

Cant, Alanna, '"Making" Labour in Mexican Artisanal Workshops', *Journal of the Royal Anthropological Institute*, 24 (2018), pp. 61-74.

Certeau, Michel de, *The Practice of Everyday Life* (Berkeley, CA: University of California Press, 1984).

Chtcheglov, Ivan, 'Formulary for a New Urbanism', in *Situationist International Anthology*, ed. by Ken Knabb (Berkeley: Bureau of Public Secrets, 2007), pp. 1-7.

Debord, Guy, 'Definitions' in *Situationist International Anthology*, ed. by Ken Knabb, (Berkeley: Bureau of Public Secrets, 2007a), pp. 51-52.

50 Finlay, *Gathering*.

Debord, Guy, 'Theory of the Derive' in *Situationist International Anthology* ed. by Ken Knabb (Berkeley: Bureau of Public Secrets, 2007b), pp. 62-67.

Emerson, Robert M., Rachel I. Fretz, and Linda L. Shaw, *Writing Ethnographic Fieldnotes* (Chicago: University of Chicago Press, 1995).

Finlay, Alec, *Gathering. A Place Aware Guide To The Cairngorms* (Zurich: Hauser & Wirth Publishers, 2018).

Florida, Richard, *The Rise of the Creative Class* (New York: Basic Books, 2002).

Ford, Simon, *The Situationist International: A User's Guide* (London: Black Dog Publishing, 2005).

Fulton, Hamish, *Mountain Time, Human Time* (New York: Charta Books, 2010).

Gell, Alfred, *Art and Agency* (Oxford: Oxford University Press, 1998).

Grasseni, Cristina, 'Skilled Landscapes: Mapping Practices of Locality', *Environment and Planning D: Society and Space*, 22 (2004), pp. 699-717.

Grijzenhout, Frans, *Vermeer's The Little Street. A View of the Penspoort in Delft* (Amsterdam: Rijksmuseum, 2016).

Harrison, Simon, 'Forgetful and Memorious Landscapes', *Social Anthropology*, 12 (2004), pp. 135-151.

Highmore, Ben, *Ordinary Lives. Studies in the Everyday* (London: Routledge, 2010).

Ingold, Tim, 'Culture on the Ground: the World Perceived through the Feet', *Journal of Material Culture*, 9 (2004), pp. 315-340.

Ingold, Tim, 'Materials against Materiality', *Archaeological Dialogues* 14 (2007a), pp. 1-16.

Ingold, Tim, 'Modes of Creativity in Life and Art. Introduction', in *Creativity and Cultural Improvisation*, ed. by Elizabeth Hallam and Tim Ingold (Oxford: Berg, 2007b), pp. 45-54.

Ingold, Tim and Elizabeth Hallam, 'Creativity and Cultural Improvisation: an Introduction', in *Creativity and Cultural Improvisation*, ed. by Elizabeth Hallam and Tim Ingold (Oxford: Berg, 2007), pp. 1-24.

Jackson, Jean, 'I am a Fieldnote: Fieldnotes as a Symbol of Professional Identity', in *Fieldnotes. The Makings of Anthropology*, ed. by Roger Sanjek (Ithaca: Cornell University Press, 1990), pp. 3-33.

Kant, Immanuel, *The Critique of Judgement*, trans. by James Creed Meredith (Oxford: Clarendon Press, 1952).

Knabb, Ken, *Situationist International Anthology* (Berkeley: Bureau of Public Secrets, 2007).

Lassiter, Luke, *The Chicago Guide to Collaborative Ethnography* (Chicago: University of Chicago Press, 2005).

Lee, Jo and Tim Ingold, 'Fieldwork on Foot: Perceiving, Routing, Socialising', in *Locating the Field: Space, Place and Context in Anthropology*, ed. by Simon Coleman and Peter Collins (Oxford: Berg, 2006), pp. 67-86.

Mauss, Marcel, *The Gift. Forms and Functions of Exchange in Archaic Societies* (London: Routledge, 1970).
Morphy, Howard, 'Art as a Mode of Action: Some Problems with Gell's Art and Agency', *Journal of Material Culture*, 12 (2009), pp. 5-27.
Morrison-Bell, Cynthia, Alistair Robinson and Mike Collier, *Walk On: From Richard Long to Janet Cardiff – 40 Years of Art Walking* (Sunderland: University of Sunderland, 2013).
Myers, Misha, 'Walking again Lively: towards an Ambulant and Conversive Methodology of Performance and Research', *Mobilities*, 6 (2011), pp. 183-201.
Rampley, Matthew, 'Creativity', *The British Journal of Aesthetics*, 38 (1998), pp. 265-278.
Rampley, Matthew, 'Art History and Cultural Difference: Alfred Gell's Anthropology of Art', *Art History*, 28 (2005), pp. 524-551.
Richardson, Tina, *Walking Inside Out. Contemporary British Psychogeography* (London: Rowman and Littlefield, 2015).
Sansi, Roger, *Art, Anthropology and the Gift* (London: Bloomsbury, 2015).
Sanjek, Roger, *Fieldnotes. The Making of Anthropology* (Ithaca: Cornell University Press, 1990).
Sanjek, Roger and Susan W. Tratner, *Fieldnotes: The Makings of Anthropology in the Digital World* (Philadelphia: University of Pennsylvania Press, 2015).
Sassen, Brigitte, 'Artistic Genius and the Question of Creativity', in *Kant's Critique of the Power of Judgement. Critical Essays*, ed. by Paul Guyer (Lanham: Rowman and Littlefield, 2003), pp. 171-180.
Scottish Arts Council (2005) *Artistic Evaluation – Visual Arts*. Unpublished report. Checked on October 2nd 2018: http://www.scottisharts.org.uk/resources/publications/Miscellaneous/Strategic%20Review%20papers/CFOs/M%20%20R/Peacock/2005%20Peacock%20Gangrel%20Bodies%20Jim%20Colquhoun.pdf.
Sinclair, Iain, *London Orbital. A Walk Around the M25* (London: Penguin, 2002).
Taussig, Michael, *The Nervous System* (London: Routledge, 1992).
Vergunst, Jo, 'Taking a Trip and Taking Care in Everyday Life', in *Ways of Walking: Ethnography and Practice on Foot*, ed. by Tim Ingold and Jo Vergunst (Aldershot: Ashgate, 2008), pp. 105-121.
Vergunst, Jo, 'Technology and Technique in a Useful Ethnography of Movement', *Mobilities*, 6 (2011), pp. 203-219.
Vergunst, Jo, 'Key Figure of Mobility: the Pedestrian', *Social Anthropology*, 25 (2017), pp. 13-27.
Wilf, Eitan, 'Semiotic Dimensions of Creativity', *Annual Review of Anthropology*, 43 (2014), pp. 397-412.

## About the Authors

**Jo Vergunst** is an anthropologist at the University of Aberdeen. His research interests are environmental anthropology, landscape and phenomenology, ethnographic research methods, walking, European rurality, farming and environmentalism, cultural heritage and art. He has conducted pioneering work on walking as a methodology.

# 9. Recreating Reconstructions: Archaeology, Architecture and 3D Technologies[1]

*Patricia S. Lulof*

**Abstract**
The application of 3D techniques and digital methodologies in visualization is by no means new for Archaeology, but relatively new to other disciplines, like heritage studies, museology, conservation and restoration, and (art)-history. This chapter focuses on the understanding of ancient built structures, taking the famous courtyard building (Zone F) from the 6th century BC in Aquarossa, Italy, as a starting point. When the path that leads to the final (3D) reconstruction of material objects, however small, monumental or intangible, is thoroughly documented, it generates a vast amount of new data otherwise never encountered. It is these data that we want to use to understand fully the concept of reconstructing ancient and historical communities.

**Keywords:** architecture, Acquarossa, technology, 3D modelling, reconstruction

1 I thank Jill Hilditch, author in this volume, for sharing her thoughts and ideas with me. We refer to each other's chapters to reinforce an archaeological understanding of agency and technology – we're using the same theoretical principles to perform our own investigations. The 3D model of the Monumental complex of Zone F has been executed in close collaboration with Mikko Kriek, 3D modeller of the Amsterdam Center of Ancient Studies and Archaeology. The Acquarossa Memory Project was funded in 2017-2018 by the Netherlands Organization for Research (KIEM-NWO): Living in the Past. Reconstructing Etruscan Houses in Acquarossa (file 314-98-103). I also thank the 4D Research Lab and Jitte Waagen in particular, for the drone applications.

Dupré, S., A. Harris, J. Kursell, P. Lulof, M. Stols-Witlox (eds.), *Reconstruction, Replication and Re-enactment in the Humanities and Social Sciences.* Taylor & Francis Group
DOI 10.5117/9789463728003_CH09

## Introduction

Reconstructing the past from material remains and traces of architecture, unearthed by archaeological excavations, has been the main focus of archaeology ever since the first explorers discovered the built remains of ancient worlds. Using the term 'reconstruction' is quite difficult without confronting the relationship between hard data and uncertainties. However, reconstruction remains part and parcel of the main task of archaeologists to explain the past. The application of 3D techniques and digital methodologies in visualization (modelling, scanning and photogrammetry) is by no means new for archaeology, but relatively new to other disciplines, like heritage studies, museology, conservation and restoration, and (art)-history. The many perspectives on the reconstruction of the spatial context, and the possibility of visualizing the chronological phases through time (the '4D' element), makes visual computing an innovative research tool for specialists in processes of construction and production. Reconstructions in 3D offer a virtual world where various kinds of experiment can be conducted by scientists from the humanities and beyond. Consequently, we have to reflect on the methods of digital research and the ways its results can be presented, including uncertainties, as well as empirically annotated. It is fundamental that reconstructions should be accessible, changeable, and correctible at any time. This is the paradigmatic difference between material models and historical models: 3D models can be changed any time. When the path that leads to the final reconstruction of material objects, however small, monumental or intangible, is thoroughly documented, it generates a vast amount of new data otherwise never encountered. It is these data that we want to use to understand fully the concept of reconstructing ancient and historical communities and the space they occupy.[2]

This chapter focuses on the understanding of ancient built structures, regarding them as complex material objects. It contributes to this volume as an analysis of the inner workings of a technological reconstruction, by following the steps of construction in the past. The theoretical ambition of this chapter is to (re-)define concepts of reconstruction and re-enactment of building procedures and to test their validity by bringing into play a specific and rich set of data, available in the stone foundations and terracotta roofs that made Archaic architecture in Italy. I would like to emphasize the great potential of the combination of comparative analysis of both ground

2 Frischer 'Digital Illustration'; Earl, 'Modelling in Archaeology'; Hermon and Nikodem '4D Modelling'.

plans and roofs of Archaic religious architecture, which effectuates the reconstruction of the entire construction of the building in such detail that it identifies originally applied building modules and building sequence. This is demonstrated by analysing the architecture and roofs of a concrete case study of one of the most important but also fiercely disputed Archaic built structures in Central Italy: the latest phase of the Etruscan Monumental Public Building in Acquarossa (Zone F), dated to the middle of the sixth century BC.[3] In light of new investigations at the site with the help of digital methodologies we unravel a system of interactions between terracotta craftsmen and architects, working together in this enigmatic period at the dawn of the Archaic period.[4]

## Methodology and aim

The best approach to exploring the socio-economic aspect of ancient architecture while performing empirical analysis of its archaeological remains, is to focus on technologies and building methods. Hence, we propose a methodology to study technology that combines *chaîne opératoire* – the framework that allows analysing the steps that unfold during the technological process of constructing a building within a social environment – and conventional (empirical) research methods and innovative research tools, such as 4D modelling. The 3D tools and theoretical concepts act together to illuminate the construction process of Archaic architecture. When used for reconstructing practice, craft and construction, 4D modelling is, today, an important research tool, as it is impossible to visualize processes through time (the 4D element) without digital tools. This approach compels the combination of otherwise separately investigated fields, such as architecture and terracotta craftsmanship, and requires the archaeologist to move beyond traditional disciplinary boundaries and integrate different datasets, such as

3 On the Acquarossa Memory project: Lulof and Sepers, 'On the Acquarossa Memory Project'; Miller, *Continuity and Change*; pp. 37-40, 116-117, 171-174; Strandberg Olofsson, *Acquarossa V*, and 'On the Reconstruction,' extensively on Zone F, Wikander and Wikander, 'The Early Monumental Complex', and Wintor, *Symbols of Wealth*, pp. 229-231, 235-237, 539-543. An overview is published by Potts, *Religious Architecture*, pp. 31-50. The importance of roofs for reconstructing the often unknown architecture is underlined in the same publication. pp. 51-61.

4 *Deliciae Fictiles* V Conference in Naples, 15-18 March 2018, aimed to discuss the network between patron elites and specialized craft communities that were responsible for the sophisticated terracotta decoration of architectural complexes in Italy between 600 and 100 BC. The Proceedings of the conference are published as *Deliciae Fictiles* V *Networks and Workshops* in 2019 by Oxbow Books.

the analysis of ground plans and roofs. In this way, digital modelling guides the interpretative processes of reconstructing architecture, generating a vast amount of new data that otherwise would remain unnoticed.[5]

Transformations of architecture – being constructed, destroyed, rebuilt or repaired – need complex and technically demanding *chaînes opératoires*.[6] The study of technology is an essential tool for understanding the materiality of the construction sequence of Archaic roofs, while stressing the importance of human *agency* in the transfer of these technologies.[7] This last concept should seek the larger social dynamics of network interactions and the types of relationships between those in practice and in power. The application of 4D modelling during research into built environments offers new insights for reconstruction and re-enactment and a new approach for analysing data by showing the constructive choices of architects and craftspeople. The process of 4D modelling creates a versatile and unique context for abductive reasoning; what are the most likely explanations for archaeological and architectural remains, and what questions do they pose in themselves? In fact, 4D modelling requires question-generated research; questions raised only during the modelling process. The process of reconstructing unravels *chaîne opératoire* and unveils *paradata* otherwise invisible. Moreover, simulation of the construction sequence of built structures provides insight into the social organization of historical workshops and the dynamics of craft communities. This social context of domestic architecture and roofs can be understood when the physical aspect of materiality is involved, by examining the construction sequence or the *chaîne opératoire* of these monuments.[8]

## Acquarossa. The monumental building in Zone F

Acquarossa lies in the heart of Etruscan territory, close to the Lago di Bolsena, but far enough from the primal coastal centres to be regarded as a hinterland town. The town is named after the red-coloured creek surrounding the

5 On the value of 3D modelling in reconstructing architecture: Barceló, 'Visual Analysis'; Hermon, 'Scientific Method'; Lulof et. al, 'The Art of Reconstruction'; Earl, 'Modelling in Archaeology'; Opgenhaffen and Sepers, '3D Modelling'; Lulof, 'Networks and Workshops' and Lulof, 'Connecting Foundations'.

6 Lounsbury, 'Architecture'; Audouze, 'Leroi-Gourhan' and Dobres, 'Technology's Links'. See, for the full explanation of *chaîne opèratoire*, the chapter by Jill Hilditch in this volume.

7 Roux, 'Technological Innovation'; Lulof, 'Networks and Workshops'; Lulof and Sepers, 'The Acquarossa Memory Project'.

8 Frischer, 'Digital illustration'; Hermon, 'Scientific Method'; Lulof, 'Connecting Foundations'.

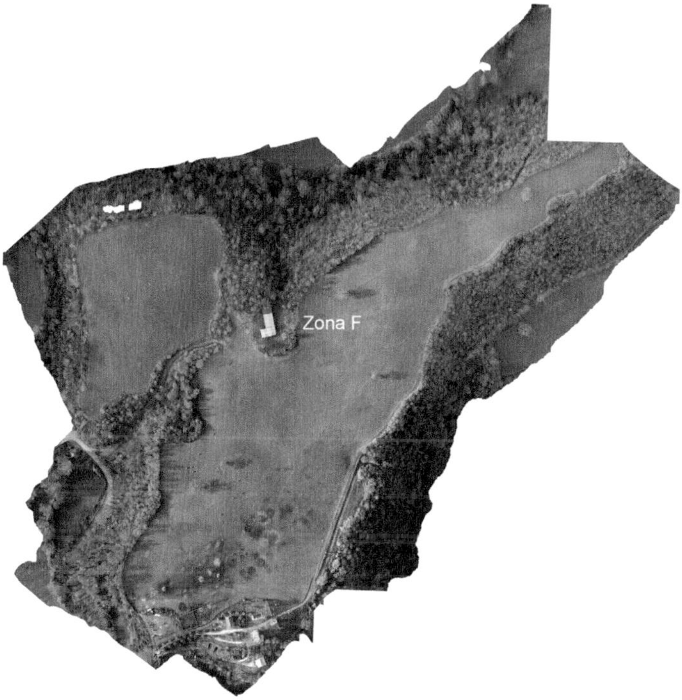


Figure 9.1: Drone view of Acquarossa, photogrammetric 3D reconstruction of site.
Photo: Jitte Waagen, 4D Research Lab, University of Amsterdam, used with permission.

site, which is situated on a tufa-plateau (Figure 9.1). Excavations carried out by the Swedish Institute in Rome (1966-1978) revealed a large series of Etruscan houses and public buildings, inhabited from the late eighth century BC until shortly after the middle of the sixth century BC, when the town was suddenly abandoned, most probably because of an earthquake.[9] The houses were left to crumble at the time and the remains of the foundations, the walls and the decorated roofs, as well as the thousands of household utensils, were all found *in situ*. It is one of the very few examples of an intact Etruscan townscape, with a unique set of family dwellings from the past. The excavations themselves were never fully published, however the

9 Until now, no explanation has been given for the sudden departure of the inhabitants; the suggestion of an earthquake comes from Prof. Örjan Wikander (personal information). Already suggested in Wikander, *Acquarossa VI:2*: 'la sua fine potrebbe risalire al 550/525 a.C. circa. Poiché mancano tracce di devastazioni belliche, si può sospettare che Acquarossa sia stata distrutta da un terremoto'.

architectural terracottas, and some specialized categories received ample attention and several exhibitions were organized. A fine presentation of the site is visible in the National Etruscan museum in Viterbo (Museo Rocca Albornoz) which shows the archaeological finds from Acquarossa.[10]

Since 2015, the 4D Research Lab and my research group has been involved in the Acquarossa Memory Project, aiming at knowledge exchange and multidisciplinary archaeological research, using digital methods as (in-)tangible heritage.[11] We will use connected data (applications) and techniques as a new intermediary in the final dissemination of the products. Re-utilization and exploitation of (archaeological) heritage is the chief goal of the project.[12] We discovered that some important clusters of foundations had been lying bare since the 1970s and were badly damaged.[13] On the basis of the 3D models, we produce we will start reconstructing real houses, for tourist purposes. We have started with domestic and public architecture to see what difficulties we will encounter and to find solutions for them. After that we will build the other houses, and they can have different roof systems, as well as differences in superstructure, to present mudbrick walls as well as *pisé* (rammed earth) walls, to mark the different construction methods. The primary demand of the architects is focused on accurate and safe construction of the built environment of the selected replica houses.

We have also focused on the reconstruction of Zone F (in the NE of the plateau, close to Pian del Sale), where a relatively complete monumental building was excavated by the Swedish team in the early 1970s. It was the first building of this size and monumentality that attracted a lot of public and scientific attention. It consisted of a then unknown type of courtyard building with one wing with columns, a kind of portico, with a series of rooms around a court. The architectural terracottas were found at the very point where the roof collapsed, in situ. The roof was practically intact and meticulously reconstructed and is now visible in the museum in Viterbo (Figure 9.2).[14]

10 The best overview of the site and its history of excavations, Wikander, *Acquarossa VI:2* and the bibliography by Wikander and Wikander in Wendt et. al, *Acquarossa VII*.

11 On the Acquarossa Memory Project: Lulof and Sepers, 'The Acquarossa Memory Project', with bibliography. The final model of Zone B has been thoroughly analysed and prepared for an annotated model, which will serve the architect in building the holiday homes.

12 The project focuses on the role of current trends in (virtual) heritage management in the scientific experience of the research group in several aspects of digital humanities and archaeology.

13 We intend to clean the site in 2019 and make high-resolution 3D scans as well as drone photogrammetry documentation. After which we will cover the remains with protective textiles, which will not allow weeds and foliage to grow into the tufa blocks.

14 The monumental building, its architecture and the roof elements have been published on several occasions, the main starting point in *Acquarossa V*, by Strandberg Olofsson. The latest

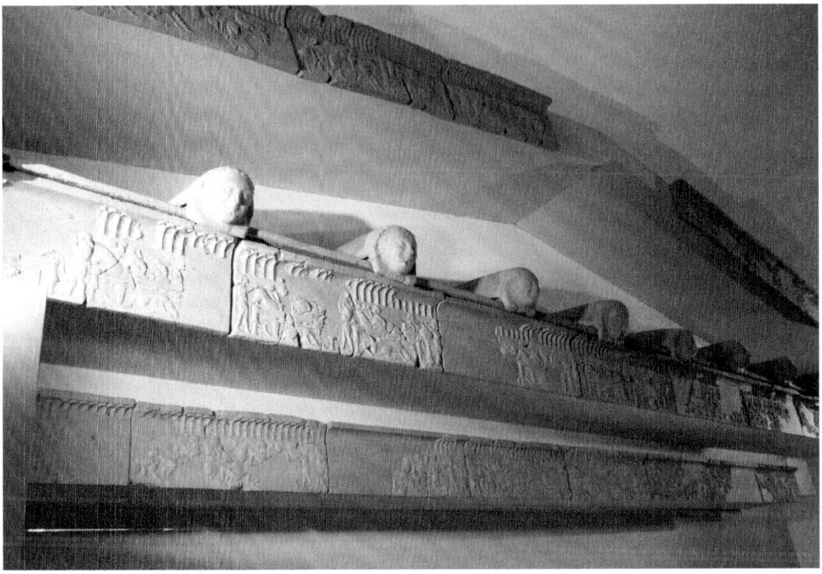

Figure 9.2: Architectural Terracottas from Zone F, Museo Etrusco Rocca Albornoz Viterbo. Photo: author.

The building consisted of several building phases and certainly had an obvious public and central function, also perhaps a sacral function, as a sacral pit was discovered, belonging to the earliest phase. The earliest phase also shows a courtyard building, with one wing ascertained and a shed- and saddle roof and an exquisite decorative roof system; it was dated to the beginning of the sixth century BC.[15] The later phase consisted, according to the excavators and later specialists, of two separate buildings that were built one after another in a short period (560-550 BC). The *portico* building A and *portico* building C, were set approximately at right angles to each other facing an open courtyard (Figure 9.3a). Both buildings consisted of rooms (two and four, respectively) fronted by a colonnade with wooden columns set at different spaces. The columns were of wood with delicately moulded bases and Etrusco-Doric capitals in *peperino* stone, two of them found *in situ*. Foundations of other spaces and rooms have been found in the complex, indicated by the excavators as separate buildings and rooms, but

comments are by Winter, *Symbols of Wealth* (Roof 4-5 and 4-9) and Chapter 9; see also Miller, *Continuity and Change*, pp. 175-179.

15 Wikander, *Acquarossa I:1* and Wikander and Wikander, 'The Early Monumental Complex'; with a completely different decorative roof system: Winter, *Symbols of Wealth*, Roof 2-15; the final publication of the early phase of Zone F will appear shortly by Ö. Wikander.

not forming part of buildings A and C. The tiled roofs of both buildings were decorated with four different types of revetment plaques showing mythical scenes as well as scenes of dancing and dining, and antefixes with female heads. The decoration was exclusively set on the façade, facing the courtyard. The buildings were destroyed shortly after 550 BC. The construction of the building complex was different from the other structures at the site, hence indicating a possible special function. Foundations were cut into the bedrock and in one case, blocks were used to build the wall (south), instead of using the *pisé* or mudbrick materials that were common at this site and elsewhere. The floors of the rooms were set higher than those of the *portico* and the slope was strongly descending towards the north, with a thick terrace wall to support the terrain. The court had pits and drains to receive superfluous rainwater and keep the courtyard dry.[16]

The first reconstructions of the later phase of Zone F were published by Carl Östenberg in 1970, and repeatedly discussed by Margareta Strandberg Olofsson, the main excavator, especially in her article in 1989, with her latest reconstruction and a full discussion because of critical attacks from Italian colleagues. In short, several options were presented and illustrated with 2D graphic reconstructions: Östenberg reconstructed the courtyard building with two wings, attached to each other and including the rooms at the corner where the two wings of the building meet, as well as the attached House D in the west (Figure 9.3b).[17] Strandberg reconstructed the east wing (building A) as a separate temple with three cellae and a gabled roof with columns in front, and a separate long stoa-like *portico* (building C) (Figure 9.3c). She based this reconstruction on the find spots of the architectural terracottas in front of the built structures.[18] The exhibited materials and models in the Museum in Viterbo are displayed according to Strandberg's reconstruction, notwithstanding the fact that this reconstruction has been fiercely debated in the literature, right from the beginning, and the reconstructions have never been adjusted (Figure 9.2).

16 Winter, *Symbols of Wealth*, Roofs 4-5 and 4-9, with previous literature and notes on the differences in chronology, pp. 230-231, 235-236.

17 Known to have been part of the earlier phase, this structure has also been regarded as 'still standing and probably in use' in the later phase, Wikander and Wikander, 'The Early Monumental Complex', pp. 197-198.

18 Östenberg, *Case etrusche*; Strandberg Olofsson, *Acquarossa V*; Torelli, 'Review of Acquarossa', carefully deconstructs the arguments in favour of the reconstruction as a gabled roof and temple, carefully explaining not only that a temple is ridiculous, but suggesting rethinking the structure in the light of the finds in Murlo and Satricum where similar courtyard buildings were discovered. In 1989 Strandberg Olofsson, 'On the Reconstruction' published a reaction to the critical notes and re-evaluated her reconstruction without changing it.

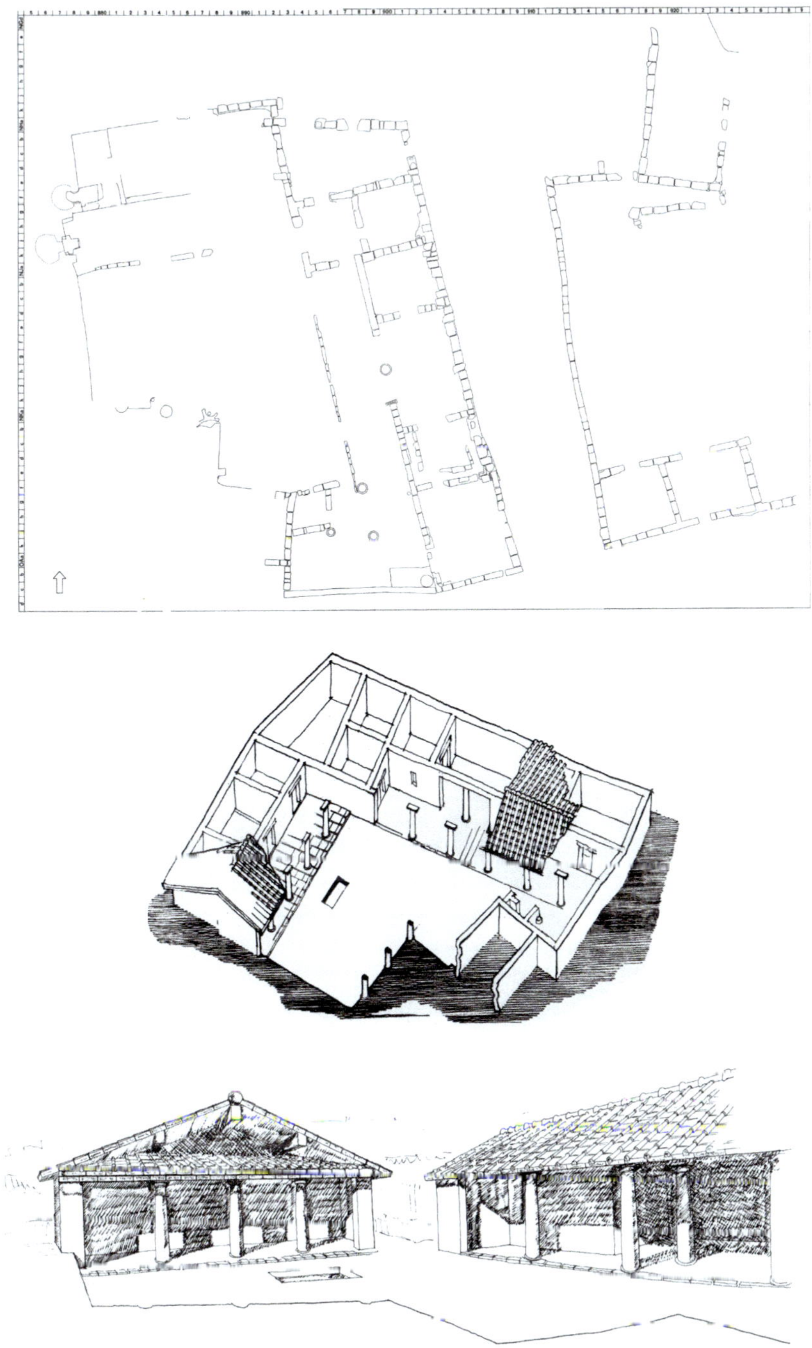

Figure 9.3 (a-c): Plans and 2D reconstructions of monumental building Zone F and adjacent buildings. (1. Re-elaborations by Maarten Sepers and Mikko Kriek after (1) Örjan Wikander, after *Acquarossa VI*, fig. 35; (2) Carl Östenberg, after 'Case Etrusche', fig. on p. 165; (3) Margareta Strandberg Olofsson, after 'On the Reconstruction', Fig. 3).

Still today, the same drawings and small-scale models are displayed in the museum. However, following recent research and careful publications, it is now well-known that the typical Tuscan three-cella temple appeared in Central Italy, e.g. Etruria at the same time as the second decorative roof system that was introduced perhaps in 500 BC at the earliest.[19] The recessed gable in Building A, as reconstructed by Strandberg Olofsson, can also be doubted, because this phenomenon originated in Campania and only appeared in Etruria in combination with the second decorative system, that is, much later.[20] A saddle-roofed façade with an open pediment and recessed gable seems very unlikely in Zone F of Acquarossa, especially since columen- and mutulus plaques (figurative relief plaques used to decorate an open pediment), raking simas (upright tile elements along the sloping roof edges), obliquely cut keystones for the apex (wedge in the roof top revetment), and acroteria (figural terracotta sculptures on the top and corners of a roof) have not been found in the collapsed *debris* of the roofs.[21]

The polemical discussion on the reconstruction of the later phase of Zone F in Acquarossa gave us the opportunity and reason to re-think and re-construct the monument from scratch, using the Swedish architects' plans and drawings, as well as the detailed reports on the stratigraphy and find circumstances of the excavators, when available.[22] We decided to make a 3D scan as well as a photogrammetric reconstruction (images set in Agisoft PhotoScan Software) of the actual remains of the site, produced by flying with an UAS (drone) under the protective roof construction that protects the foundation walls of Zone F today (Plate 9.1). With this 3D reconstruction of the actual state of the remains, following the data presented in the

19 On the second decorative system and the introduction of the Tuscan three-cella temple: Lulof, 'Reconstructing Rome's Golden', pp. 119-123, with references.

20 Winter, 'The Origin', p. 45, regards the reconstruction of the recessed gable of Strandberg Olofsson as 'anomalous' and chronologically unjustified.

21 Strandberg Olofsson, 'On the Reconstruction', in reaction to Torelli, 'Review of Acquarossa', Winter, *Symbols of Wealth*, disregards the dispute. However, she does speak out on the rareness of the gabled roof in this period in Winter, 'The Origin', p. 45. The technical terms in describing terracotta roof elements that were implemented in an Archaic roof are best explained in the figure on p. 6, Winter, *Symbols*.

22 Östenberg, *Case etrusche*; Wikander, *Acquarossa I:1*; Strandberg Olofsson, *Acquarossa V* and 'On the Reconstruction', published the basic drawings and photos of the excavations. A preliminary report on the first phase of Zone F (Wikander and Wikander, 'The Early Monumental Complex') presents a lot of information referring to the site, fundamental to the understanding of the later monumental complex. The final excavations report on the so-called second phase of Zone F has yet to be published.

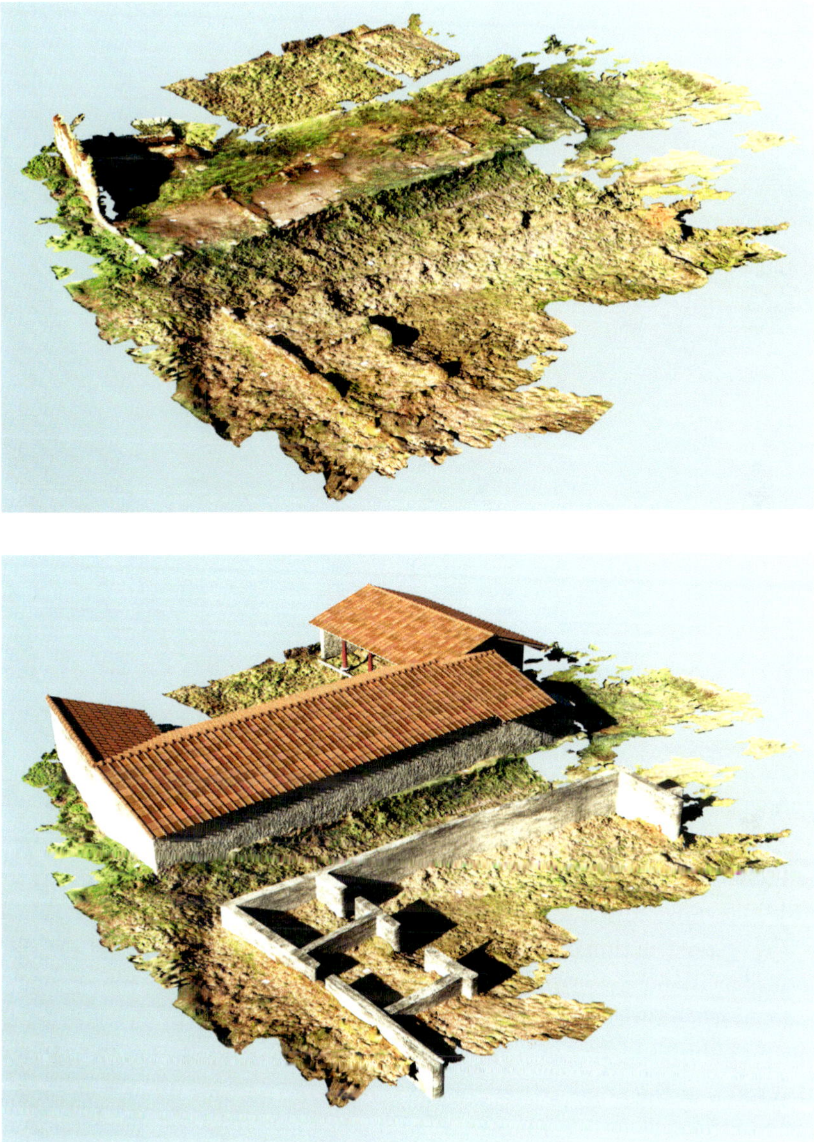

Plate 9.1 (a-b): Photogrammetrical (1) model of the site of Zone F produced with drones and (2) walls and roofs of Zone F in the first 3D model. Photo: Jitte Waagen and Mikko Kriek, 4D Research Lab, University of Amsterdam, used with permission.

publications and covered by results from recent research into Archaic architecture and newly excavated parallels, we set to work on the 3D reconstruction of this building, entering images and plans in a digital model, using 3D Studio Max software, departing from the concept that Zone F

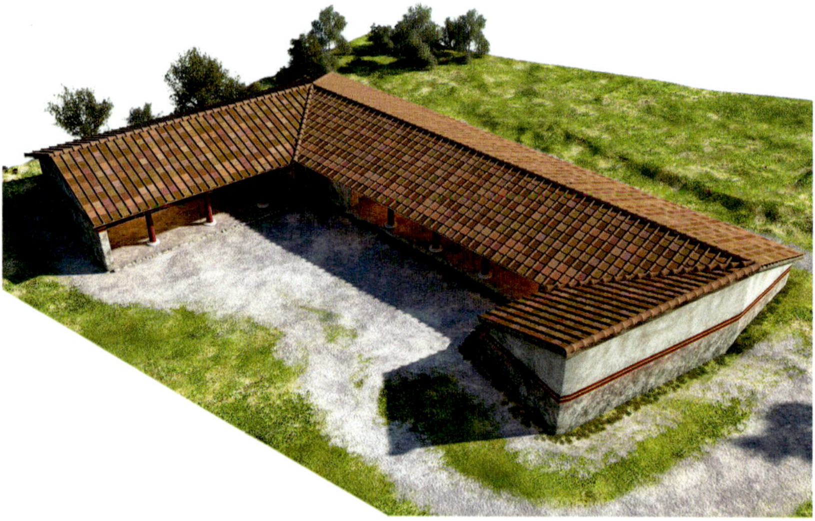

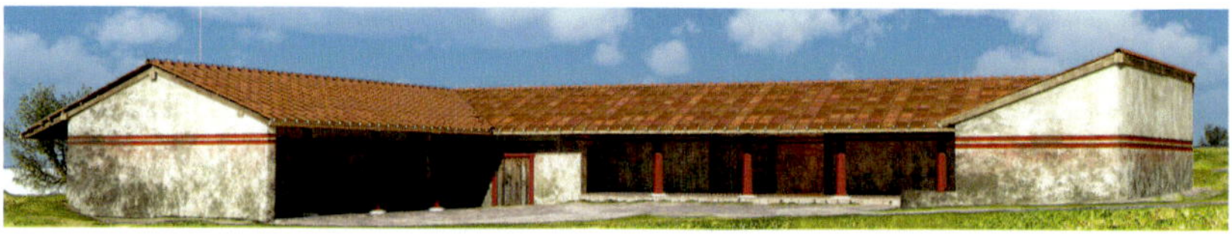

Plate 9.2 (a-b): Roof and walls of monumental courtyard building F in 3D in (1) bird-eye view and (2) from the south-west. Model: Mikko Kriek, 4D Research Lab, University of Amsterdam, used with permission.

had one building with three wings surrounding the courtyard, similar to buildings now known to have existed in the same period (Plates 9.1-3).[23]

The reconstruction of the whole area of Zone F has been included: On the basis of available data from parallel sites we have taken all built structures as a starting point, and set off to re-create a built structure as a unity: a courtyard building including all available spaces, such as House D, and the fifth room at the angle where the two *portico* structures meet.[24] Perhaps some of these structures belonged to the earlier phase, but foundations and walls could possibly have been integrated in the later monumental

23 Fieldwork was executed in May 2018. I thank Mikko Kriek and Jitte Waagen for the technical preparation of the model with the help of drones and photogrammetry models, and the ACASA Research Fund for making the project possible. Part of the 3D Model discussed and presented in this chapter was funded within the NWO-KIEM project, see note 1.

24 The drains running along the foundations just outside the row of columns meet in the angle between structure A and C, which means again that these wings were connected. Torelli, 'Review of Acquarossa' sees this as proof of 'strutture porticate ad un solo piovente'.

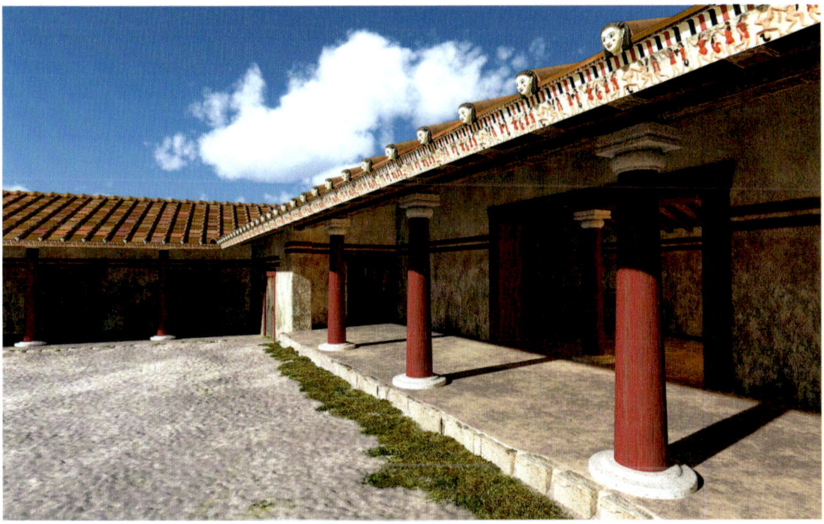

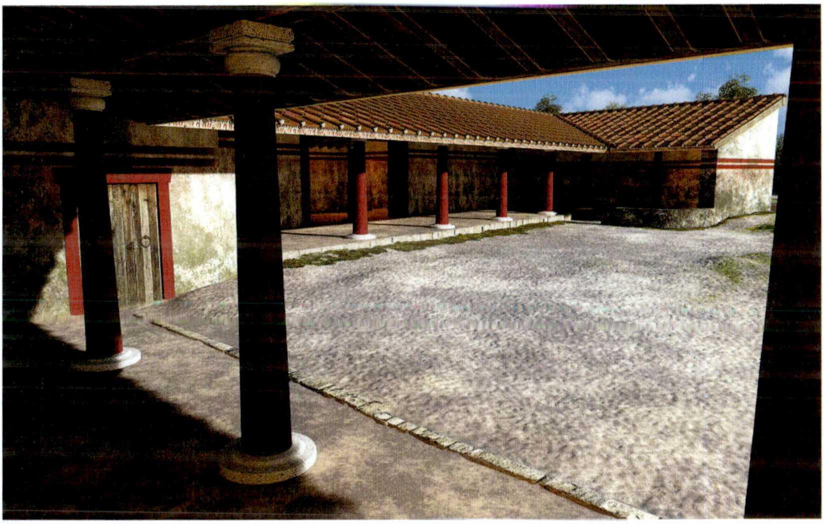

Plate 9.3 (a-b): View of (1) interior of monumental building F from the south and (2) into courtyard from the north. Photo: Mikko Kriek, 4D Research Lab, University of Amsterdam, used with permission.

complex, as was already suggested by the excavators.[25] We have also altered the idea that all columns should have the same height, as well as that all spaces should have the same height. We had to consider the strong slope of the terrain towards the north and adjusted the height of the walls as well as the columns to meet with the roof. We see that materials and elements

25 Wikander and Wikander, 'The Early Monumental Complex', p. 197.

could be adapted to serve the final product, in this case a built environment. Therefore, the columns in Building A could have been higher than in C, so that the wings with the rooms could have met at the corner, covered by the same roof. Including House D would make it possible to enclose a public space with a built structure surrounding a court. The western part, unfortunately, was completely destroyed, so it would have been equally possible to have a four-winged monument, just like contemporary Murlo, and suggested also for Satricum.[26]

We suggest to return to Östenberg's original reconstruction and regard Zone F as one building and a public place where the people of Acquarossa gathered for games, festivals and symposia, suggested by the figurative decoration of the revetment plaques (Plate 9.3b).[27] The annotated 3D reconstruction with elements of augmented reality has been finished only very recently, and was obviously based on thorough research and discussions. It can be altered and adjusted when new data reveal new information. A QR code enables the public to walk around inside Zone F.[28]

Reconstructing the roof of the 'new' monumental complex was complicated, notwithstanding the fact that the roof elements and their position have been known from the first publications onwards, that is, the tiles and imbrices of type II, the female head antefixes and the revetment plaques.[29] Terracotta roofs need a sophisticated and solid sub-construction, like the timber truss. With help and advice from modern architects we came up with constructive solutions where we were in doubt. Here archaeological research and architectural expertise are indispensable.[30] We therefore suggest saddle roofs running in east-west (wing A) and in north-south (wing C) direction with antefixes and revetment plaques facing the courtyard. The roof of structure D could have been a shed or single-sloped and intersected with the slope of the saddle roof of wing C with diagonally cut tiles as a compluviate roof. The same solution could be suggested for the connecting roofs at the N-E corner, at the front, facing the courtyard. The back of the N-E corner of the

26 On the contemporary monumental courtyard buildings discovered after Zone F: Potts, 'Religious Architecture', pp. 141-143 (general), p. 142 (Murlo with bibliography) and Van 't Lindenhout, 'Constructing Urban Landscapes', pp. 131-133 (Satricum, with bibliography).

27 Torelli, *Fictilia Tecta*, pp. 5-7.

28 The reconstruction in 3D will be projected when the visitor uses the application at the site itself, and it will be made available on the website we plan to develop for the Archeological Park of Acquarossa; http://www.2vr.in/V-1KZL (checked on March 14th, 2019).

29 See above, note 15.

30 Hodge, *The Woodwork*, pp. 17-40; see also Hopkins, *The Genesis*, pp. 55 and 104. As in tomb architecture, Naso, *Architetture dipinte*, p. 286, figure 2, 293, figures 218-219.

roof proved to be the most difficult to construct. Several options have been proposed, but no definite proof has been given by the typology of the tiles. The presence of tiles belonging to a hipped roof type could prove decisive for one option in our reconstruction (Plate 9.2a). We had already decided to include as many spatial structures as possible, and we wished to include the N-E corner B structure as well, as a kind of connecting set of rooms between wing A and C. The small space in the far north with the diagonal thick wall that was interpreted as a terrace wall to protect the structure from sliding along the lower slope in particular needs clarifying (Figure 9.3a). It seems unlikely that this large space was left uncovered.[31]

Fragments to prove compluviate or hipped roofs (two connected roofs, V-shaped or the opposite concave or convex) have not been found during excavation. However, the roof definitely had round sky-light tiles, some of which have indeed been found in excavation. These permitted light into the rooms below and perhaps made windows at the eastern side of the building superfluous.[32] The subsequent layering of the roof construction with light rafters of wood on top, which may have been covered with wattled mats of reed, provides an even surface for the rows of tiles.[33] The reconstruction of the interior of the monumental complex of Zone F cannot be certain, since no archaeological data are available that show the details of the inner rooms. Proof of entrances at several points have been adequately argued for, mostly on the basis of the plans of the foundation walls. In the final 3D model, different hypotheses are visualized as variations, leaving the actual remains and other data intact and accessible. The birds-eye view of the area shows how the latest phase of Zone F was constructed (Plate 9.2a b).

Reconstruction of roof-building is indispensable when one studies architectural terracottas. We aim to create an innovative interdisciplinary approach to source and reconstruct roof production, focusing on the distribution of moulds, as well as roof construction techniques, in order to study the later phase in the production process (the actual roof decoration), production tools (*matrix* or mould), and building techniques, and using Zone

31 The 'terrace' wall already formed part of the earliest phase of the monumental complex, but also is well-connected to the back wall of wing C with structure B, Wikander and Wikander, 'The Early Monumental Complex', Fig. 9. Solutions will be discussed in the near future, together with the Swedish specialists. I thank Prof. Wikander for sharing his thoughts with us.

32 Wikander, *Roof-Tiles*, p. 145 (compluviate and hipped roofs) and p. 195 (skylight tiles).

33 Suggested by Turfa and Steinmayer, 'The Comparative Structure', pp. 19-20, Figure 3; Sapirstein, 'How the Corinthians', pp. 313-314 suggests this method for the Protocorinthian rooftiles. See also Miller, *Continuity and Change*, pp. 198-200 and pp. 173-174 for the rejection of this system.

F as a case study, the workshops manufacturing architectural terracottas in Acquarossa between 680 and 530 BC.[34]

Focusing on the accurate, closest to a true and real, construction of the monumental building of Zone F (and its roof) at the site, we made a set of 3D scans in the National Etruscan Museum in Viterbo (Museo Rocca Albornoz). We implemented these 3D scans into the model (Figure 9.2 and Plate 9.3a-b). For the dynamic 4D model, showing the different phases of the structure in monumental Zone F, we worked with a vast data collection that is needed for a refereed virtual model. These data should become visible, available and controllable. Although some work has been done on the conceptual aspects of this process (the so-called *London Charter*), it remains fundamental in 3D modelling to create a new publication format that enables us to follow each step and argument, as well as annotate them, for reconstructing built heritage from Antiquity. The model of Zone F will be published in an annotated system that has been developed recently for the 4D Research Lab at the University of Amsterdam.[35] Finally, it remains important that 3D models should be open-access and correctible. This is exactly what makes 3D models and 3D reconstructions innovative: they can be changed and deleted at the same time.[36]

## Conclusions

Investigating craft communities, building technologies and processes has never been thoroughly undertaken for the Archaic period and the region of Central Italy, or for this exceptionally rich category of materials, or for the craftspeople producing architectural terracottas. Reconstructing ancient architecture and re-enacting building processes with 4D modelling as a research tool, has proven elucidating. The socio-economic dynamics of craft production in sacred areas and networks between sanctuaries have

34 Acquarossa is an exemplary site to study the concept of workshops and their impact. C. Wikander, *Acquarossa I:1*, *Acquarossa I.2* and Ö. Wikander, *Acquarossa VI:1*, Winter, *Symbols of Wealth*, chapter 8, esp. pp. 513-514: Around 560 BC., there is a major shift in practice, where only public buildings receive decoration and domestic structures carry undecorated or modestly decorated roofs.

35 The 4D Research Lab of the University of Amsterdam already has an initial system in place for storing built heritage information digitally (Noordegraaf et. al, 'Cinema Parisien 3D'). An interactive website has been activated with access to other media. The models will be completely accessible and dealt with as a scientific publication, as is argued for in a wider context, Ryan, 'Documenting and validating', Huvila, 'The Unbearable Complexity'.

36 In this project, dynamic digital reconstructions are to be preferred above static replications, like the exhibited replication of the façades of Zone F in the Etruscan Museum in Viterbo.

only recently been addressed in a wider context; connections have been made with the phenomenon of the urbanization of the region.[37] With the exception of a project on the petrography of architectural terracottas from the earliest phase, technical, science-based research and computer-based methods, like petrography, x-ray imaging, and 3D GIS or 4D modelling, have not yet been used for Archaic architecture and this material group to address the reconstruction of production. 4D modelling as a research tool enables us to re-enact or perform the building process, thus encountering construction issues and production problems, solving them or discussing them while entering the actual production chain. It reveals *chaîne opératoire* while processing *paradata* and emphasizes new data and intense research questions about what can be reconstructed and what is actually lost and can never be retrieved. Moreover, analysis of the *chaîne opératoire* may reveal the different social groups involved in the building process, completion, maintenance, destruction and rebuilding of these objects according to the developments in construction technology. Reconstructing the roof with all its elements within a 3D model, is no less than the actual production of a replication, which forces us, during the process, to look at details of technical adaptions and rethink the actions of the craftsmen at work. This combination of reconstruction, re-enactment and replication creates a synthesis that is needed to understand and clarify construction methods and architecture in the period under study in this chapter, to reveal actions of specialized craftspeople and, when compared with other groups, eventually shed light on their influence on society. This interdisciplinary approach combined innovative methodologies with traditional research tools, creating a unique opportunity that will result in insights into chronologies, production, and consumption of Ancient architecture and its terracotta roofs between circa 700 and 550 BC. The archaeological research group that operates in the field of re-enactment, replication and reconstruction has set its aim on the preparation of data for the reconstruction, as a research experiment which transforms traditional data (ranging from archaeologically excavated foundations and numerous (terracotta) roof decorations), drawings and previous reconstructions, and has built advanced digital 3D models, taking this case study as a starting point. The digital reconstructions have been used as a research tool, the usage of which necessitates scholars to test hypotheses in a much more detailed way (i.e. architectural and structural in this case-study) than with research based on descriptions or two-dimensional

37 See the recently published volume P.S. Lulof et al. *Deliciae Fictiles V, Networks and Workshops in Italy*.

reconstructions only. It has been, in fact a true Experimental Archaeology project, using 3D software programs and computers to literally re-build a vanished building, block after block. We closely collaborate with modern architects and engineers, both in the past and at this very moment. Stone foundations, columns of wood and stone, stone entablature and wooden roof structures, terracotta roof decoration, all materials and their function in the building structures, have been studied in detail in order to argue for a specific reconstruction. Finally, it has been very important to be able to show hypotheses and assumptions in the model that was created. We have developed a specific database in which the 3D models can be stored together and connected with their data. In this we introduce a new publication format for digital models, complete with variations and annotations. The results of the architectural investigation into data complexes that once belonged to the well-known Monumental Building in Acquarossa and to Archaic structures in Pre-Roman Italy in general, have been processed to publish dynamic 3D models that can be discussed, peered at and referred to as an archaeological publication, with annotations and levels of uncertainties. It is stressed that in performing actions to create models, to visualize choices and decisions in this built space, a wide variety of new data and knowledge has been produced that would otherwise have gone unnoticed.

We hope that in this chapter we have shown that it is possible to encounter the impossible: return to the past and relive the experience of ancient building technology and learn from it by understanding the choices and processes. From today on, studying ancient architecture cannot be done without reconstructing the lost built environment, recreating and replicating the technology of architecture step by step, re-enacting and visualizing the process in lucid and transparent digital models and finally, understanding the magnificent monuments and their architectural and environmental context.

## Bibliography

Audouze, Françoise, 'Leroi-Gourhan, a Philosopher of Technique and Evolution', *JArchaeolRes*, 10 (2002), pp. 277-306.

Barceló, Juan A., 'Visual Analysis in Archaeology. An Artificial Intelligence Approach', in *Morphometrics for Nonmorphometricians*, ed. by Ashraf M.T. Elewa (Berlin/Heidelberg: Springer Verlag, 2010), pp. 93-156.

Dobres, Marcia-Anne, 'Technology's Links and Chaînes, The Processual Unfolding of Technique and Technician', in *Contemporary Archaeology in Theory, The New Pragmatism*, ed. by Robert W. Preucel and Stephen A. Mrozowski, (Malden/ Oxford: Wiley-Blackwell, 2010), pp. 124-146.

Earl, Graeme, 'Modelling in Archaeology, Computer Graphic and other Digital Pasts', *Perspectives on Science*, 21 (2013), pp. 226-244.

Edlund-Berry, Ingrid, 'The Language of Etrusco Italic Architecture: New Perspectives on Etruscan Temples', *AJA*, 112 (2008), pp. 441-447.

Frischer, Bernard, 'From Digital Illustration to Digital Heuristics', in *Beyond Illustration, 2D and 4D Digital Technologies as Tools for Discovery in Archaeology*, ed. by Bernard Frischer and Anastasia Dakouri-Hild (Oxford: BAR International Series 1805, 2008), pp. v-xxii.

Hermon, Sorin and Joanna Nikodem, '4D Modelling as a Scientific Research Tool in Archaeology', in *Layers of Perception. Proceedings of the 35th International Conference on Computer Applications and Quantitative Methods in Archaeology* (CAA, Berlin, Germany, April 2-6, 2007), ed. by Axel Posluschny et al. (Bonn: Dr. Rudolf Habelt, 2008), pp. 140-146.

Hermon, Sorin, 'Scientific Method, *chaîne opératoire* and Visualization: 3D Modelling as a Research Tool in Archaeology', in *Paradata and Transparency in Virtual Heritage*, ed. by Anna Bentkowska-Kafel and Hugh Denard (Farnham: Ashgate, 2012), pp. 13-22.

Hodge, A. Trevor, *The Woodwork of Greek Roofs*, (Cambridge and New York: Cambridge University Press, 1960).

Hopkins, John N., *The Genesis of Roman Architecture* (New Haven: Yale University Press, 2016).

Huvila, Isto, 'The Unbearable Complexity of Documenting Intellectual Processes: Paradata and Virtual Cultural Heritage Visualisation', *Human IT*, 12 (2012), pp. 97-110.

Lindenhout, Elizabeth van 't, 'Constructing Urban Landscapes in Latium Vetus in the Archaic Period', in *Current Approaches to the Archaeology of first Millennium B.C. Italian Urbanism*, ed. by Elizabeth C. Robinson (Portsmouth Rhode Island: *JRA Supplements* 97, 2014), pp. 127-144.

Lounsbury, Carl R., 'Architecture and Cultural History', in *The Oxford Handbook of Material Culture Studies*, ed. by Dan Hicks and Mary C. Beaudry (Oxford: Oxford University Press, 2010), pp. 484-501.

Lulof, Patricia S., 'The Art of Reconstruction and the Image of Power', in *Interpretando l'antico. Scritti di archeologia offerti a Maria Bonghi Jovino*, ed. by Cristina Chiaramonte Treré et al. (Milano: *Quaderni di ACME*, 134, 2012), pp. 111-130.

Lulof, Patricia S., et al., 'The Art of Reconstruction. Documenting the Process of 4D Modeling, some Preliminary Results', in *Proceedings of the 2013 Digital*

*Heritage International Congress* (*DigitalHeritage*), ed. by Alonzo C. Addison et al. (Marseille: IEEE Xplore 2013), Vol. 1, pp. 333-337.

Lulof, Patricia S., 'Reconstructing Rome's Golden Building Age. Temples from the last Tarquin to the Roman Republic (530-480 BC)', in *Current Approaches to the Archaeology of First Millennium B.C. Italian Urbanism*, ed. by Elizabeth C. Robinson (Portsmouth Rhode Island: *JRA Supplements* 97, 2014), pp. 113-126.

Lulof, Patricia S., 'Networks and Workshops. Construction of Temples at the Dawn of the Roman Republic', in *Arqueología de la Construcción V. Man-made materials, Engineering and Infrastructure* (*5th International workshop on the Archaeology of Roman Construction, Oxford 11-12 April 2015*), ed. by Janet Delaine and Stefano Camporeale (Madrid: *Anejos de AespA* 77, CSIC Press 2016), pp. 331-342.

Lulof, Patricia S., and Christopher Smith, 'The Age of Tarquinius Superbus. Ancient History and Archaeology: an Introduction', in *The Age of Tarquinius Superbus. Central Italy in the late Sixth Century BC. Proceedings of the Conference* (*November 7-9, 2013, Rome*), ed. by Patricia S. Lulof and Christopher Smith (Leuven: *BABesch* Supplements 29, 2017), pp. 3-13.

Lulof, Patricia S., and Maarten H. Sepers, 'The Acquarossa Memory Project. Reconstructing an Etruscan Town', *Giornale di Archeologia e Calcolatori*, 28 (2017), pp. 69-78.

Lulof, Patricia S., Ilaria Manzini, and Carlo Rescigno, *Deliciae Fictiles V: Networks and Workshops. Architectural Terracottas and Decorative Roof Systems in Italy and Beyond*. Proceedings of the Fifth International Conference held at the University of Campania "Luigi Vanvitelli" and the National Archaeological Museum in Naples (March 15-17, 2018). (Oxford and Philadelphia: Oxbow Books 2019).

Lulof, Patricia S., 'Connecting Foundations and Roofs. The Satricum Sacellum and the Sant'Omobono Sanctuary', in *Etrusco-Italic Architecture in its Mediterranean Setting*, ed. by Charlotte Potts (Oxford: Oxford University Press, pending).

Miller, Paul M., *Continuity and Change in Etruscan Domestic Architecture* (Oxford: Access Archaeology Archaeopress, 2017).

Naso, Alessandro, *Architetture dipinte. Decorazioni parietali non figurate nelle tombe a camera dell'Etruria meridionale* (*VII-V secolo a.C.*) (Roma: *Bibliotheca archaeologica* 18. L'Erma di Bretschneider, 1996).

Noordegraaf, Julia, et al., 'Cinema Parisien 3D: 3D Visualization as a Tool for the History of Cinemagoing', *Journal of Film and Screen Media*, 11 (2016), pp. 45-61.

Nylander, Carl, *Architettura etrusca nel Viterbese: Ricerche svedesi a San Giovenale e Acquarossa 1956-1986*: Catalogue of the Exhibition at Viterbo (Rome: De Luca,1986).

Opgenhaffen, Loes, and Maarten H. Sepers '3D Modelling: Crossing Traditional Boundaries between Different Research Areas', in *The International Archives of the Photogrammetry, Remote Sensing and Spatial Information Sciences* (2015), Vol. XL-5/W4, pp. 411-14.

Östenberg, Carl E., *Case etrusche di Acquarossa* (Rome: Multigrafica, 1975).

Potts, Charlotte R., *Religious Architecture in Latium and Etruria c. 900-500 BC* (Oxford: Oxford University Press, 2015).

Roux, Valentine, 'Technological Innovation and Developmental Trajectories, Social Factors as Evolutionary Forces', in *Innovations in Cultural Systems, Contributions from Evolutionary Anthropology*, ed. by Michael J. O'Brien and Stephen Shennan (Cambridge Mass: The MIT Press 2009), pp. 217-234.

Russel, Ben, and Elisabeth Fentress, 'Mud-Brick and *pisé de terre* between Punic and Roman', in *Arqueología de la Construcción V. Man-Made Materials, Engineering and Infrastructure (5th International Workshop on the Archaeology of Roman Construction, Oxford 11-12 April 2015)*, ed. by Janet Delaine and Stefano Camporeale (Madrid: *Anejos de AespA* 77, CSIC Press, 2016), pp. 131-145.

Ryan, Nick, 'Documenting and Validating Virtual Archaeology', *Giornale di Archeologia e Calcolatori*, 12 (2001), pp. 254-273.

Rystedt, Eva, *Acquarossa IV: Early Etruscan Akroteria from Acquarossa and Poggio Civitate (Murlo)* (Stockholm: Paul Åströms Förlag, 1983).

Sapirstein, Philip, 'How the Corinthians Manufactured Their First Roof Tiles', *Hesperia*, 78 (2009), pp. 195-229.

Staccioli, Romolo A., *Modelli di edifici etrusco-italici. I modelli votivi* (Firenze: Sansoni, 1968).

Strandberg Olofsson, Margareta, 'On the Reconstruction of the Monumental Area of Acquarossa', *Opuscula Romana*, XVII:12 (1989), pp. 163-183.

Strandberg Olofsson, Margareta, *Acquarossa V. 1. The Head Antefixes and Relief Plaques: a Reconstruction of a Terracotta Decoration and its Architectural Setting* (Stockholm: Paul Åströms Förlag 1984).

Torelli, Mario, 'Review of Acquarossa vols. I-V', *Gnomon*, 58 (1986), pp. 259-267.

Torelli, Mario, '*Fictilia Tecta*. Riflessioni storiche sull'arcaismo etrusco e romano', in *Deliciae Fictiles IV. Architectural Terracottas in Ancient Italy. Images of Gods, Monsters and Heroes*, ed. by Patricia Lulof and Carlo Resigno (Oxford: Oxbow Books, 2011), pp. 3-15.

Turfa, Jean M., and Alwin G. Steinmayer , 'The Comparative Structure of Greek and Etruscan Monumental Buildings', *Papers of the BSR*, 64 (1996), pp. 1-39.

Ward-Harvey, Ken, *Fundamental Building Materials* (Boca Raton Florida: Universal Publishers, 2009).

Wendt, Leni, et al., *Acquarossa VII. Trial Trenches, Tombs and Surface Finds* (Stockholm: Paul Åströms Förlag 1994).

Wikander, Charlotte, *Acquarossa I: The Painted Architectural Terracottas. 1. Catalogue and Architectural Context* (Stockholm: Paul Åströms Förlag, 1981).

Wikander, Charlotte, *Acquarossa I: The Painted Architectural Terracottas. 2. Typological and Decorative Analysis* (Stockholm: Paul Åströms Förlag 1988).
Wikander, Charlotte, and Örjan Wikander, 'The Early Monumental Complex at Acquarossa', *Opuscula Romana*, XVIII:11 (1990), pp. 189-205.
Wikander, Örjan, 'Acquarossa' in *Enciclopedia dell'Arte Antica* (1993).
Wikander, Örjan, *Acquarossa VI: The Roof-Tiles. 1. Catalogue and Architectural Context* (Stockholm: Paul Åströms Förlag 1986).
Wikander, Örjan, *Acquarossa VI. The Roof-Tiles. 2. Typology and Technical Features* (Stockholm: Paul Åströms Förlag 1993).
Wikander, Örjan, *Roof-Tiles and Tile-Toofs at Poggio Civitate (Murlo). The Emergence of Central Italic Tile Industry* (ACTA Rom-4, Stockholm: Paul Åströms Förlag, 2018), Vol. 63.
Winter, Nancy A., 'The Origin of the Recessed Gable in Etruscan Architecture' in *Deliciae Fictiles III, Architectural Terracottas in Ancient Italy: New Discoveries and Interpretations*, ed. by Ingrid Edlund-Berry et al. (Oxford: Oxbow Books, 2006), pp. 45-48.
Winter, Nancy A., *Symbols of Wealth and Power: Architectural Terracotta Decoration in Etruria and Central Italy, 640-510 B.C.* (Supplement to the *Memoirs of the American Academy in Rome* 9. Ann Arbor: University of Michigan Press, 2009).
Winter, Nancy A., 'Traders and Refugees: Contributions to Etruscan Architecture', *Etruscan Studies*, 20 (2017), pp. 123-157.

## About the Author

**Patricia Lulof** is Associate Professor of Mediterranean Archaeology at the University of Amsterdam. She specializes in Pre-Roman Archaeology, Ancient Architecture, and (digital) building techniques. She is scientific director of the interdisciplinary 4D Research Lab of the Faculty of Humanities at the University of Amsterdam, founded in 2012; and was PI of 'Archaeology of Architecture' (2013-2014, Faculty Research Priority Area Creative Industry-Digital Humanities), 'Biographies of Buildings. Virtual Futures for our Cultural Past' (2014-2015, Netherlands Institute for Advanced Study), the KIEM/NWO project 'Living in the Past. Reconstructing Etruscan Houses in Acquarossa' (2017-2018), and 'Van Klooster tot Kwartier' (2019-2020, Amsterdam University Fund).

# 10. Science and the Knowing Body: Making Sense of Embodied Knowledge in Scientific Experiment

*H. Otto Sibum*

**Abstract**
The historiographical approach called 'Experimental History of Science' explores embodied knowledge of the past through reworking of historical experiments. The actual performance of a 19th-century experiment (James Joule's paddle-wheel experiment) with a replica has served as the key tool to unlock tacit dimensions of past scientist's working knowledge. The narrative discloses an embodied knowledge (gestural knowledge) that was crucial for the performance of this experiment and is intimately entangled with already existing sites of knowledgeable labour to which Joule had access and which served him as informal spaces of learning.

**Keywords:** experimental history of science, embodied knowledge, gestural knowledge, reworking, paddle-wheel experiment

## Introduction

In this chapter I am going to investigate science and the knowing body (Plate 10.1). By means of a historiographical approach called 'Experimental History of Science', I will explore embodied knowledge of the past through reworking of historical experiments. However, before we can make sense of embodied knowledge, we have to be clear about the role things play in our understanding of the human past. In history and sociology of science this question has been tackled at least since the 1980s when the authors of

Dupré, S., A. Harris, J. Kursell, P. Lulof, M. Stols-Witlox (eds.), *Reconstruction, Replication and Re-enactment in the Humanities and Social Sciences*. Taylor & Francis Group
DOI 10.5117/9789463728003_CH10

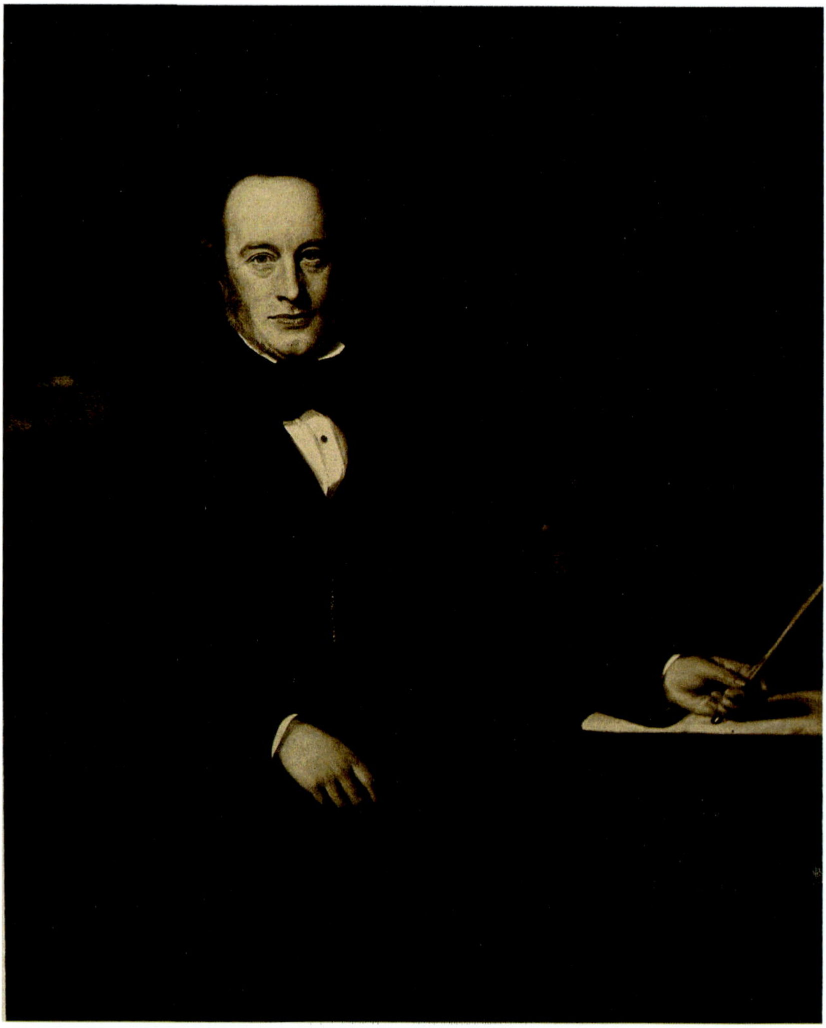

Plate 10.1: Portrait of James Prescott Joule, ca. 1913 after G. Patten,1863. To the right in the back the depiction of the paddle-wheel set-up. In his left hand Joule is holding his new thermometer especially made for the experiment by the Manchester instrument maker John Benjamin Dancer. Photo: Wellcome Collection, CC BY.

*Leviathan and the Air-Pump* 'ballasted epistemology with an unknown new actor, a leaky, pieced-together, handmade air pump'. But as the cultural anthropologist Bruno Latour remarked:

> 'We possess hundreds of myths describing the way subjects (or the collective or intersubjectivity, or epistemes) construct the object – Kant's Copernican Revolution is only one in a long line of examples. Yet we

> have nothing that recounts the other aspect of the story: how objects construct the subject. Shapin and Schaffer have access to thousands of archival pages on Boyle's ideas, and Hobbes's, but nothing about the tacit practice of the air pump or on the dexterity it required. The witnesses to this second half of history are constituted not by texts or languages but by silent, brute remainders such as pumps, stones and statues.'[1]

Prominent parts of these silent representatives of the past, this second half of history, are stored in museum collections. However, things (and not just those in museum collections) require a thorough practical as well as theoretical investigation in order to understand their role in our attempts of making sense of the world – past and present. For Latour, the state we are in we owe to our modernist thinking that perpetuates the great divide which exists between humans and things and hence tends to silence the latter. Scholars in archaeology, the master discipline of material culture studies, bemoan an asymmetry between things and texts in our reflections about studying the past and in particular how little is done to better understand how objects construct the subject.[2] Since the 1980s historians and sociologists of science have produced remarkable accounts on the investigation of material culture and quite recently general historians have invited a fresh discussion about the potential and limits of studying this second half of human history.[3] 'Experimental History of Science' (EHS) is a program developed in the early 1990s in order to give the endeavour of reworking past experimental practices and their silent representatives of the past a voice.[4] It started off from the still unsatisfactory but already much earlier bemoaned insight that

> 'When talking of natural sciences, one often forgets that there exists a living scientific practice, parallel to this, an official *Gestalt* of science on paper. These two worlds, however, are frequently as different from one another as are the practice of democratic government and its official theory.'[5]

These two worlds of science, the shiny, polished, enduring one on paper and the rather ephemeral lived practice of science, exist to a large degree still today, and that despite the many efforts to overcome this divide. Hence,

1 Latour, *We Have Never*, p. 82.
2 Olsen, 'Material Culture'; Ingold, 'Materials against Materiality'.
3 See for example Auslander, et al., 'AHR Conversation'.
4 Sibum, 'Experimentelle Wissenschaftsgeschichte'.
5 Fleck, 'Zur Krise', p. 50.

EHS aims at studying past science at work. With its approach of building replicas of old scientific instruments and the performance of past trials, it drew attention away from the textual remains of science towards these silent actors, the material culture as well as the scientist's body.

The project began as a pedagogical one in the physics department at Oldenburg University. We wanted to shift physics students' attention away from the polished textbook images and descriptions of canonical experiments to the often messy daily work of a scientist. By replacing modern table-top set-ups of canonical experiments with replicas of the historical ones, we made these students experience physics at work and even prompted them to raise historical questions about their own discipline. This approach soon became labelled the replication method. However, as a historian of science I developed this approach further to a historiographic method emphasizing the explorative character of experimentation for historical studies. This diagnostic adventure of past experimental practice I described as reworking rather than replication or re-enactment or reconstruction.[6] When coining the term 'Experimental History of Science' I did so partly in dependence on 'experimental archaeology'.

For the sake of distinction, we may analytically divide this approach into two parts. Firstly, building replicas of historical experiments. The term replica should not be mixed up with the concept of replication, and I will come to that later on. Producing a good copy of an eighteenth or nineteenth-century instrument requires a lot of knowledge and skills in scientific and non-scientific fields. For example, one has to contact craftsmen requesting their expertise about techniques in producing a precision scale or a particular mixture of glass etc. Often these experts are challenged and initially refused to start working on that, because quite often even they themselves did not know any longer how to make such an old device. But after a while these workshop jobs became their most beloved ones, because they were challenging them in a refreshing way, opening up lost dimensions of their own craft and thereby even improving their own field of expertise. In most cases, these craftsmen performed tremendously in producing replicas of the original.[7]

Art curators are equally challenged: For example, Venetian glass vases are difficult to reproduce. By means of using the materials of the historical

6 Sibum, 'Experimental History'.

7 For example, a leading German museum asked us whether we would be willing to paint our replica of one their objects in a strange colour, otherwise they would fear that our copy would ruin the market value of the original.

epoch and his own expertise as a glassblower, the leading expert William Gudenrath was able to obtain new information about the production techniques involved. These acquired skills made him produce replicas that won world recognition, but it should be emphasized that these will never be identical to the originals; they are unique pieces of art.[8]

Secondly, Experimental History of Science comprises performing past experiments with a replica. In science, performing an experiment is understood as replication as scientists try to achieve the same experimental results that were produced in the original experiment. However, it is important to note here, that replication in science does not necessarily mean to use a replica of the original experimental set up. Replication implies first of all to reproduce the same experimental results. Achieving the same results by means of using a very different experimental set up (not a replica of the original device) would even be considered as a much stronger and hence more persuasive replication.[9] Historians of science have taken up the question of replication in a number of different ways. Steven Shapin and Simon Schaffer write for example: 'We had read much about experiment; we had both even performed a few as students; but we did not feel that we had a satisfactory idea of what an experiment was and how it yielded scientific knowledge.'[10] Like anthropologists who perform fieldwork, they studied Robert Boyle's eighteenth-century experiments with an air-pump through the rich textual remains that reveal important dimensions of a famous controversy between Boyle and Thomas Hobbes thereby bringing out excellently the different technologies of knowledge production that determined the course of this controversy, the troubles with replication and how experimental philosophers succeeded in convincing an academic public of the existence of the vacuum as a matter of fact. Peter Galison took up the problem of replication in a different way; he sought to answer historically the question of how an experiment ends. Since there is no unambiguous way to determine when one can consider the replication of an experiment as satisfactory, it is of great importance to learn under which conditions a scientist or a collective decided not to carry out further investigations.[11] David Gooding followed a third way of exploring an experiment, analysing Michael Faraday's extensive laboratory notebook entries line by line in order

8 Gudenrath, *Master Class Series*, II: https://renvenetian.cmog.org. Checked on March 4th, 2019.

9 On replication see Collins, *Changing Order.*

10 Shapin and Schaffer, *Leviathan*, p. 16.

11 Galison, *How Experiments End.*

to derive from them the cognitive processes that led to the development of the electromotor. It must be made clear however that laboratory notebook entries like those of Faraday represent a retrospective consideration of the course of a research day. That is, Faraday's very expressive account indeed offers excellent possibilities for understanding the course of his experimentation, but it should not be forgotten that the textual coherence of the experimental narrative only corresponds in part to the course of past real-time actions.[12]

Performing James Joule's early nineteenth-century paddle-wheel experiment to determine the mechanical equivalent of heat serves as another example. In particular, taking seriously the many painstaking breakdowns in these performances, I have shown that even minute notebook entries of this past scientist do not necessarily yield traces of those experimental techniques I came to recognize as most important in carrying out the experiment. Furthermore, performing the experiment led to the insight that the choreography of actions derived from the publication did not harmonize with the course of action experienced with the replica. Even the sense-economy of the experimenter alone demanded the carrying out of simultaneous working rhythms; an assistant, unnamed in the text, was necessary to manage the mechanical operation of the instrument (Plate 10.2).[13]

Furthermore, a replica constructed according to the scale drawings in Joule's publication did not allow for a repetition of the experiment that corresponded with the results he recorded there. Only further studies on the surviving historical instrument in the Science Museum in London yielded the information that was required to achieve this. Although James Joule himself prepared the illustrations of his apparatus and experimental set-up, these were not sufficiently precise to carry out a successful performance of the experiment. These experienced differences between textual and performative coherence of experimental actions was most annoying but at the same time the source of new discoveries, one of which is that there is a lived practice of science beyond the laboratory notebook narrative which can only be disclosed through performing – or better reworking – the experiment with a replica. I have chosen to speak of 'reworking' experiment rather than 're-enacting' because I want to indicate that every performance of an experiment is a unique event and certainly different from the original one. Hence even if 'reworking' is striving for re-enactment of a

12 Gooding, *Experiment*; Sibum, 'Narrating by Numbers'.

13 Joule, 'On the Mechanical Equivalent'; Sibum, 'Reworking the Mechanical Value'; Sibum, 'Experimental History of Science'.

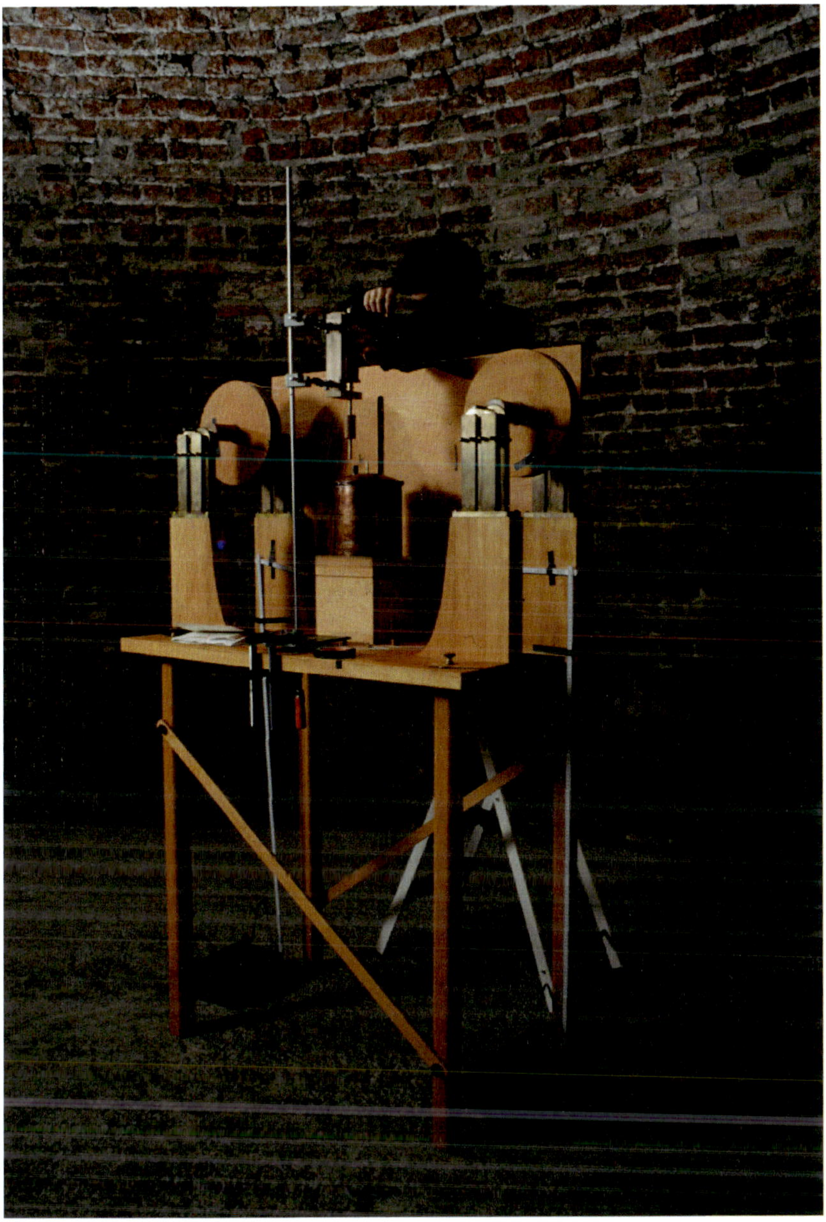

Plate 10.2: Performing James Joule's paddle-wheel experiment with a replica: By means of employing a hand gear the experimenter turns a wooden spindle to which strings are attached. In this way he is winding up the weights that are placed on the floor. He then connects the spindle with the top end of the axle of the stirrer that is placed in the copper vessel. After disconnecting the hand gear the weights will slowly descent to the floor thereby churning the water in the vessel. After twenty repetitions the temperature of the water will have increased slightly. Joule could not have done the experiment on his own, already the mechanical performance of the experiment requires support by an assistant.

past experimental performance, the term 'reworking' serves as a reminder that repetition, replication or re-enactment is impossible. At the most, by means of performance of experiment we are able to reconstruct past events.

In the early 1990s, various historians regarded the performance of a historical experiment with a replica as an ill-conceived approach because this investigation of things could never be acknowledged as a historical source, in fact we were practicing an anachronism.[14] Only gradually they came to realize that they were confusing reworking with replication in the scientific sense. This conflict is still to a certain degree virulent in scholarly discussions. But it is important to understand that reworking past experiments by means of performance is not an attempt to find out 'how it really was'. It is a complementary method to the existing modes of historical exploration that will reveal dimensions of past practice of knowledge-making that are not captured in any literary record of the past and that lead to insights about matters hitherto unrecognized.

In other fields like music or the arts we find similar considerations. The aforementioned glass artist who is today probably the most knowledgeable person about Venetian glass blowing techniques would never claim that he could perform the one and only manufacturing technique as used by Renaissance Venetian glassblowers. Similarly, the director of Academy of Ancient Music and authority in historically informed performance of music, Christopher Hogwood, always denied that performing historical master pieces of music with historical instruments is a certain way of achieving an authentic historical performance.[15]

## Reworking embodied knowledge

What then can reworking of historical experiments be good for? What can it achieve? Although replication is a key methodological component of modern experimental science and broadly understood as a warranty for truth and objectivity, historical studies have shown quite clearly that scientists themselves take rather controversial positions on the meaning of replication.[16] Without going into detail about this aspect here, I would like to stress, however, that employing the performance of past experiments

14 This critique was voiced predominantly orally, but to my knowledge, was never substantiated in print.

15 Personal communication by Christopher Hogwood to the author.

16 See for example Collins, 'The Seven Sexes'.

as an instrument for research in the history of science requires a distancing from an understanding of the scientific term 'replication'. Performing historical experiments with a replica is much more than that. It certainly provides deep insight into the process of scientific replication, but in the first instance it is an explorative approach to past practices of knowledge-making that reveals dimensions of the lived practice of science that have not been recorded in our literary remains. In our predominantly literary culture, academic approaches to the lived practice of knowledge-making remain to a large degree within this literary domain. As long as these narratives are exclusively based on observational practices of literary, visual and material sources, these historical accounts may still be insufficient. Only the performance of experiment in combination with the conventional methodological repertoire of textual analysis can unlock and provide full access to a working knowledge that existed in the scientific laboratories, artisanal workshops and studios of the past.

Within conventional history of science, historical investigations of lived practice of past science have merely pointed to the existence of invisible assistants and their embodied knowledge conventionally described as personal or tacit knowledge. Generally speaking, historians know quite a lot about the construction, representation and the disciplining of the human body, but still too little about the productive role of the human body, i.e. the cognitive implications of their bodily disciplining through their engagement in the material world. Reasons for this unsatisfactory situation are manifold and I will just mention the two most relevant ones: Firstly, historians and historians of science have traditionally developed their analytical tools exclusively through engagement with the practices of a literary culture. Despite their increasing interest in the hardware of science or the body techniques of their users, they often lack appropriate diagnostic tools and procedures. Performing past experiments with a replica was initially considered by many as anachronistic and the history of tacit knowledge as a nearly 'impossible archaeological feat because it is usually impossible to go back and find or operate the original instruments in their original settings and setup' – as a historian of science once stated.[17]

Secondly, academic research is still to a large degree part of an influential tradition that defines knowledge as being rational, cerebral, ubiquitous and communicable through texts. Consequently, the implied divide between epistemology and practice leads to a denigration of forms of knowledge, which are embodied, and local. The approach of reworking past experimental

17 Biagioli, 'Tacit Knowledge', pp. 69-81, 73.

practice (EHS) has been developed with the aim of making the impossible archaeological feat possible and to challenge the mentioned epistemological divide. But it is important to mention that this diagnostic adventure is not just drawing the reader's attention away from the software to the hardware of science, rather it uses performance of historical experiments as a technique complementary to conventional historical scholarship. Providing access to these silent representatives of the past and using them in these performances allows for the study of the self-evident gestures at work in the laboratories.

Historians of gesture have pointed to the same methodological problem that they 'cannot walk through the streets of the past in order to observe ordinary situations [...] usually [they] work with incomplete documentation which is more often biased towards the exceptional than to the normal events of everyday life'.[18] Therefore they watch out for sites of social frictions where norms get questioned and self-evident human behaviours become objects of attention and debate.

Complementary to this approach, reworking historical experiments, including the performance of experimental trials, is not dependent on documents of the exceptional. On the contrary, it allows for the investigation of the mundane, self-evident gestures at work in the laboratories of the past. We can study key aspects of the knowing body of the scientist and the changing self-evidence of gestures even of those experiments that were performed in the absence of witnesses.[19]

By means of performing historical experiments with replicas of scientific apparatus and studying their technical breakdowns – the failures – we reveal dimensions of past practice hitherto unrecognized even by the actors themselves or their contemporaries. *Doing* an experiment and recognizing the troubles encountered in getting it to work, creates an awareness of the behaviour of the historical experimenter and the unarticulated techniques which were indispensable for the performance of the experiment. Not even in the scientists' notebooks do we find hints about crucial working techniques because these were most often self-evident.[20]

18 Bremmer and Roodenburg, *A Cultural History*, in particular chapter 6 by Muchembled, 'The Order of Gestures', pp. 129-151, 131.

19 In contrast to the eighteenth-century understanding of experiment as public spectacle, early nineteenth-century experiments were more and more performed at specially protected locations. In particular, the sensitivity of instruments used in these experiments required special forms of protection and did not allow the presence of witnesses.

20 James Joule's notebook pages record immense data about the performance of the paddle-wheel experiment. However, as the reworking of the experiment has demonstrated, there is not a single hint to be found on these pages that would reveal knowledge about the most important technique employed in this experiment, i.e. how to perform temperature measurements. See also Sibum, 'Narrating by Numbers', pp. 141-158.

Moreover, reworking past experiments depends above all on the improvisational work and knowledge of the researchers, and the material objects (as well as accompanying texts) serve as a kind of choreography for this performance because they provide partial direction for our thinking and acting. This locally created working knowledge I have termed *gestural knowledge*.[21] The unfamiliar concept will break from the traditional, static concept of disembodiment of knowledge in favour of a conception of knowledge united with the actor's performance of work which changes according to the specific kinds of performance, for example in the manipulation of an instrument or the use of mathematical tools, and in ever new historical circumstances. Therefore, repeating past experiments should not be regarded as showing that they function reliably; reworking aims at exploring the forms of knowing involved in these performances. While in traditional studies knowledge achieved through the use of instruments is simply understood as skills, competences or practices, with the conception of gestural knowledge we succeed in overcoming the fundamental, systematic separation of epistemology and practice which itself has a long historical development. Skills or manual capabilities are now to be considered a constitutive part of the knowledge of the experimenter. What the physical chemist Michael Polanyi once designated as 'tacit' or 'personal knowledge' can be described as the expression of a historically located gestural knowledge, which is historically embodied in the form of the performative actions of the experimenter or in part in the artefacts. And most importantly, their cultural origins can be retrieved. With this dynamic conception of embodied knowledge that is bound to the performative actions of the researcher or collective, we are able to grasp those cultural repertoires of action which are essential for the formation of experimental knowledge but which usually escape the historian's attention because they belong to different worlds of sense, often described as non-verbal or oral knowledge traditions.[22]

## Gestural knowledge and scientific change: The case of James Prescott Joule

In order to deepen these reflections, let me refer to an old study of mine, the experimental work of the Victorian brewer James Prescott Joule. His experiment on the determination of the mechanical equivalent of heat is

21 See for example Sibum, 'Reworking the Mechanical'; Sibum, 'Les gestes'.

22 Polanyi, *Personal Knowledge*; Polanyi, *The Tacit Dimension*.

today considered as one of the canonical experiments in physics, because it demonstrated for the first time the existence of a constant relationship between mechanical force and heat that soon became the building block for the science of energy, and energy conservation in particular. Today, in every physics class room, students repeat this experiment and museums display table-top versions in order to illustrate this great achievement in science. And, indeed, visual representations infer that in principle the experiment should work rather easily: You take a pot of water that is stirred through a mechanical device, thereby the liquid gradually warms up. The surprising find is that the ratio of the mechanical work used in order to stir the water and the temperature difference achieved always amounts to the same constant value, called the mechanical value of heat. However, as soon as you perform this experiment with the replica you will notice tremendous troubles in performing it. One of the key experiences in performing this famous experiment is that it requires outstanding thermometrical skills and the construction of extraordinary precision instrumentation (Plate 10.3). No textual remains of this important moment in history of science provide the necessary clues that explain how this was possible. Not even in Joule's notebooks will one find any hint, and hence previous generations of historians of science declared Joule's extraordinary experimental achievement a gift of providence. Donald Cardwell, Joule's biographer and outstanding historian of technology concluded: 'In the final analysis I must concede that these particular gifts (Joule's immense skill and precision in performing experiment) were bestowed by Providence and not by his fellow men.'[23]

By studying the material culture of this early Victorian science, we can, however, do better than perpetuating this well known, but highly problematic construction of a genius. The detailed list of troubles that I encountered in practically reworking this early nineteenth-century experiment allowed me to identify hitherto unrecognized connections James Joule was entangled in, that provide explanations as to why he became such a singularly outstanding investigator of nature. We know now that in his work Joule was implicitly drawing on cultural repertoires of action, like his close collaboration with the instrument maker John Benjamin Dancer as well as his upbringing and active participation in the brewing business. Being the son of the wealthiest brewer in Manchester and having worked in that business for more than twenty years (eight hours a day) became crucial in building up thermometrical skills unheard of outside the brewer's world. One should know that in the 1830s brewing had undergone

23 Cardwell, *James Joule*.

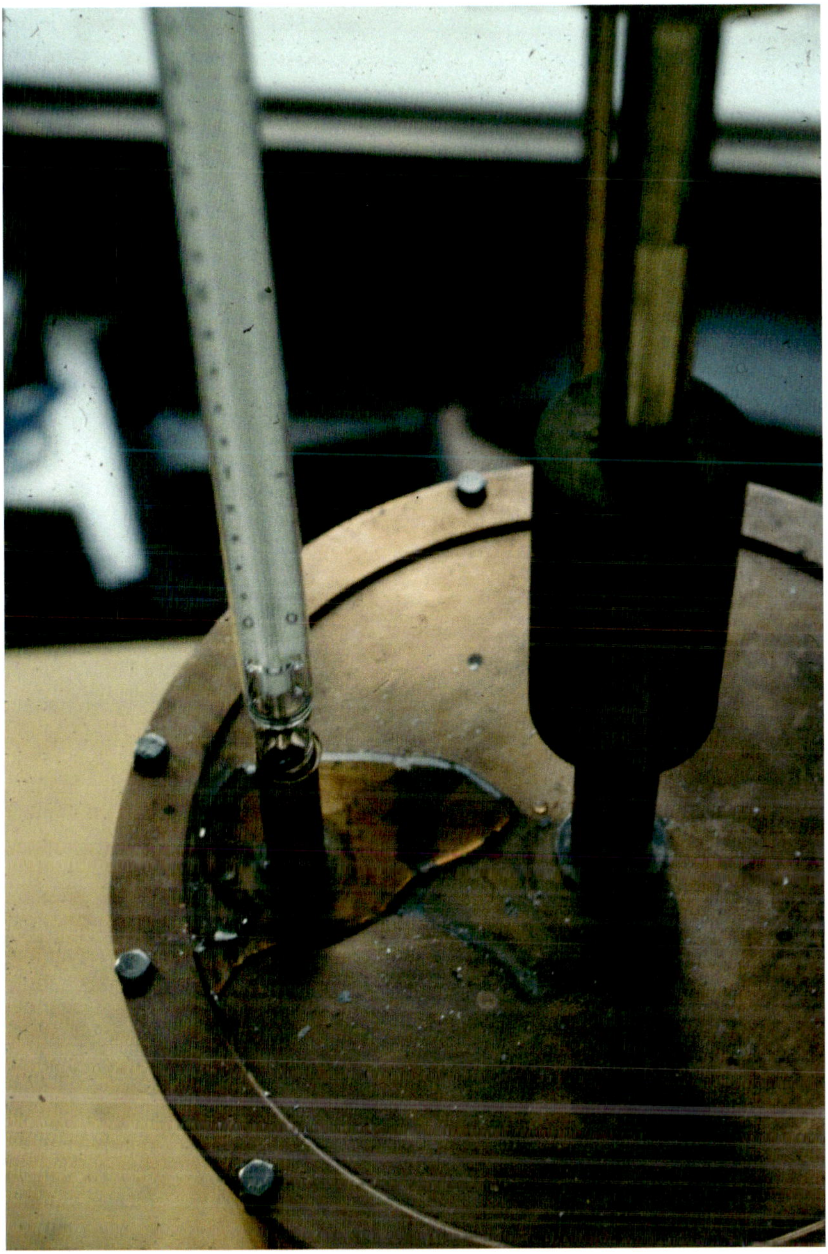

Plate 10.3: The thermometer has to be placed in the copper vessel to determine the exact temperature of the water at the beginning and the end of the experiment. Furthermore a second thermometer has to be placed in the laboratory to record permanently the room temperature. We have no literary traces left from Joule that would allow us to learn when, for how long, how deep the thermometer has to be placed in the vessel, nor at which place, how often the air temperature has to recorded.

massive transformations in Victorian England, from being a craft tradition to becoming industrial production. This was the place where extraordinary precision techniques, especially in heat measurement, were developed and, more generally, where you learned to trust in numbers.

Similarly important was Joule's close interaction with the leading instrument maker John Benjamin Dancer, with whom he designed and built the most precise thermometer in the country. Joule's active participation in these hitherto unrecognized Manchester subcultures led him to develop this unique calibration technique and his skills in taking heat measurements, which became key in his scientific experiment on the determination of the mechanical equivalent of heat. As I have shown elsewhere, in these informal spaces of learning, Joule acquired a gestural knowledge that would become crucial in the performance of his scientific experiment.[24]

However, this lived scientific practice did not find its representation in the related publication. On the contrary, being a brewer and collaborator with an instrument maker would have made him appear more a member of the Manchester community of skilled artisans rather than one of the established scientific authorities of natural philosophy. Hence, in writing, Joule carefully crafted the official 'Gestalt' of science (as he understood it), because he wanted to be recognized as an authority in natural philosophy. And, indeed, the final publication is a masterpiece in creating this ideal of science: the disembodied observer of natural phenomena and a master of precision measurement. In the published narrative, he made the established fact appear as a constant of nature by effacing all the immense human labour that had gone into the experimental performance. The unskilled working class represented by the unknown brewing mate that had assisted Joule in the mechanical performance of the machine was not mentioned at all. The instrument maker J.B. Dancer, as a representative of the skilled working class, was given the status of a valuable servant. Only Joule himself remained, but now as the disembodied observer of nature who displayed herself. He distanced himself from any kind of sensuous perception other than those gestures of accuracy which demonstrated remote control of the object of research: Reading off temperature scales from the best instrument by means of a 'practiced eye' (Plate 10.4). This would fit the assumed manners of the reader of the *Philosophical Transactions* and was precisely the kind of science that the Cambridge doyen of natural philosophy, William Whewell, who coined the word 'scientist' at this time, helped to promote, the ideal of scientific practice in which sight and hearing were privileged above the

24 Sibum, 'Les gestes'.

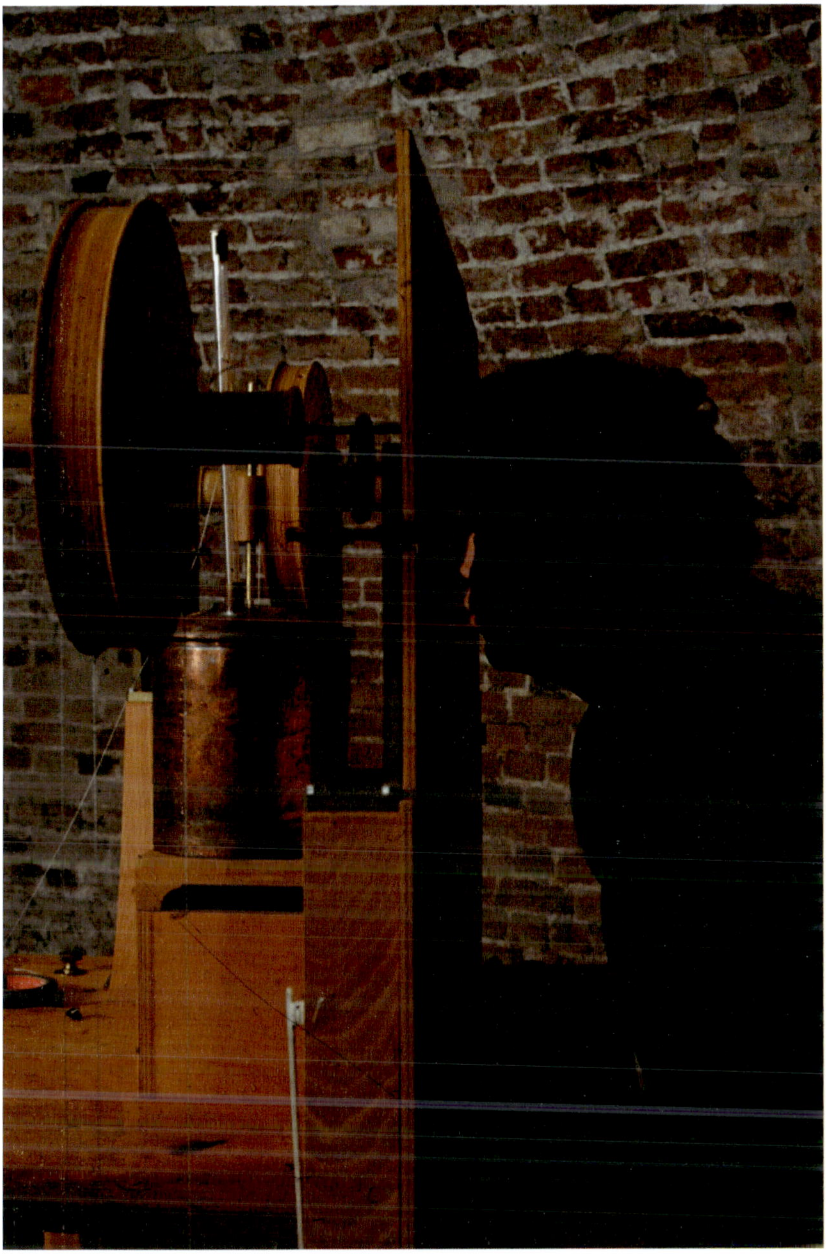

Plate 10.4: Reading of temperature values of a thermometer that is placed in the water of the copper vessel. A wooden board is shielding the experimenter from the vessel and the thermometer in order to prevent body heat radiation

other senses. 'The other senses have not any peculiar prerogatives, at least none which bear on the formation of science.'[25]

Moreover, in order to get permission for this text to be printed in the *Philosophical Transactions* Joule even took out the most important theoretical conclusion originally drawn from the experiment, after his referees of the Royal Society questioned its validity. By means of his experiment Joule was convinced that he had deduced that 'friction is the conversion of force into heat' which implied that heat would be an entity that could be transformed into mechanical force and vice versa. This was violating the then accepted theory of heat. Joule did take this conclusion out of the manuscript, not because he was convinced by the arguments of the reviewers, but because he wanted to become an author in this most prestigious journal. The letter from Joule to the Cambridge mathematician G.G. Stokes from the same year of the publication leaves no doubt about this: 'I beg your acceptance of the enclosed paper in which I have endeavoured to determine the mechanical equivalent of Heat with accuracy. The result at which I conceived I had arrived was that Friction consists in the conversion of Force into Heat; but the Committee of the R.S. having disapproved of such a deduction from the experiments I thought it best to withdraw it, although I think this view will ultimately be found to be the correct one.'[26]

## Conclusion

In 1934 the French ethnographer Marcel Mauss published an important essay called 'Les techniques du corps', 'Body Techniques', in which he emphasized that 'the physical training of all ages and both sexes is made up of masses of details which pass unobserved; we must undertake to observe them.' As mentioned above, historians and social scientists have taken up his advice and studied the disciplining of the human body in cultural settings. The programme exemplified in this essay is equally inclined to Mauss' work but wants to go a step further. We are not regarding body techniques as representing whole social classes or gender as the notion of *habitus* implies. In our historical anthropological approach, we aim at identifying body techniques of past individuals, but most importantly with the aim to understand the cognitive implications of this bodily formation. The actual performance

25 Whewell, *The Philosophy*, Vol. 1, p. 280.

26 Joule to Stokes, July 3rd, 1850, Cambridge University Library, ADD.

of a nineteenth-century experiment with a replica has served as the key tool to unlock tacit dimensions of James Joule's experimental practice. Particularly through the various technical breakdowns, this trial became the site of acquiring a gestural knowledge which provided clues to search for hitherto unrecognized historical connections. Through this exploration, the conventional image of this experiment has been challenged. The narrative has provided important insights into hitherto neglected but already existing sites of knowledgeable labour to which Joule had access and which served him as informal spaces of learning. Therefore, this micro study of Joule's experiment provides an explanation of scientific change which acknowledges his singularity and at the same time shows that Joule's mundane self-evident gestures of work and the related embodied knowledge is entangled in a much broader structural development of a web of nineteenth-century practitioners' knowledge. From this perspective, Joule's tacit or personal knowledge is better described as a historically embodied gestural knowledge drawing on cultural repertoires of actions that can be retrieved. In order to gain credit for his achievement, Joule himself shaped his scientific experiment in such a way that his embodied knowledge appeared as an achievement of a disembodied genius. In doing that, he himself reinforced the traditional divide between epistemology and practice and perpetuated the distinction between philosophical authorities and skilled artisans.

## Bibliography

Auslander, Leora, Amy Bentley, Leor Halevi, H. Otto Sibum, and Christopher Witmore, 'AHR Conversation: Historians and the Study of Material Culture', *American Historical Review*, 114 (2009), pp. 1355-1404.

Biagioli, Mario, 'Tacit Knowledge, Courtliness, and the Scientist's Body', in *Choreographing History*, ed. by Susan Leigh Foster (Bloomington/Indianapolis: Indiana University Press, 1995), pp. 69-81.

Bremmer, Jan, and Herman Roodenburg, *A Cultural History of Gesture* (Ithaca: Cornell University Press, 1992).

Cardwell, Donald S.L., *James Joule: A Biography* (Manchester: Manchester University Press, 1989).

Collins, Harry M., 'The Seven Sexes: A Study in the Sociology of a Phenomenon, or the Replication of Experiments in Physics', *Sociology*, 9 (1985), pp. 205-224.

Collins, Harry, *Changing Order: Replication and Induction in Scientific Practice* (London: Sage Publ., 1985).

Fleck, Ludwik, 'Zur Krise der "Wirklichkeit"', Die Naturwissenschaften 17 (1929), trans. by G.H. Schalit and Yehuda Elkana as 'On the Crisis of Reality', in *Cognition and Fact – Materials on Ludwik Fleck*, ed. by R.S. Cohen and T. Schnelle (Dordrecht: Reidel Publishing Company, 1986).

Fors, Hjalmar, Lawrence M. Principe, H. Otto Sibum, *From the Library to the Laboratory and Back Again: Experiment as a Tool for Historians of Science, Ambix*, 63 (2016), pp. 85-97.

Galison, Peter, *How Experiments End* (Chicago: Chicago University Press, 1987).

Gooding, David, *Experiment and the Making of Meaning: Human Agency in Scientific Observation and Experiment* (Dordrecht: Kluwer Academic Press, 1990).

Gudenrath, William, *Master Class Series II: Introduction to Venetian Techniques with William Gudenrath* (Corning: The Corning Museum of Glass, 1998).

Gudenrath, William, *The Techniques of Renaissance Venetian Glassworking* (Corning: Corning Museum of Glass, 2016).

Ingold, Tim, 'Materials against Materiality', *Archaeological Dialogues*, 14 (2007), pp. 1-16.

Joule, James Prescott to Stokes, July 3rd, 1850, Cambridge University Library, ADD.

Joule, James Prescott, 'On the Mechanical Equivalent of Heat', *Philosophical Transactions of the Royal Society of London*, 140 (1850), Part. 1, pp. 61-82.

Latour, Bruno, *We Have Never Been Modern* (Cambridge, MA: Harvard University Press, 1993).

Olsen, BjØrnar, 'Material Culture after Text. Re-Membering Things', *Norwegian Archaeological Review*, 36 (2003), pp. 87-104.

Polanyi, Michael, *Personal Knowledge: Towards a Post-Critical Philosophy* (Chicago: Chicago University Press, 1962).

Polanyi, Michael, *The Tacit Dimension* (London: Routledge, 1967).

Shapin, Steven, and Simon Schaffer, *Leviathan and the Air-Pump: Hobbes, Boyle, and the Experimental Life* (Princeton: Princeton University Press, 1985).

Sibum, H. Otto, 'Reworking the Mechanical Value of Heat: Instruments of Precision and Gestures of Accuracy in Early Victorian England', *Studies in History and Philosophy of Science*, 26 (1995), pp. 73-106.

Sibum, H. Otto, 'Les gestes de la mesure: Joule, les pratiques de la brasserie et la science', *Annales Histoire Sciences Sociale*, 53 (1998), pp. 745-774.

Sibum, H. Otto, 'Experimental History of Science', in *Museums of Modern Science*, ed. by Svante Lindqvist, Nobel Symposium 112, (Canton, MA: Science History Publications, 2000), pp. 77-86.

Sibum, H. Otto, 'Experimentelle Wissenschaftsgeschichte', in *Instrument – Experiment: historische Studien* / im Auftr. des Vorstandes der Deutschen Gesellschaft für Geschichte der Medizin, Naturwissenschaft und Technik hrsg. von Christoph

Meinel (Berlin/Diepholz: Verlag für Geschichte der Medizin, Naturwissenschaft und Technik, 2000), pp. 61-73.
Sibum, H. Otto, 'Narrating by Numbers. Keeping an Account of Early 19th Century Laboratory Experiences', in *Reworking the Bench*, ed. by Frederick L. Holmes, Jürgen Renn and Hans-Jörg Rheinberger, Archimedes: New Studies in the History and Philosophy of Science and Technology (Dordrecht: Springer, 2003), Vol. 7, pp. 141-158.
*The Techniques of Renaissance Venetian Glassworking.The Techniques of Renaissance Venetian Glassworking* (Corning: Corning Museum of Glass, 2016).
Whewell, William, *The Philosophy of the Inductive Sciences, Founded Upon Their History* (London: John W. Parker, 1840), Vol. 1.

## About the Author

**Heinz Otto Sibum** is a historian of science, Hans Rausing Professor of History of Science and Director of the Office for History of Science at the University of Uppsala. His research covers the history of the physical sciences from the eighteenth century until present. His studies are dedicated to exploring new approaches in social and cultural history of the sciences. He has recently worked on laboratory practices, instruments, experiment and embodied knowledge in scientific research.

# Index of RRR Terminology

# Index of Keywords

For Product Safety Concerns and Information please contact our EU representative GPSR@taylorandfrancis.com Taylor & Francis Verlag GmbH, Kaufingerstraße 24, 80331 München, Germany

Batch number: 10722100

Printed by Printforce, the Netherlands